ADVANCES IN MEMBRANE SCIENCE AND TECHNOLOGY

ADVANCES IN MEMBRANE SCIENCE AND TECHNOLOGY

TONGWEN XU
EDITOR

Nova Biomedical Books
New York

For permission to use material from this book please contact us:
Telephone 631-231-7269; Fax 631-231-8175
Web Site: http://www.novapublishers.com

Library of Congress Cataloging-in-Publication Data
Available upon request

ISBN: 978-1-60692-883-7

Published by Nova Science Publishers, Inc. ✢ New York

Contents

Preface		**vii**
Chapter I	A Random Walk through Membrane Science—From Water Desalination and Artificial Kidneys to Fuel Cell Separators and Membrane Reactors *H. Strathmann*	**1**
Chapter II	Preparation and Application of Ion Exchange Membranes: Current Status and Perspective *Tongwen Xu*	**21**
Chapter III	Proton Exchange Membranes and Fuel Cells *J. Kerres and F. Schönberger*	**75**
Chapter IV	Organic/Inorganic Hybrid Membranes: Overview and Perspective *Cuiming Wu and Tongwen Xu*	**119**
Chapter V	Pervaporation Membranes for Organic Separation *Xiangyi Qiao, Lan Ying Jiang, Tai-Shung Chung and Ruixue Liu*	**163**
Chapter VI	Membrane Bioreactor: Theory and Practice *In S. Kim, Namjung Jang and Xianghao Ren*	**201**
Chapter VII	Membrane Integration Processes in Industrial Applications *Hong-Joo Lee and Seung-Hyeon Moon*	**243**
Chapter VIII	Membrane Controlled Release *Liang-Yin Chu*	**279**
Index		**295**

Preface

Even though membrane-related technologies can be traced back to 250 years ago, the first laboratory-scale membrane synthesis became available only 60 years ago. Since then, membranes and related technologies have gained more technical and commercial relevance. Their applications have extended to environmental, chemical, medical, food, and energy industries. Since membrane-related technologies will play a key role in solving the problems in environmental protection and energy supply, we can not have a happy and sustainable life without membranes. As stated by Prof. Richard Bowen, "If you are tired of membranes, you are tired of life".

The academia and industry also attach importance to these technologies because of their characteristics, such as high energy efficiency, environmental benignity, integration flexibility, and economical competence. To date, many books on membranes have been published to awaken reader's interest in this field. This book, however, is intended not to make a summary of the literature in these areas, but to focus on the current status of some advanced membrane technologies which are well related to human life. Eight chapters were contributed by the selected researchers and professors in the corresponding fields. Chapter 1 takes a roam through membrane science and technology—from desalination and artificial kidneys to fuel cell separators and membrane reactors, informing the reader of what kind of membrane technologies have come true, or might or might never come true. Chapter 2 concentrates on the current science and technology using electro-membranes. Such topic was selected because electro-membrane related processes are playing a great role in cleaner production and separation due to ionic selectivity of membranes. Chapter 3 treats of the application of membranes to energy supply, which has been a hot issue for sustainable development of our earth. Inorganic-organic hybrid membranes, which combine the advantages of inorganic and organic membranes, were recently developed and expected to be applied in such harsh conditions as high temperature and strongly oxidizing circumstances. So an overview and perspective of such domain is included in this book as Chapter 4. The improvements on traditional chemical processes using membrane technologies are summarized in Chapter 5 and exemplified with pervaporation for organic separation. Chapter 6 covers a hot issue in our daily life: recycling of municipal waste water using membrane bioreactors. Another distinctive characteristic about membrane technologies is integration flexibility, which is crucial to the realization of multiple functions needed for a specific complex industrial application. Chapter 7 deals with such technique integrations. Membrane controlled release, the focus of Chapter 8, is an emerging membrane technology that might come true and has proved important in medical and pharmaceutical applications. The main purpose of editing this book is to provide valuable information and enlightening ideas to all the scientists and researchers in membrane research area. Besides, dabblers can find food for thought in some chapters, too.

As an editor, I appreciate the cooperation and active participation from all contributors, who, despite their extreme busy schedule of teaching and research (and even administration), found time to make this book a reality. Sincere thanks are extended particularly to Ms. Maya Columbus from Nova Science Publishers for her help and encouragement to publish this book.

Tongwen Xu, Prof.& Dr.
Laboratory of Functional Membranes
School of Chemistry and Materials Science
University of Science and Technology of China
Hefei, Anhui 230026
P.R. China
Tel:+86-551-3601587
Email:twxu@ustc.edu.cn

In: Advances in Membrane Science and Technology
Editor: Tongwen Xu
ISBN: 978-1-60692-883-7

Chapter I

A Random Walk through Membrane Science—From Water Desalination and Artificial Kidneys to Fuel Cell Separators and Membrane Reactors

*H. Strathmann**
Institute for Chemical Technology, University of Stuttgart,
Boeblinger Str 72, D-70199 Stuttgart, Germany

Abstract

When during the middle of the last century the first synthetic membranes became commercially available, a multitude of potential applications were identified and the expectations for the future technical and commercial relevance of membranes were very high. Membrane processes have indeed several intrinsic properties which make them especially suited for certain applications such as the separation, concentration or purification of molecular mixtures. They are energy efficient, do not generate hazardous by-products and can be operated at ambient temperature. However, in spite of their obvious advantages in many applications, membranes have not met the needs of modern industrial production processes. In some applications, such as in hemodialysis and water desalination, membranes indeed play a key role today. But in other applications such as the treatment of industrial effluents or the production of chemicals and drugs, membranes find it difficult to compete with conventional separation processes. The reason for the poor performance of membranes in certain applications is the unsatisfactory properties of today's available membranes. However, with the development of new membranes with improved properties, the membrane-based industry is responding to the market needs. An assessment of today's membrane products and their application illustrates the importance but also the limitation of membrane science and technology. Examples for successful applications of membranes are presented and emerging membrane processes and

* Telephone: +49-(0)-7071-687323; Email: heiner.strathmann@t-online.de

potential future applications as well as research needs to meet the future challenges of the membrane market are indicated.

Keywords: Membrane sciences, membrane processes, membrane application, membrane market, water desalination; artificial kidneys; fuel cells; membrane reactors; controlled release.

Introduction

When during the middle of the last century the first synthetic membranes became available on a laboratory scale, the expectation for their technical and commercial relevance was very high and a multitude of potential applications were identified in the chemical, the food and the pharmaceutical industry as well as in energy conversion systems, in artificial organs and medical devices. Membranes and membrane processes are characterized by high energy efficiency and transport selectivity. They can be operated at ambient temperature and do not generate hazardous by-products. These properties make membrane processes very attractive for a cost effective separation of aqueous mixtures and for the concentration or purification of temperature-sensitive biological materials. Integrated in conventional processes, membranes often provide more efficient and cleaner production techniques and better quality products.

However, in spite of their obvious advantages in many applications, membranes have not always met the needs of modern industry. In some applications, such as in hemodialysis or in reverse osmosis sea water desalination and in the production of high quality industrial water, membranes indeed play an important role today. But in other applications, such as the recycling of valuable products from industrial effluents and waste streams or the efficient production of chemicals, petrochemicals and drugs, etc., membranes find it difficult to compete with conventional separation processes. The reason for the poor performance of membranes in certain applications is due to the unsatisfactory properties of today's available membranes, lack of basic process know-how and long-term practical experience. With the development of new membranes with improved properties in recent years, a number of new potential applications have been identified and the membrane-based industry is rapidly exploiting these applications responding to the market needs for clean production processes and sustainable technologies with a maximum production efficiency and recycling of valuable materials and a minimum of toxic by-products and energy consumption [1]. An assessment of today's membrane products and their application indicates the research needs to meet the requirements of the future market.

2. The Structures of Synthetic Membranes and Their Function

The term "membrane" covers a large variety of different structures made from different materials with different transport properties. The same is true for membrane processes which can be very different in terms of the applied membrane structures and the driving force utilized for the transport of different components. Before discussing the present state of membrane science and technology in more detail, some basic terms concerning membranes and membrane processes shall be described.

2.1. Membrane Materials and Structures

Typical membrane structures used today in the various applications consist of solid dense polymer, ceramic or metal films, and porous symmetric or asymmetric polymer and ceramic structures, liquid films with selective carrier components and electrically charged barriers.

The key properties determining the membrane performance are:

- High selectivity and permeability for the transport of specific components
- Good mechanical, chemical and thermal stability under operating conditions
- Low fouling tendency and good compatibility with the operation environment
- Cost-effective and defect-free production

Membranes are manufactured in different geometrical forms such as flat sheets or tubes with a diameter of 1 to 2 cm. For many applications membranes are produced as capillaries with a diameter of 200 to 500 μm or as hollow fibers with a diameter of less than 80 μm. Capillary membranes generally have an asymmetric structure with the selective layer facing the feed solution which can be on the inner or outer surface of the capillary. Hollow fibers generally have the feed stream on the outside and the permeate stream on the inside of the fibers. Different membrane structures, materials and configurations are illustrated in the schematic drawing of Figure 1.

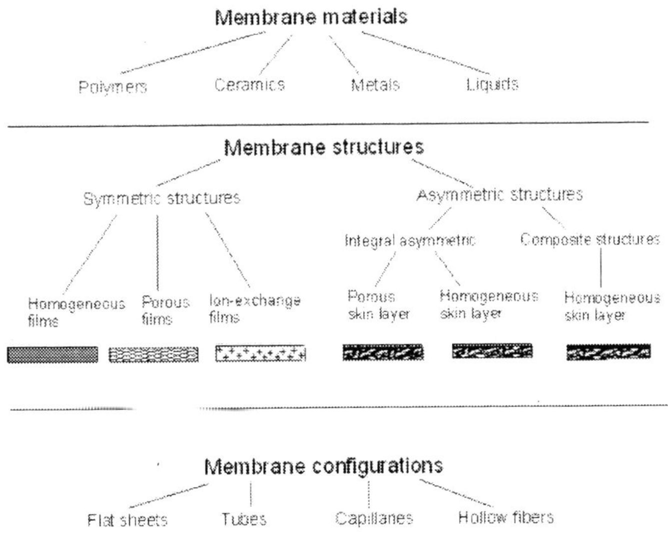

Figure 1. Schematic diagram illustrating the materials, the structures and the configurations of most of today's synthetic membranes

2.2. Membrane Modules

For the efficiency of a membrane process in a certain application the geometry of the membrane and the way it is installed in a suited device, i.e. the so-called membrane module, is also of importance. Widely used membrane modules are the pleated cartridges generally applied in dead-end filtration, the spiral-wound module, used mainly in reverse osmosis but also in gas separation and pervaporation and the capillary and hollow fiber module used mainly in hemodialysis and ultrafiltration and to a lesser extent in reverse osmosis and gas separation. Membrane modules with tubular and plate-and-frame devices are used for the treatment of industrial effluents or in chemical or biochemical production processes. For special applications, devices such as the rotating cylinder, etc., are used. The key properties of efficient membrane modules are:

- High packing density
- Good control of concentration polarization and membrane fouling
- Low operating and maintenance costs
- Low costs

Most of today's available modules fulfill one or more of the above listed requirements and all module types have found an area of application in which they provide the technical and economical best design. The costs of the membrane modules depend to a large extent on the market size. The costs of modules which have a large market such as the spiral-wound module or the capillary hemodialyser are an order of magnitude lower per installed membrane area than the costs of special plate- and-frame devices or tubular systems.

2.3. Membrane Processes and Their Application

Some of the membrane processes such as reverse osmosis, ultra- and microfiltration or electrodialysis are today state-of-the-art and utilized on a large industrial scale with a significant technical and commercial impact. Other processes such as membrane contactors, membrane emulsifiers or catalytic membrane reactors are still the subject of intensive research and development efforts. However, all membrane processes have one feature in common, i.e. they can perform the separation of certain molecular mixtures effectively, energy efficiently, and economically.

Even more diversified than the membrane structures and membrane processes are their applications. Presently the application of membranes is concentrated on four main areas, i.e. separation of molecular and particulate mixtures, the controlled release of active agents, membrane reactors and artificial organs, and energy storage and conversion systems.

In these applications a large variety of processes and membrane structures are used which are tailor-made for the specific application as indicated in table 1.

Table 1. Application of the state-of-the-art membranes

Separation of molecular and particulate mixtures	Controlled release of active agents	Membrane reactors and artificial organs	Energy conversion and energy production systems
Microfiltration and ultrafiltration	Therapeutic drug delivery systems	Catalytic membrane reactors	Fuel cell separators
Reverse osmosis	Pesticide depots	Bioreactors	Battery separators
Dialysis and electrodialysis	Fertilizer delivery systems	Diagnostic devices and sensors	Electrochemical synthesis
Gas separation and pervaporation	Air conditioning systems	Bio-hybrid and artificial organs	Pressure retarded osmosis
Membrane distillation and pertraction	Packing material and protective coatings	Integrated membrane system	Reversed electrodialysis

The applied driving forces to achieve a certain mass transport in membrane processes are gradients in the electrochemical potential of individual components induced by concentration gradients causing a diffusion of molecules or an electrical potential gradient resulting in a migration of ions and hydrostatic pressure differences causing a convective flux.

2.4. Historical Development of Membranes and Membrane Processes

Semipermeable membranes are not a recent invention. As part of the living organism, membranes have existed as long as life has existed on earth. Synthetic membranes and their technical utilization, however, are a more recent development which began with the studies of Nollet [2], who discovered that a pig's bladder preferentially passes ethanol when it was brought in contact on one side with a water-ethanol mixture and on the other side with pure water. Nollet was probably the first to recognize the relation between a semipermeable membrane, osmotic pressure, and volume flux. More systematic studies on mass transport in semipermeable membranes were carried out by Graham [3], who studied the diffusion of gases through different media and discovered that rubber exhibits different permeabilities to different gases. The theoretical treatment and much of the interpretation of osmotic phenomena and mass transport through membranes are based on the studies of Fick [4] and van't Hoff who gave a thermodynamic explanation for the osmotic pressure of dilute solutions [5]. With the introduction of the flux equation for electrolytes by Nernst and Planck [6] and the description of the membrane equilibrium and membrane potentials in the presence of electrolytes by Donnan [7], the early history of membrane science ends with most of the basic phenomena satisfactorily described and theoretically interpreted.

In the early membrane studies natural materials such as pig's bladders or rubber were used as selective barrier materials. In the first half of the last century, work was concentrated on the development of synthetic membranes with defined mass transport properties. Based on a patent of Zsigmondy, a series of nitrocellulose membranes with various pore sizes became commercially available [8]. These membranes were mainly a subject of scientific interest with only a very few practical applications in microbiological laboratories. This changed drastically from 1950 on when the practical use of membranes in technically relevant applications became the main focus of interest, and a significant membrane-based industry developed rapidly. Progress in polymer chemistry resulted in a large number of synthetic polymers which ultimately became available for the preparation of new membranes with specific transport properties plus excellent mechanical and thermal stability. Membrane transport properties were described by a comprehensive theory based on the thermodynamics of irreversible processes by Staverman and Kedem [9,10]. The development of the first successfully functioning hemodialyser by Kolff and Berk [11] was the key to the large-scale application of membranes in medicine.

A milestone in membrane science was the development of a reverse osmosis membrane based on cellulose acetate which provided high salt rejection and high fluxes at moderate hydrostatic pressures by Loeb and Sourirajan. This was a major advance toward the application of reverse osmosis membranes as an effective tool for the production of potable water from the sea. The membrane developed by Loeb and Sourirajan had an asymmetric structure with a dense skin at the surface which determined the membrane selectivity and flux and highly porous substructure providing its mechanical strength [12]. Soon other synthetic polymers such as polyamides, polyacrylonitrile, polysulphone, etc., were used as basic material for the preparation of synthetic membranes. However, cellulose acetate remained the dominant material for the preparation of reverse osmosis membranes until the development of the interfacially polymerized composite membrane by Cadotte became available [13]. These membranes showed significantly higher fluxes and rejection than the cellulose acetate membranes and better chemical and mechanical stability. Even earlier than the large-scale use of reverse osmosis for sea and brackish water desalination was the industrial-scale application of electrodialysis which became a practical reality with the work of Juda and McRae [14] who developed the first reliable ion-exchange membranes having both good electrolyte conductivity and ion permselectivity.

Parallel to the development of membranes was the development of appropriate membrane housing assemblies, called modules, in which the membrane could be installed for large-scale applications. The criteria for the design of such modules included high membrane packing density, reliability, ease of membrane or module replacement, control of concentration polarization, and low cost. A key development was the so called spiral-wound module to be used in reverse osmosis and the hollow fiber membrane module mainly applied in hemodialysis. In addition to reverse osmosis and ultra- and microfiltration a number of other membrane processes such as pervaporation, vapour separation, membrane distillation, continuous electrodeionization were commercialized during the last two decades of the last century [15].

3. Assessment of Today's State of Membrane Science and Technology

In the early stage of membrane science and technology, main effort was concentrated on the understanding of membrane transport properties and the preparation of membrane structures with tailor-made properties. During the second half of the last century a large number of membranes with highly improved fluxes and separation properties were developed and their applications studied extensively in the laboratory. The large-scale utilization of membrane processes and the development of a significant membrane-based industry began during the last 25 years of the 20th century. During this time, the expectations were extremely high as far as the technical and commercial relevance of membranes is concerned. Today, some of the early enthusiasm is replaced by a more realistic view. Nevertheless, some of the expectations have come true and others still might come true in the future. However, in many applications the performance of membranes have been disappointing and dreams have turned into nightmares.

3.1. Membrane Technology Dreams Come True

One of the many anticipated membrane applications that has indeed met the optimistic expectations is reverse osmosis sea water desalination. In the middle of the last century it became increasingly difficult in many areas of the world to supply a rapid growing population with potable water of sufficient quality. Sea water desalination seemed to be the only feasible solution to the problem and reverse osmosis was considered to be the most energy efficient and economic process for this task. Under the slogan "let's make the deserts green" a large research and development program was started all over the world, and 50 years later the dream came true. Today, reverse osmosis sea water desalination is providing potable water at affordable costs for millions of people. The road to success in reverse osmosis sea water desalination is illustrated in Figures 2 a) and b) which show the reduction of reverse osmosis sea water desalination costs and the improvement in membrane properties achieved during the last twenty years. When in 1980 the first successful sea water desalination plants were installed, the production costs of potable water were several US \$ per m^3. The membranes used in these plants were based on cellulose acetate. They had product water fluxes between 0.3 to 0.6 m^3 per m^2 and day when operated at a pressure between 60 and 80 bars with sea water. The cellulose-acetate-based membranes had a relatively low water flux and poor chemical and thermal stability. Around 1990 the thin film composite membranes became available which had a much higher water flux of 0.5 to 0.7 m^3 per m^2 and day, and a salt retention well above 99.3% which was considered necessary for single stage reverse osmosis sea water desalination. But not only did the properties of the membranes improve significantly over the years, their production cost also decreased drastically. With further improved properties, lower membrane costs, longer membrane life and by recovering the energy of the compressed brine with a turbine, the total costs of reverse osmosis sea water desalination decreased to 0.6 to 0.8 US \$ per m^3 in the year 2006.

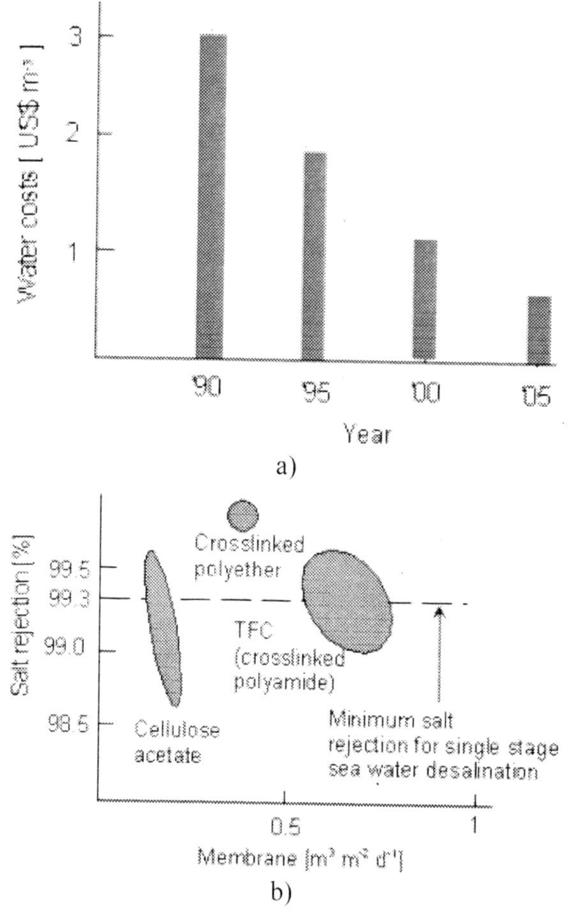

Figure 2. Schematic diagram showing a) the decrease in reverse osmosis sea water desalination costs between 1990 and 2005 from more than 2 to less than 0.7 US $ per m^3 of potable water, and b) the salt retention as a function of the flux of various membranes with sea water at an applied pressure of 60 bar

While sea water desalination is today the biggest single application for reverse osmosis it is not the only one. Reverse osmosis is also widely used for the production of high quality water to be used in the electronic and metal surface treating industry, in biotechnology, in analytical and medical laboratories and in surface- and wastewater treatment. Reverse osmosis is today the most important water treatment process and a real success story in the utilization of membrane technology.

Another area in which membranes have become an indispensable tool is hemodialysis. More than 1.4 million people suffering from acute and chronic renal failure must be treated three times per week by a so-called artificial kidney which contains as a key component a membrane through which certain toxic low molecular weight components such as urea are removed extra-corporally from a bloodstream by dialysis and filtration. One of the first recorded treatments of renal failure with an artificial kidney, which the patient survived, took place in the Netherlands in the winter of 1945. A photo of the kidney machine used in this treatment by Kolff and Berk is shown in Figure 3 together with a modern hemodialyser illustrating impressively the progress made over the years in hemodialysis. The artificial kidney developed by Kolff and Berk consisted of a cellophane tube wrapped around a wooden rack which was then rotated in an open vessel containing the stripping solution. Blood from the patient was pumped through the tubing and toxic components were removed by dialysis. The total effective surface area of the membrane facing the blood stream was ca. 1 m^2 and the volume of the extra-corporal blood required to fill the dialysis tubing and pumping system was ca. 4 to 5 liters. The treatment time was about 7 hours. The first successfully applied artificial kidney was still a crude apparatus. Due to the large required extra-corporal blood volume, the low clearance and low flux of the tubular membrane, the treatment was a strenuous affair for a patient. The development of a new hollow fiber or capillary membranes based on polyacrylonitrile and polysulphone with significantly improved flux and clearance during the following years led to a very compact and efficient device which also has a surface area of ca 1 m^2, but it requires an extra-corporal blood volume of only 0.2 to 0.3 liters and the treatment time is reduced to ca. 3 hours. Today, the artificial kidney provides for more than 1.5 million patients a bearable life with an additional life expectancy of more than 15 years. Hemodialysers based on polysulphone can be reused several times which is done in the USA and several Asian countries, while in Europe and Japan reuse is, for safety reasons, quite rare. In total, more than 180 million kidneys are used per year. This makes hemodialysis by far the largest single market for synthetic membranes which is well in excess of 3×10^9 US$ per year. The artificial kidney is certainly one of the most successful membrane applications where today a dream has come true not only for the membrane-based industry but more so for patients suffering from renal failure.

Reverse osmosis desalination of sea water and hemodialysis are today the most prominent single applications of membranes. However, there are many more membranes and membrane processes such as electrodialysis and micro- and ultrafiltration which have been very successful in serving smaller and more diversified markets. These processes have replaced conventional techniques such as distillation, ion-exchange, freeze-drying and adsorption in many applications because they are very often more efficient, yield higher product quality and have a smaller environmental impact because there are no toxic or hazardous by-products in membrane separation processes [16].

The main application of electrodialysis is the demineralization of brackish water. A completely different use of electrodialysis was envisioned in Japan. Here, electrodialysis is used to concentrate sodium chloride from seawater for the production of table salts. More recently electrodialysis is also used for treating heavy metal containing industrial effluents and in the production of boiler feed water from surface and well water [17]. The chlorine/alkaline electrolysis with fluorine-carbon polymer based ion-exchange membranes is a well established technique today. However, the market for electrodialysis is rather small and served by a few companies.

Controlled release of active components through flux-controlling membranes is used today in chemotherapeutic devices, and pesticide and pheromone applications. However, the impact on membranes in these applications has been rather low [18]. The same is true for membranes used in sensors and diagnostic devices where the market for membrane products is very small compared to applications in water treatment and hemodialysis.

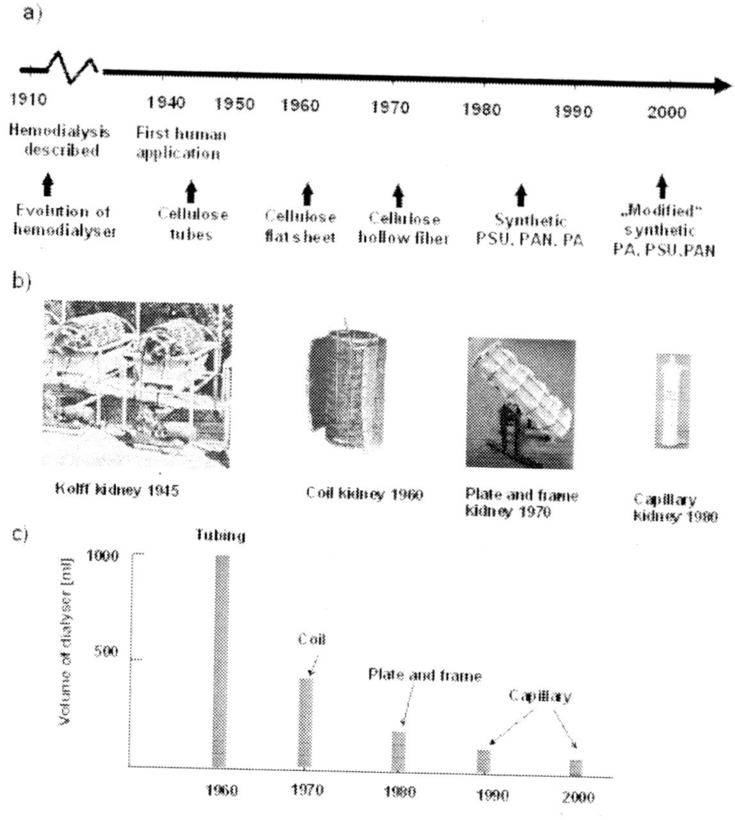

Figure 3. Schematic diagram showing a) the historical development of the artificial kidney, b) various dialysers used between 1945 and today, and c) the volume of different dialyser concepts

3.2. Emerging Membrane Processes and Their Applications

Membrane processes and applications which have been evaluated extensively on a laboratory and a pilot plant scale but have had little commercial impact so far, are gas and vapor separation, pervaporation, membrane contactors, membranes in fuel cells, etc. But due to the development of better membranes, increasing energy costs, declining raw material resources and increasing environmental awareness, these processes will most likely be of increasing importance in the future.

In the separation of gas and vapor mixtures, membranes have met the high expectations only in a limited number of applications. In the recovery of hydrogen from off-gas of ammonia production plants or the enrichment of nitrogen in air, membranes are now used commercially, but still on a small scale [19]. The same is true for the separation of CO_2 from natural gas or biogas and the recovery of organic solvents from air. De-gassing of water by membranes is another application which is well established today. But in the separation of gases and vapors, membrane technology is competing with cryogenic or adsorption processes and membrane processes are generally only competitive in a certain range of required product concentration and a given gas feed mixture composition. The production of oxygen and nitrogen from air is a typical example for the probabilities and limitations of gas separation membranes. Both nitrogen and oxygen are needed in large quantities in various industrial applications and produced mainly by cryogenic distillation from air, and membrane separation would be an interesting alternative. However, the separation factor of today's available membranes for O_2/N_2 is between 2 and 10 depending on the fluxes. To produce oxygen of high purity from air with these membranes, however, is not economic. This is illustrated in Figure 4 a) which shows the oxygen concentration in the permeate using air as a feed stream for different membrane selectivity assuming infinite high pressure ratio and counter current flow of feed and permeate. The maximum oxygen concentration in air that can be achieved by membrane separation in a single stage is 30 to 70 % depending on the selectivity of the membrane, the stage-cut and the ratio of feed to permeate pressure. The market for oxygen enriched air of this quality, however, is very limited. To produce higher oxygen concentration requires a multi-stage process since oxygen is the preferentially permeating component in the separation of air by membranes. A large number of stages require large membrane areas and additional energy which then makes the process uneconomic for the production of oxygen enriched air with high oxygen content. Since nitrogen is enriched in the retentate of the membrane, the situation is better. Even with membranes having a relatively low selectivity, nitrogen enriched air of ca 98 to 99 % nitrogen can be obtained economically. This is illustrated in Figure 4 b) in which the nitrogen recovered from the air feed stream is shown as a function of the nitrogen concentration in the retentate for different membrane O_2/N_2 selectivity assuming counter-current flow of feed and permeate and a high pressure ratio. The graph indicates that the nitrogen concentration in the retentate is inverse proportional to the nitrogen recovery rate and is increasing with the membrane selectivity. Higher recovery rates mean lower required membrane area and energy consumption, i.e. lower production costs.

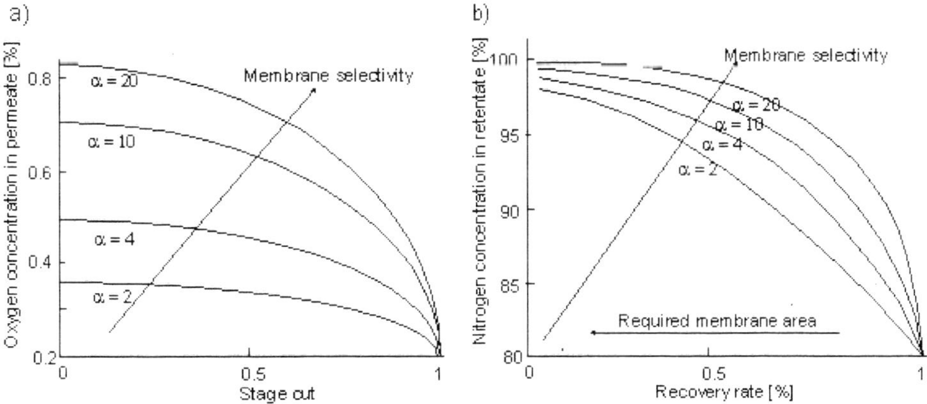

Figure 4. Schematic diagram showing a) the oxygen concentration in oxygen-enriched air as a function of the stage cut calculated for membranes with different O_2/N_2 selectivity, b) the nitrogen concentration in the retentate as a function of the nitrogen recovery rate calculated for membranes with different selectivity assuming counter-current flow and an infinitely high ratio of feed to permeate pressure

The graph of Figure 4 b) also indicates that with membranes with an O_2/N_2 selectivity of 4 to 6, nitrogen concentrations of 98 to 99 % nitrogen can be achieved with a recovery rate of 30 to 50%, which makes the membrane process competitive with other separation techniques.

Since there is a substantial market for nitrogen enriched air of this quality, membrane gas separation can be applied efficiently and economically and has met the expectations. In the production of oxygen-enriched air, the use of membranes has so far been disappointing and membranes with significantly higher O_2/N_2 selectivity are required.

The recovery of volatile organic components from waste air streams of the chemical and petrochemical industry is another potential application for membranes. Although the market is quite large, membrane processes have so far covered only a small portion of it due to the inefficiency of today's available membranes. The development of membranes with higher selectivity and better stability under operating condition and higher raw material costs could increase the importance of membrane technology in applications of gas and vapor separation drastically.

3.3. Membrane Application Dreams that Might Not Come True

Another group of membrane processes and applications that have been extensively studied in the laboratory are membrane distillation, liquid membranes with facilitated transport properties, affinity membranes, implanted membrane devices for immunoisolation of cell tissues producing drugs such as insulin or dopamine used in long-term therapy. Membrane distillation offers some interesting advantages in water desalination compared to reverse osmosis and conventional distillation. In membrane distillation, a salt solution is separated from pure water by a hydrophobic membrane which is permeable to water vapor but not to liquid water. The driving force for the transport of water across the membrane is the water vapour pressure difference between the salt solution and the pure water. This water vapour pressure difference is the result of a temperature difference between the two liquids. The water flux is affected very little by the osmotic pressure of the solution, and brine salt concentrations of more than 20 % salts can be obtained. The feed solution temperature can be kept relatively low and often waste heat can be used to establish the required temperature gradient across the membrane. The main problem is the water vapour flux through the membrane which is relatively low, i.e. ca. 0.2 to 0.3 m^3 per m^2 and day. Thus, a large membrane area is required for a given capacity plant. Furthermore, due to organic trace components in the feed solution which act as a surfactant, the membrane may become permeable to aqueous solutions which will affect the performance and reduces the life of the membranes in practical applications. Affinity membranes having functional components such as certain ligands fixed to the membrane matrix have been developed and tested for various biomedical applications. So far, however, their large-scale practical application has been rather limited. The use of zeolites in membrane structures have shown very promising results in experimental laboratory studies showing extremely high separation factors for the separation of alcohol water/mixtures and for the separation of low molecular weight hydrocarbons, and a series of interesting applications in catalytic membrane reactors are identified. But difficulties in preparing defect-free zeolite membranes have so far limited their large-scale use.

3.4. Membrane Technology Dreams that Might Never Come True

There are several membrane processes such as piezodialysis, pressure-retarded osmosis, and reversed electrodialysis, that have been extensively studied in the laboratory and that are still under investigation. However, it is very questionable if these processes will ever reach the expected technical and commercial relevance. Piezodialysis, for example, can be applied to remove and concentrate salts from an electrolyte solution using a hydrostatic pressure as a driving force through a mosaic membrane, which consists of macroscopic domains containing positive and negative ions in a neutral polymer matrix. The process is technically feasible. However, there is presently no commercially interesting application for this process and also in the future it will be difficult to find a large-scale commercial application of piezodialysis.

The production of energy by mixing sea water with river water through a reverse osmosis membrane provides a clean and sustainable energy source and is certainly very interesting. However, the maximum amount of energy that can be recovered is the Gibbs' free energy of mixing fresh water and sea water which is ca. 0.7 kWh when $1m^3$ of fresh water is mixed with sea water having an osmotic pressure of 25 bars. The maximum driving force for the water transport through the membrane which separates the sea water from the fresh water is the osmotic pressure difference which is about 25 bars. With today's membranes this pressure would result in a flux of ca. $1m^3$ per m^2 membrane area and day. In praxis, however, the osmotic pressure difference would be significantly lower because of the dilution effect of the sea water while passing through the membrane module. Concentration polarization effects at the membrane surface will further reduce the effective osmotic pressure, and finally energy is needed to pump the sea and the fresh water through the membrane device. Therefore, it seems not unreasonable that the net energy obtained by mixing sea and fresh water by pressure retarded osmosis is in the range of 0.3 to 0.5 kWh per 1 m^2 of installed membrane area per day. Considering today's investment and operating costs without energy costs of ca. 0.3 to 0.4 US $ per $1m^3$ product water, it becomes evident that membrane fluxes must be significantly higher and membrane costs significantly lower for pressure-retarded osmosis to become economically feasible [19]. For mixing river and sea water by permeation through ion-exchange membranes, i.e. by reversing electrodialysis, the situation is similar. Also in this case, the costs of membrane must be significantly reduced and their ion flux increased to be competitive with other energy sources used today. [20]. If retarded osmosis and reversed electrodialysis will ever be competitive with other primary energy sources will depend very much on the development of better suited membranes and also on energy cost increase in the future.

4. The Membrane-Based Industry, Its Structure and Market Strategy

With the large-scale application of membranes and membrane processes, a membrane-based industry developed. The structure of this industry is very heterogeneous as far as the size of the companies and their basic concept towards the market is concerned [21]. The membrane market is characterized by a few rather large market segments such as sea and brackish water desalination or hemodialysis and a large number of small market segments in the food and chemical industries, in analytical laboratories, and in the treatment of industrial waste water streams. The larger markets are dominated by a relatively small number of large companies. A multitude of small companies are active in market niches providing service to industrial sectors. The structure of the membrane industry is illustrated in the schematic diagram of figure 5 which shows a selected number of membrane processes and the different technology levels which can be distinguished during the development of a membrane process from basic research to the final application.

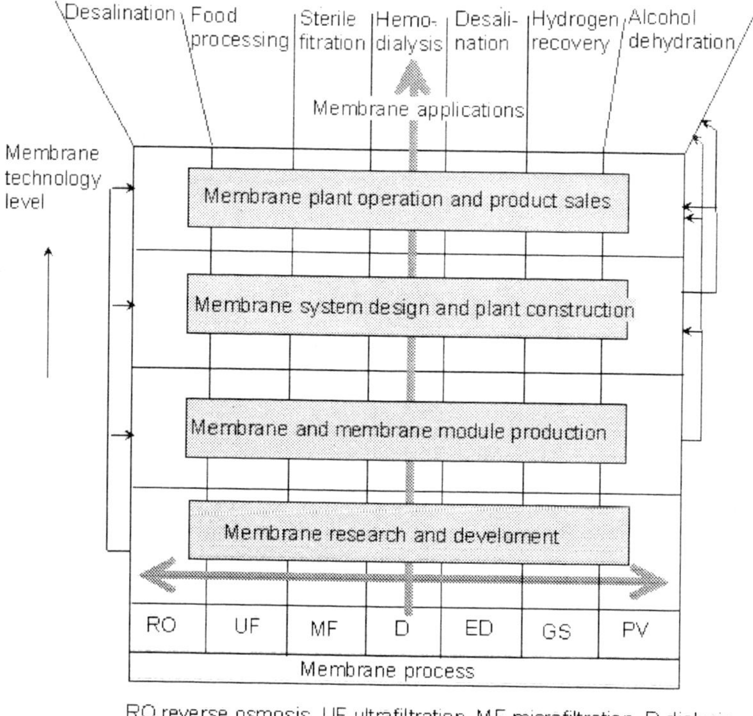

Figure 5. Schematic diagram illustrating the structure of the membrane-based industry indicating the various levels of technology for various membrane processes from research and development to applications and the different market strategies of companies

Research and development is the basis of all further development and therefore indicated as the first technology level. The next level is the membrane and membrane module production followed by membrane system design and plant construction. The last level is the membrane plant operation and sales which then go into various applications. Some companies have concentrated on one technology level such as membrane research and development. This is the case of many public or private research institutions. They supply products, processes and knowledge to the next level of technology which is concentrating on the production of membranes and membrane modules. Membrane producers are often divisions of major chemical companies or medium-size enterprises. They usually produce a series of different membranes and modules such as flat sheets, hollow fibers or capillaries for various applications from reverse osmosis, ultrafiltration and dialysis. In the third technology level, mostly small or medium size companies are active. These companies buy membranes or membrane modules from one or several membrane manufacturers and design and construct plants for various applications. They often concentrate on certain areas of application where they have a specific know-how such as water purification, gas separation, or food and drug processing. They generally provide a solution to a customer's separation problem which might result in a combination of separation processes such as ion-exchange, carbon adsorption, or chemical and biological treatment procedures in addition to various membrane processes. Several of these companies do not only build the plant but also own and operate it, and guarantee the customer a certain product of a given quality for a fixed price. Their membrane sales are often not very large since these companies usually concentrate on small market niches. However, they are an important link in the membrane industry because of their specific application know-how in different fields and markets.

There are also companies which follow the vertical line shown in the diagram of Figure 5. These companies concentrate on one process. They have their own research activities, membrane and module production and plant design and operation, and focus on one large market which is big enough to generate the adequate profit to support a relatively large organization involved in membrane research, manufacturing and marketing. Typical membrane applications with a large market are sea water desalination or hemodialysis. The total annual sales in these markets are several billion US $. On the water market several companies have adapted the so-called BOOT (build, own, operate, transport) strategy. They no longer sell membranes or membrane plants but water which they supply to the end-user. Hemodialysis companies not only build and sell artificial kidneys today but also operate dialysis stations in hospitals.

5. Research Needs in Membrane Science and Technology

Many of the basic aspects of membrane structures and mass transport properties were extensively studied during the last century. But there are still many open questions concerning the basic understanding of the relation between membrane material and membrane transport properties as well as the interactions between membranes and permeating components and their effect on membrane transport properties, etc.

But there is also a substantial need for applied research to improve membrane properties and to develop tailor-made membranes and membrane processes for solving certain separation problems more efficiently and economically.

5.1. Basic Research Needs in Membrane Science

Due to the availability of increasingly powerful computers, simulation of mass transport and chemical reactions in homogeneous or heterogeneous structures has become possible.

Molecular dynamic simulations can provide valuable information about the solubility and diffusivity of molecular components in polymers or glass transition temperature of polymers. This information is very helpful in selecting polymers for the preparation of membranes with specific, problem-oriented transport properties. The phenomena of concentration polarization and membrane fouling need to be further elucidated by developing mathematical models to describe the mass transport at the membrane surface and the adsorption of components by the membrane as a function of various parameters such as the chemical nature of the components, their size and shape, and the hydrodynamic situation at the membrane feed side surface. Mathematical modeling of membrane processes in different applications to be used for designing and constructing membrane plants is needed. The phenomena of the overlimiting current density and electroconvection which are observed in electrodialysis and continuous electrodeionization need further elucidation.

5.2. Research Needs in Applied Membrane Processes

Although membranes have been available for more than 20 years as efficient and versatile tools in many applications in industry, energy conversion, and life science, it seems that membrane science is still in the beginning of utilizing the total potential that membranes offer considering the functions membranes fulfill in the living organism. To utilize functions observed in biological membranes also in synthetic membranes, a substantial research and development effort is needed. But also the state-of-the-art membranes and membrane processes must be further improved in terms of their efficiency, and their introduction into new areas of application must be evaluated. This also requires a substantial commitment to membrane related research.

Research Needs in the-State-of-the-Art Membranes

In reverse osmosis for example, the membrane performance has reached a level where only marginal improvements can be expected as far as selectivity and flux is concerned. One research target is the development of membranes with better temperature stability which can be steam sterilized for applications in the food and drug industry. The chemical stability especially towards oxidizing agents and high or low pH-values is another research target that needs to be addressed. An even more severe problem in reverse osmosis and most other membrane separation processes is membrane fouling due to adsorption of organic compounds or precipitation of salts with low solubility. In sea water desalination by reverse osmosis, costs for the pretreatment of the feed water to avoid membrane scaling and fouling is a substantial fraction of the total water production costs. More efficient and less costly feed water pretreatment can decrease the water production cost significantly. The development of asymmetric microfiltration membranes with a minimum of internal fouling would be desirable to obtain a long useful life of the membranes under operating conditions. Since the practical application of membrane processes is extremely diversified, there is often a serious lack of application know-how and long-term experience. This requires a close cooperation between the membrane scientists and experts in the field of application such as food engineers, microbiologists, and medical doctors.

Research Needs in Emerging Membrane Processes and Their Applications

The research needs in the developing processes such as gas separation, pervaporation, bipolar membrane electrodialysis and especially catalytic membrane reactors and membrane contactors involve the optimization of processes in their various applications which requires substantial application know-how. Process intensification by integration of membrane processes in conventional production processes to increase efficiency and economics as well as reducing production hazard by-products is another area requiring substantial research and development efforts. The application of existing and to be newly developed membranes in regenerative and diagnostic medicine does not only provide a significant future market for the membrane industry but are also an important tool for a better and more efficient health care system. But further research is needed in this area. Membranes used today in energy storage and energy conversion systems such as batteries and fuel cells are far from being perfect and need to be improved in terms of the overall properties. Pressure-retarded osmosis and reverse electrodialysis do not look very promising when they are applied to mixing sea and surface water. The situation, however, might change in the future when more efficient membranes become available and when highly concentrated brines from industrial production processes can be mixed with surface water. Thus, under certain conditions reverse electrodialysis especially might be an option for the production of "clean" energy.

6. The Future of Membrane Science and Technology

The development of membrane science and technology is quite different in established processes, developing processes and to be developed processes.

In the established processes, progress will be marginal. Polymer membranes with higher fluxes, higher selectivity and better chemical and thermal stability for reverse osmosis and ultra- and microfiltration will probably be developed in the future. Ceramic membranes with the same transport properties as polymer membranes will most likely be commercially available at competitive costs in the very near future. Membrane modules with improved control of concentration polarization and membrane fouling at lower energy input will probably be developed using new spacer and turbulent promoter concepts. Ion-exchange membranes with lower electric resistance and better permselectivity will become available at significantly lower production costs. As far as the different processes are concerned, reverse osmosis will probably more extensively be applied in the food and drug industry and in the treatment of industrial effluents and even municipal waste water. Furthermore, reverse osmosis will most likely also be used in the future for separating organic compounds in organic solvents. Ultrafiltration has found a multitude of applications in the treatment of industrial effluents, and it is used in the food and chemical industry as well as in biotechnology to separate, concentrate, and purify molecular mixtures. These applications will certainly be extended into new areas such as the treatment of municipal drinking water obtained from surface waters. The use of membrane bioreactors treating industrial and municipal waste water will probably increase significantly due to decreasing fresh water sources. Finally, ultrafiltration will be used in processing organic solvents when better solvent resistant membranes become available. Microfiltration is presently used mainly as a depth filter to remove particles or bacteria and viruses from water used in the pharmaceutical and electronic industry and in analytical laboratories and various medical applications. These applications are growing steadily and are extended into new areas. In the future, cross-flow microfiltration with asymmetrically structured membranes which would have a significantly longer useful life under operating conditions than a depth filter will also be used to treat surface water and effluents that have much higher loads of solid materials. A very important role in the future of microfiltration will be played by the ceramic membranes. Because of their good mechanical and chemical stability, they can easily be cleaned by back-flushing and aggressive chemicals such as strong acids or bases or oxidizing media. Production costs of these membranes will most likely be reduced when the market for their application is increasing. The growth of electrodialysis in brackish water desalination will only be marginal in the future due to increasing competition of nanofiltration. However, its relevance in the treatment of industrial effluents generated by galvanic processes or in the production of pulp and paper as well as its application in the chemical and food processing industry and in biotechnology will most likely increase. Thus, in the state-of-the-art membrane processes, a steady growth can be expected in the near and midterm future.

With the developing processes the situation is quite different. Some processes such as gas and vapor separation and the use of membrane contactors and catalytic membrane reactors will most likely grow significantly in the years to come. Especially in the treatment of natural gas and in the petrochemical industry, membranes have large opportunities when a structure with the required properties eventually based on ceramic materials becomes available. The market for catalytic membrane reactors is very large but also very heterogeneous and requires in addition to the proper membranes a significant amount of specific application know-how. The same is true for membrane contactors and integrated membrane processes [21, 22]. Here process development is a key factor for success. For other processes such as facilitated transport with fixed and mobile carriers and the production of acids and bases by electrodialysis with bipolar membranes, the future development will be more difficult to estimate, since no real large markets for these processes have been developed so far because of the shortcoming of today's available membranes.

In addition to the membranes and the membrane applications, that are available or under development today, completely new membranes and a large number of new possible applications will be identified in the future. Considering the versatility of the biological membranes which not only have very energy efficient and specific mass transport properties but which are also capable of information transfer in living organisms, today's membrane dreams are even more ambitious than those 50 years ago. Today's membrane dreams include more efficient production processes in the chemical industry, better use of our natural resources, the development of artificial organs and controlled drug delivery systems due to membranes and membrane processes. The list of dreams is long and some may come true in the near future, some will take a little longer and some might not come true at all.

Nevertheless, membrane technology is an important tool for sustainable industrial development, and its impact will most likely increase further in the future with diminishing natural resources and increasing energy costs.

References

[1] H. Strathmann, In: M.K. Turner (Ed.), *Effective Industrial Membrane Processes: Benefits and Oppertunities*, Elsevier Applied Science, London, 1991.

[2] J.A. Nollet, Recherches sur les Causes du Bouillonnement des Liquides, Histoire de l'Academie Royale des Sciences, Paris Annee MDCCXLVIII, 57, 1752.

[3] T. Graham, On the absorption and dialytic separation of gases by colloid septa, *Philos. Mag.* 32 (1866) 401-420.

[4] A. Fick, Über Diffusion, Poggendorff's Annalen der Physik und Chemie. 94 (1855) 59-81.

[5] J. H. van't Hoff, Die Rolle des osmotischen Druckes in der Analogie zwischen Lösungen und Gasen, *Z. Phys. Chem.* 1(1887) 481-508.

[6] M. Planck, Ueber die Potentialdifferenz zwischen zwei verduennten Loesungen binaerer Electrolyte, *Ann. Phys. Chem.* N.F. 40 (1890) 561-576.

[7] F.G. Donnan, Theorie der Membrangleichgewichte und Membranpotentiale bei Vorhandensein von nicht dialysierenden Elektrolyten, Z. für Elektrochemie und angewandte physikalische *Chemie.* 17 (1911) 572-581.

[8] R. Zsigmondy, US Patent 1421341, 1922.

[9] A.J. Staverman, Non-equilibrium thermodynamics of membrane processes, *Trans. Faraday Soc.* 48 (1952) 176-185.

[10] O. Kedem, A. Katchalsky, A physical interpretation of the phenomenological coefficients of membrane permeability, *J. Gen. Physiol.* 45 (1961) 143-179.

[11] W.J. Kolff, H.T. Berk, The artificial kidney: a dialyzer with great area, *Acta Med. Scand.* 117 (1944) 121-134.

[12] S. Loeb, S. Sourirajan, *US Patent* 3133132, 1964.

[13] J.E. Cadotte, R.I. Petersen, Thin film reverse osmosis membranes: origin, development, and recent advances, in: A.F. Turbak(Ed.), Synthetic Membranes, *ACS Symposium Series* 153, Vol. I, 1981, pp. 305 -325.

[14] W. Juda, W.A. McRae, Coherent ion-exchange gels and membranes, *J. Amer. Chem. Soc.* 72 (1950) 1044-1053. (Also: W. Juda, W.A. McRae, US Patents 2636851 and 2636852, 1953.)

[15] R.W. Baker, *Membrane Technology and Applications*, J. Wiley & Sons, Chichester, 2004.

[16] W.J. Koros, Evolving beyond the thermal age of separation processes: membranes can lead the way, *AIChE J.* 50 (2004) 2326-2334.

[17] H. Strathmann, *Ion-exchange membrane separation processes*, Elsevier, Amsterdam, 2004.

[18] R.W. Baker, Future directions of membrane gas separation technology, *Ind. Eng. Chem. Res.* 41 (2002) 1393-1411.

[19] S. Loeb, Energy production at the dead sea by pressure-retarded osmosis: challenge or chimera? *Desalination.* 120 (1998) 247-262.

[20] J.W. Post, J. Veerman, H.V.M. Hamelers, G.J.W. Euverink, S.J. Metz, K. Nymeijer, C.J.N. Buisman, Salinity–gradient power: evaluation of pressure-retarded osmosis and reverse electrodialysis, *J. Membr. Sci.* 288 (2007) 218-230.

[21] E. Drioli, M. Romano, Progress and new perspectives on integrated membrane operations for sustainable industrial growth, *Ind. Eng. Chem. Res.* 40 (2001) 1277-1300.

[22] E. Drioli, E. Curcio, Membrane engineering for process intensification: a perspective, *J. Chem. Technol. Biotechnol.* 82 (2007) 223-227.

In: Advances in Membrane Science and Technology
Editor: Tongwen Xu
ISBN: 978-1-60692-883-7

Chapter II

Preparation and Application of Ion Exchange Membranes: Current Status and Perspective

Tongwen Xu *

Laboratory of Functional Membranes,
School of Chemistry and Materials Science,
University of Science and Technology of China, Hefei,
Anhui 230026, People's Republic of China

Abstract

During the last 50 years, ion exchange membranes have evolved from laboratory tools to industrial products with significant technical and commercial impact. Their successful applications include desalination of sea and brackish water, treatment of industrial effluents, and concentration or separation of the food and pharmaceutical products containing ionic species. The membrane technology can not only make the process cleaner and more energy-efficient but also recover resources from waste discharges and thus make the development of society sustainable.

Therefore, this chapter treats of the development of ion exchange membranes' preparation and application. The focuses are preparation of anionic, cationic, and bipolar membranes, and application using diffusional dialysis, conventional electrodialysis, and electrodialysis with bipolar membranes. Not only aqueous systems but also organic or aqueous-organic systems are discussed in the case of meeting specific needs. The most relevant literature is surveyed, and some case studies, including some research programs conducted in the author's laboratory, are presented.

Keywords: Ion exchange membranes; bipolar membrane; mosaic ion exchange membranes; hybrid ion exchange membrane; electrodialysis; diffusional dialysis; water dissociation; alcohol splitting.

---

* Correspondence concerning this chapter should be addressed to Tongwen Xu at twxu@ustc.edu.cn

1. Development of Ion Exchange Membranes and Related Processes

1.1. Classification of Ion Exchange Membranes

An ion exchange membrane is generally defined as an ion exchange resin in membrane shape. This definition leads to the classification analogous to that of ion exchange resins. For example, traditional ion exchange membranes are classified as anionic and cationic membranes according to the type of ionic groups attached to the membrane matrix. Cationic exchange membranes contain negatively charged groups, such as $-SO_3^-$, $-COO^-$, $-PO_3^{2-}$, $-PO_3H^-$, and $-C_6H_4O^-$, which are fixed to the membrane backbone and allow the passage of cations but reject anions. In contrast, anionic membranes contain positively charged groups, such as $-NH_3^+$, $-NRH_2^+$, $-NR_2H^+$, $-NR_3^+$, $-PR_3^+$, and $-SR_2^+$, which are fixed to the membrane backbone and allow the passage of anions but reject cations [1-2]. Notably, there are three additional types of ion exchange membranes: (1) amphoteric ion exchange membrane, one containing both negatively and positively fixed ionic groups but randomly distributed; (2) bipolar membrane or ion exchange composite membrane, one consisting of a cationic and an anionic layers which are laminated together; (3) mosaic ion exchange membrane, one composed of macroscopic domains of polymers with negatively fixed ions but with positively fixed ions randomly distributed in a neutral polymer matrix.

For an ion exchange membrane of any type, it at least consists of three parts: the neutral polymer matrix, the functional groups (also called fixed ion groups), and the moving ions (also called counter ions) attached therein. For example, a hydrocarbon cationic membrane as shown in Figure 1 often consists of polystyrene backbone, sulfuric acid groups, and moving cations (e.g., proton, and sodium ions).

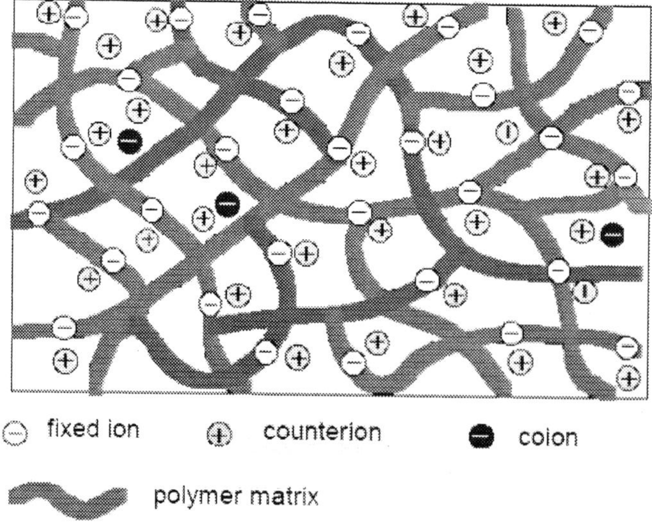

Figure 1. Schematic of the structure of a cationic membrane, taken from the reference [2].

The separation performances of ion exchange membranes rely on not only the type of above-mentioned fixed groups but also the membrane structure, i.e., the connection manner of fixed groups with the neutral polymer matrix. Accordingly, ion exchange membranes are further classified as homogenous and heterogeneous membranes, in which charged groups are chemically bonded to and physically mixed with the membrane matrix, respectively. Nonetheless, the completely homogeneous and the macroscopically heterogeneous ion exchange membranes are two extremes [2]. Most ion exchange membranes show a certain heterogenity on the microscopic scale. In particular, the heterogenity of ion-exchange membranes increases according to the following order [2-3]:

- homogeneous ion-exchange membranes,
- interpolymer membranes,
- microheterogeneous graft- and block-polymer membranes,
- snake-in-the-cage ion exchange membranes,
- heterogeneous ion-exchange membranes.

Besides the type and concentration of the fixed ionic moiety, the basic materials to prepare ion exchange membranes, i.e., the membrane matrix, determines to a large extent the mechanical, chemical, and thermal stability of the membranes [4]. These basic materials for the ion-exchange membranes are often classified as follows [2]:

- hydrocarbon or partially halogenated hydrocarbon polymers,
- perfluorocarbon polymers,
- inorganic materials,
- inorganic ion exchange material and an organic polymer,
- inorganic-organic hybrids.

Most of the commercial ion exchange membranes are rather homogenous and composed of either hydrocarbon or perfluorocarbon polymer films hosting the ionic groups [3]. However, recent studies show that the ion exchange membranes prepared from inorganic-organic hybrids possess high thermal stability and mechanical strength [5]. On some occasions, such membranes can be developed into free-standing ones and replace polymeric ion exchange membranes for application under some critical circumstances [6].

1.2. Development of Ion Exchange Membrane-Based Processes [2,7]

The earliest ion exchange membrane-based process started in 1890 with the work of Ostwald [8] who studied the properties of semipermeable membranes. He discovered that a membrane could be impermeable for any electrolyte if it is impermeable either for its cations or anions. To illustrate that, the so-called "membrane potential" at the boundary between a membrane and its surrounding solution was postulated as a consequence of the difference in concentration. In 1911, Donnan [9] confirmed the existence of such boundary and developed a mathematical equation describing the concentration equilibrium: the so-called "Donnan exclusion potential". However, the actual basic studies related to ion exchange membranes firstly begun in 1925 and were carried out by Michaelis and Fujita using the homogeneous, weak-acid collodium membranes [10]. In the 1930s, Sollner put forward the idea of a charge-mosaic membrane or amphoteric membrane containing both negatively and positively charged ion exchange groups and showed distinctive ion transport phenomena [11]. Around the 1940s, more interest in industrial applications led to the development of synthetic ion exchange membranes using phenol-formaldehyde-poly-condensation [12]. Simultaneously, Meyer and strauss proposed an electrodialysis process in which anionic and cationic membranes were arranged in alternating series to form many parallel solution compartments between two electrodes [13]. But a large industrial scale electrodialysis was not available until the 1950s when the stable, highly selective ion exchange membranes of low electric resistance were developed by Juda and McRae of Ionics Inc. in 1950 [14] and Winger et al at Rohm in 1953 [15]. From then on, the electrodialysis based on ion exchange membranes rapidly became an industrial process for demineralizing and concentrating electrolyte solutions. Afterwards, both ion exchange membranes and electrodialysis were greatly improved and used in many fields. For examples, in the 1960s, first production of salt from sea water was realized by Asahi Co. using monovalent ion permselective membranes [16]; in 1969, the invention of electrodiaylsis reversal (EDR) realized a long-term run without salt precipitation or deposition on membranes or electrodes [17]; in the 1970s, a chemically stable cationic membrane based on sulfonated polytetra-fluorethylene was first developed by Dupont as Nafion®, leading to a large scale use of this membrane in chlor-alkali industries and energy storage or conversion systems (fuel cells) [18]; in 1976, Chlanda et al [19] combined cationic and anionic layers into a bipolar membrane, which brought many novelties in electrodialysis applications for cleaner separation and production [20]. Also stimulated by the development of new ion-exchange membranes with higher selectivity, lower electrical resistance, and improved thermal, chemical, and mechanical properties, other applications of ion exchange membranes, apart from desalination of brackish water, have recently extended to food, drug, and chemicals processing and waste water treatment [20-27].

The combination of electrodialysis with conventional ion exchange technology and the use of conducting spacers, which was suggested in the 1970s [28-29], is the first integration process for ion exchange membranes. The process is now referred to as continuous electrodeionization (EDI) and is commercialized on a large scale for production of ultra-pure deionized water at the end of last century. Since then, the integration of electrolysis with chemical unit operations or other membrane units have received special attentions. These integrated processes are classfied as follows:

- The integration of electrodialysis with traditional unit operations, such as adsorption, extraction, complexation, striping, absorption, and distillation, which is widely used in chemical separation and environmental protection [21,27,30-31].
- The integration of electrodialysis with pressure-driven membrane processes, such as microfiltration, untrafiltration, nanofiltration, and reverse osmosis, which is often used in water treatment, feed pretreatment, fraction of biological species, concentration and desalination of sea water, etc. [27, 32-42]

- The integration of electrodialysis with biological unit operations, such as fermentation, and membrane biological reactors, which is widely used in separation of biochemicals, conversion of organic acid from the fermentation broth, and treatment of waste water, etc.[43-44]

The integration allows diverse applications and demonstrates the respective technological advantage of electrodialysis and chemical or other membrane units. A detail description of some integration processes can be found in Chapter 7 of this book.

In most cases, electro-membrane processes, specifically electrodialysis, have been mainly applied in aqueous media. The use of electro-membrane processes, more generally electrochemical applications of membranes, in aqueous-organic media or pure organic media seems to be a new research field for development of ion exchange membranes (IEMs) [45]. Although the use of membranes in organic media is considerably restricted as compared to the aqueous solution, the solvent resistance of membrane materials allows operating membrane processes in organic solvents. In the 1990s, the application of ED processes in this domain induced some basic research on ionic behaviors across charged membranes, such as membrane potential, ionic mobility, and electrical resistance[46-47].. The application of electrodialysis in aqueous-organic media achieved a success for demineralization of salt solutions containing a high amount of solvent (40–60 vol.%), and conversion of linear organic acid salts ($C_nH_{2n+1}COONa$, with $n = 7–15$) into their corresponding acids by using organic chemicals (such as alcohols) to increase the solubility [45,48]. An earlier report of electrodialysis in pure organic media is the methanol dissociation with bipolar membranes to produce corresponding alcoholates in the late 1990s by Shridhar et al [49-50]. A few years later, a conventional electrodialysis was used in organic media for conversion of organic salts into organic acids, which have a very low solubility in water [45].

Just the same as the development of electro-membrane processes, materials used to prepare ion exchange membranes were also diversified out of the scope of organic polymers. Since the 1960s, inorganic materials, such as zeolite, betonite, or phosphate salts, have been employed for membrane preparation [51-53]. These membranes, however, are rather unimportant due to their high cost and other disadvantages such as unideal electrochemical properties and too large pores, though they can endure higher temperatures than organic membranes [54]. It can be expected that ion exchange membranes prepared from polymers can possess both chemical stability and excellent conductivity if the membranes are incorporated into inorganic components, such as silica. So, inorganic-organic ion exchange membranes were developed in the late 1990s by sol-gel method for application in such severe conditions as higher temperature and strongly oxidizing circumstances [54-57]. Combining the advantages of inorganic and organic membranes, these hybrid membranes deserve special attentions [57], so an extension of such domain is included as Chapter 9 of this book. To date, various ion exchange membranes, including inorganic-organic (hybrid) ion exchange membranes, amphoteric ion exchange membranes, mosaic ion exchange membranes, and bipolar membranes (ion exchange composite membranes), are available for application in aqueous, aqueous-organic, and orgainic media. As a summary, Figure 2 demonstrates the chronology of ion exchange membranes development and the relevant, important events

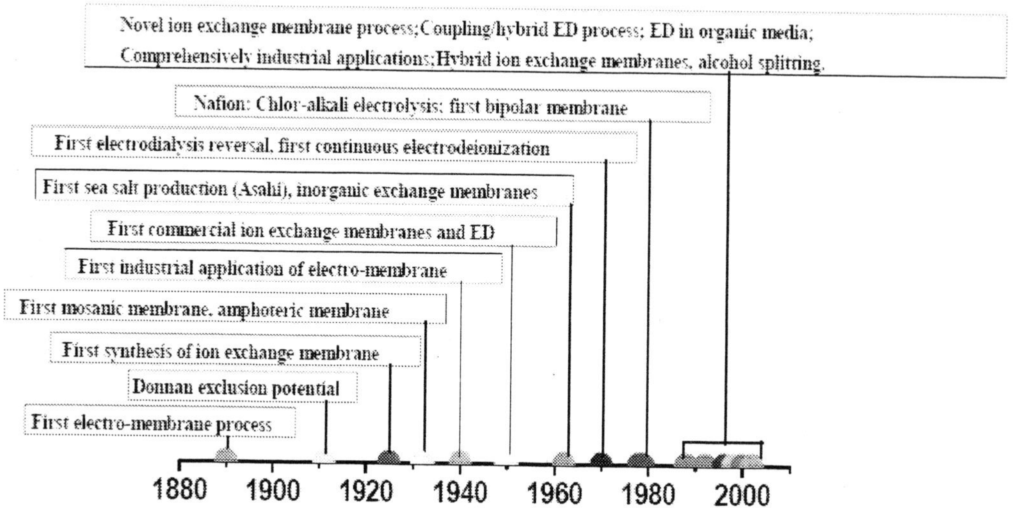

Figure 2. Development of ion exchange membranes and their related processes [7].

2. Preparation of Ion Exchange Membranes

Ion exchange membranes are key components in the afore-mentioned electro-membrane processes. Their properties determine to a large extent the technical feasibility and the process economics. For a practical ion exchange membrane, it should meet the criteria as follows [2]:

- high permselectivity--an ion exchange membrane should be highly permeable for counter-ions but impermeable to co-ions,
- low electrical resistance--the permeability of an ion exchange membrane for the counter-ions under an electrical potential gradient should be as high as possible,
- good mechanical and dimension stability - the membrane should be mechanically strong and should have a low degree of swelling or shrinking in the transition from dilute to concentrated ionic solutions,
- high chemical stability--the membrane should be stable over the entire pH-range and in the presence of oxidizing agents.

The preparation of an ion exchange membrane is very similar to that of normal ion-exchange resins: preparation of the membrane matrix and introduction of ionic groups. In general, the method to introduce ionic groups can be classified as three basic categories according to the starting materials. [2,7]

a. Starting with a monomer containing a moiety that is or can be made anionic or cationic, followed by copolymerization with non-functionalized monomer to eventually form an ion exchange membrane;

b. Starting with a polymer film which can be modified to have ionic groups, either directly by grafting of a functional monomer or indirectly by grafting non-functional monomer followed by functionalizing reaction.
c. Starting with polymers or polymer blends which have anionic or cationic moieties, followed by polymer dissolving and casting.

During the last 50 years, a significant amount of work conducted in academic institutions and industrial research centers has been concentrated to develop efficient membranes with high permeability, good permselectivity for certain ionic components, and a long usage period under operating conditions. These have been summarized as Figure 3 in the categories of cationic, anionic, and bipolar membrane preparation. These three kinds of ion exchange membranes were selected as examples because they are representatives in most applications using electro-membrane processes, such as diffusion dialysis, electrodialysis, and the water/alcohol splitting. Their main preparation procedure as well as the examples will be described in the following sections.

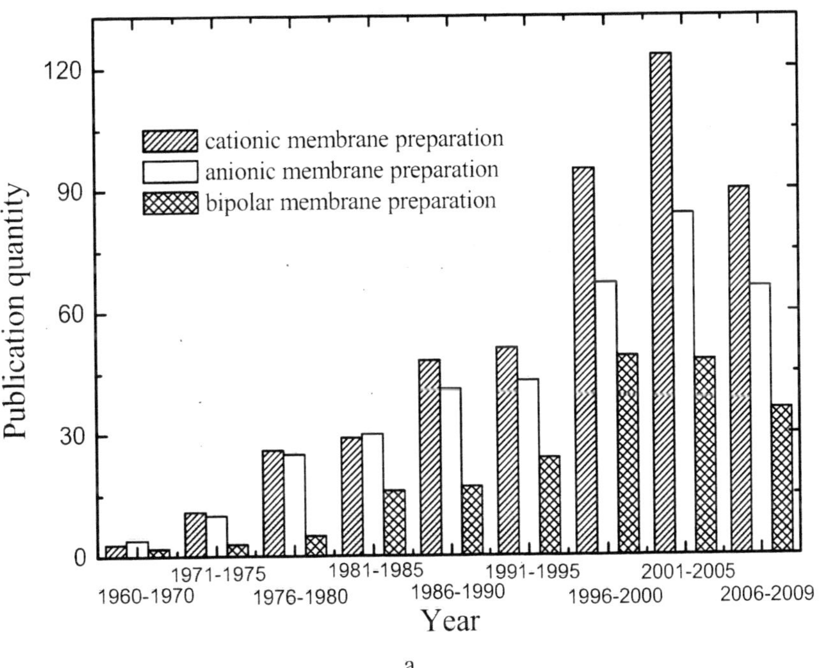

a

2.1. Cationic Membranes

Following the general procedure and concept of ion exchange membranes, cationic membranes, including proton conductive membranes, have been prepared by the following methods [2, 58].

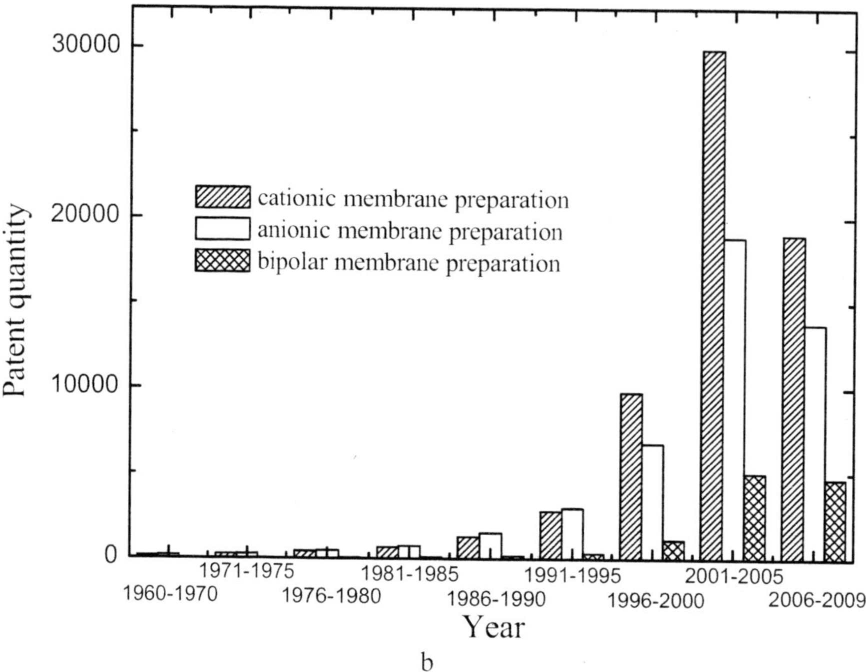

Figure 3. Chronology of ion exchange preparation documents: (a) publications; (b) patents. Source: www.scopus.com [Search settings: TITLE-ABS-KEY(cationic membrane/anionic membrane/bipolar membrane preparation). Search date: Jan 31, 2009].

2.1.1. Direct Polymerization of Monomers

This is a conventional method of preparing ion exchange membranes [20]. The membranes are prepared directly by polymerization of possible monomers, such as St and DVB, and subsequent sulphonation. For example, traditional polystyrene membranes were prepared from styrene and divinylbenzene according to the following reaction scheme [2]:

$H_2C{=}CH$–C$_6$H$_5$ + $H_2C{=}CH$–C$_6$H$_4$–$CH{=}CH_2$ —(polymerization, sulfonation)→ cross-linked polymer with –CH–CH_2–CH– backbone and SO_3H groups on the phenyl rings

In the first step, styrene is partially polymerized with divinylbenzene at about 60°C, using benzoyl peroxide as an initiator for the polymerization. The polymer is partially cross-linked and obtained as a block and then is cut into slices before sulfonation in the second step using concentrated sulfuric acid at room temperature. The obtained membranes show high ion-exchange capacity and low electrical resistance, but their mechanical strength is insufficient to be used without a proper support material.

The monomer polymerization has been developed into many novel procedures for cationic membranes. One of them is directly polymerization of a monomer that contains a cationic moiety with a neutral polymer,

$$\left[\underset{R^{+,-}}{\overset{|}{M}} \right]_n + m\,M \xrightarrow{\text{polymerization}} \left[\left[\underset{R^{+,-}}{\overset{|}{M}} \right]_n \left[M \right]_m \right]_k$$

or preparation from block polymers or polymer mixtures of a neutral polymer with the polymer containing cationic moieties:

$$\left[\underset{R^{+,-}}{\overset{|}{M}} \right]_n + \left[M \right]_m \xrightarrow{\text{common solvent}} \left[\left[\underset{R^{+,-}}{\overset{|}{M}} \right]_n \left[M \right]_m \right]_k$$

In this way, the subsequent sulphonation can be cancelled. Furthermore, by co-polymerization of a monomer containing negative charge with a neutral monomer the density of fixed charges can be easily controlled by adjusting the ratio of two monomers.

In addition, polymerization of monomers often takes place in an inert matrix during monomer soaking or pore filling, as discussed below.

2.1.2. Pore Filling/Soaking/Grafting Method [59-70]

In this method, monomers soak, fill, or graft onto a porous matrix, followed by the same polymerization and sulphnonation. A great deal of work has been dedicated to preparing ion exchange membranes in this way. Generally, these porous matrixes are insoluble in any solvents, such as polymer films of hydrocarbon (PE and PP) or fluorocarbon origin (PTFE, FEP, PFA, ETFE, and PVDF). In such route, two functional acidic groups were mainly identified as fixed ionic groups that confer the membrane its cationic character, namely, carboxylic (weakly acidic) and sulfonic acid (strongly acidic) groups. The former can be prepared by (1) direct grafting/filling/soaking of acrylic monomers like acrylic, methacrylic acids, and their mixtures with acrylonitrile and vinylacetate, or (2) grafting of epoxy acrylate monomers(such as glycidyl acrylate or glycidyl methacrylate) onto polymer films, followed by conversion of the epoxy group into carboxylic group (such as iminodiacetate groups) using post-grafting ring-opening reaction [59-62]. Strongly acidic membranes are usually prepared by grafting/soaking/filling of styrene, and the resulted graft copolymer films are subsequently sulfonated [63-69]. Note that in some pore filling processes, sulphonation can be cancelled by using a mixture of some functional monomers (*e.g.*, sodium vinylsulfonate) and DVB [70]. An example will be given here for preparation of semi-interpenetrating polymer network polystyrene cationic membranes using pore soaking method [67]. The concept of the method is schematically illustrated in Figure 4. The supporting material PVC swells while the monomers are absorbed. The monomer-absorbing PVC film permits an enlarged free volume due to the solvent effect, and the monomers can be polymerized therein. Interpenetrating polymer network (IPN) is formed between the polymer chains of St–DVB and the supporting material. The results show that two polymers—poly(styrene–DVB) and PVC—become very miscible. In this method, the membrane morphology and mechanical property can be controlled by the polymerization time and cation-exchangeable sites introduced to the base membrane by a conventional sulfonation process. As shown in Table 1, the resulting membrane exhibits good mechanical, physical, and electrochemical properties.

Table 1. Comparison between the membranes prepared by soaking method and some commercial cationic membranes (data collected from [67])

Membrane*	IEC (meq/g dry)	W_R (g.H_2O/g dry)	R_m (Ω.cm^2)	Transport number	Tensile strength MPa	Elongation at break %	LCD (mA/ cm^2)
Soaking M	2.44	0.38	2.11	0.998	13	52	1.97
CMX	1.67	0.27	2.98	0.987	29	15	1.94
CMB	3.11	0.43	3.42	0.980	-	-	1.91
HQC	1.79	0.45	4.69	0.915	-	-	1.70

*Soaking M: Membrane prepared by soaking method; crosslinking degree, 1%; polymerization time, 5-7h at 80 °C; sulphonation time, 1h.

CMX and CMB: Homogeneous; strongly-acidic and cation-permeable (sulfonic acid group); from Tokuyama Corp. (Japan).

HQC: Heterogeneous; Hangzhou Qianqiu Chemical Co. Ltd (China).

2.1.3. Phase Inversion Method [71-73]

This method derived from the conventional preparation of asymmetrical membranes. The sulphonated polyaromatic polymers are dissolved in solvents and then cast onto a substrate. The membranes are obtained by immersing the casting film in a gelatin medium or evaporating the solvent in the atmosphere. Recently, the membranes are prepared from modified sulphonated polymers, including SPES, SPEK, SPEEK, and SPPO, as referred to some recent reviews [71-73]. This is a very simple method to introduce electrical charges into a polymer matrix and to prepare an ion exchange membrane. However, the degree of functionalization is often difficult to control. In most cases, there is limitation on charge density due to the high swelling degree of sulphonated materials. So the method is often used to prepare cationic membrane for application in fuel cells. To control the methanol crossover, such membranes are often prepared from blends of sulphonated and non-sulphonated polymers [74-75].

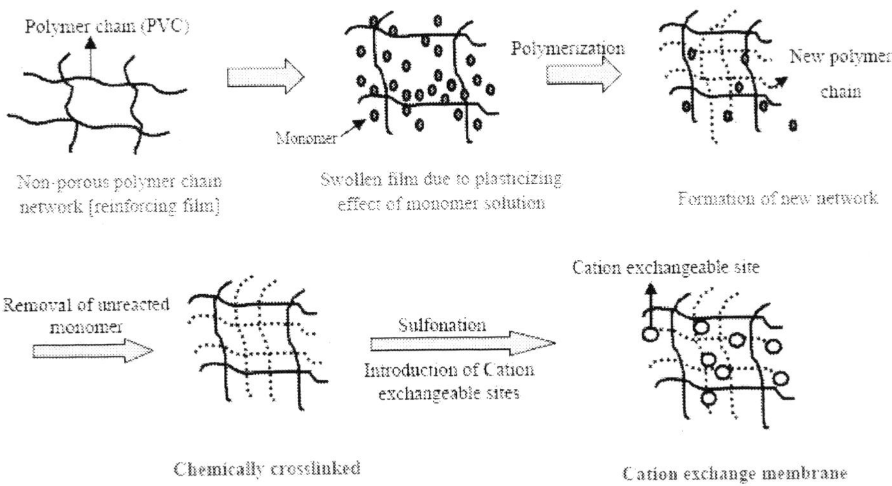

Figure 4. Concept of the proposed preparation method (taken from [67]).

2.1.4. Paste Method

In this method, the monomers are polymerized in the same insoluble polymer, which endows the membrane with flexibility and toughness. The typical components include (1) the monomer styrene, and the crosslinking agent divinylbenzene, which form the membrane matrix and crosslinking structure, (2) the fine powder of poly(vinyl chloride), which increases the membrane strength and flexibility, (3) dioctyl phthalate, which acts as a plasticizer, and (4) benzoyl peroxide which is the catalyst or initiator [76]. These components are mixed thoroughly and deaerated under reduced pressure. Since PVC can not be dissolved into monomer, the final mixture is pasty and easy to be coated on a reinforcing material, and then a base membrane forms in a hot press. The final cationic membrane can be obtained using the sulfonation as described above. The membrane in this way constructed a three-dimensional resinous structure due to the introduction of reinforcing material, the crosslinking by DVB, and the fusion of PVC into a continuous film. Nonetheless, PVC can not be sulphonated to obtain the functional groups, so the final membrane is semi-homogenous.

2.1.5. In-Situ Polymerization

There are disadvantages with the phase inversion method and the paste method; i.e., the former needs to treat the residual solvents while the latter can not obtain a homogeneous structure. In our recent studies, a new method, namely in-situ polymerization, was proposed for preparing novel cationic membrane [77]. In essence, the procedure is the same as the paste or phase inversion method; however, we use soluble aromatic polymer to replace PVC of the former and monomer to replace the solvent of the latter. After dissolving the aromatic polymer completely in the monomer, the homogeneous solution was cast onto a reinforcing material and then underwent polymerization in the same way as the paste method.

The method is exemplified with a system of bromomethylated poly(2,6-dimethyl- 1,4-phenylene oxide) (BPPO, polymer)-styrene (monomer)-DVB (crosslinker). Figure 5 shows the IEC, water content, and proton conductivity of the membranes with different crosslinking degrees [77].

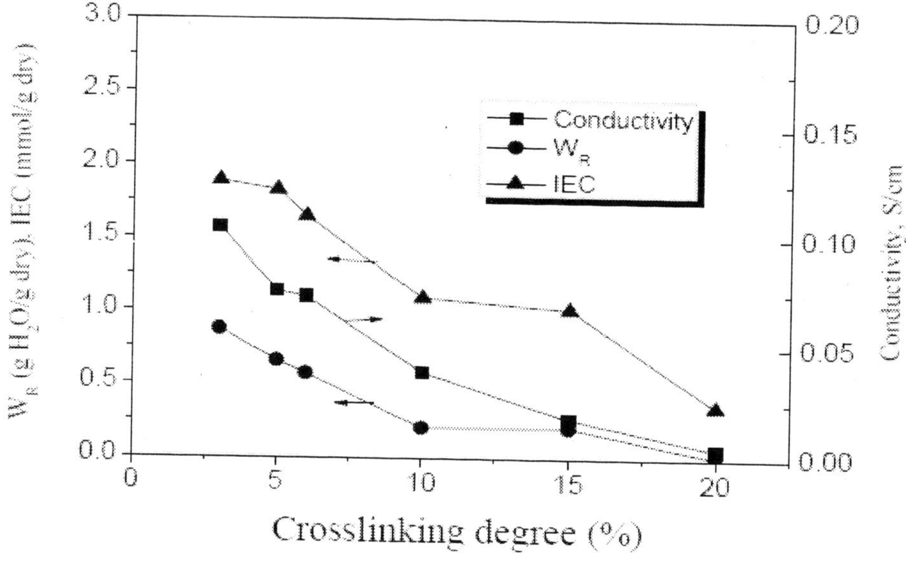

Figure 5. Effect of crosslinking degree on membrane conductivity, IEC, and W_R. Sulphonation conditions: 0.5% chlorosulfonic acid (v/v) and 99.5% 1,2-dichloroethane at room temperature (20±1 °C) for 9 min; base membrane: 3g of BPPO in 10 ml of Styrene+DVB [77].

2.2. Anionic Membranes

The base membranes for anionic membrane preparation are the same as those for cationic membranes. So the above methods can be directly used for preparing anionic membranes except that solufonation is replaced by chloromethylation-quaternary amination using a tertiary amine or direct quaternization with methyl iodide if vinylpyridine monomers are used instead of styrene.

It should be noted that, as compared to cationic membranes, preparation of an anion exchange membrane seems to be more complicated and costly because the agent commonly used for chloromethylation is chloromethyl methyl ether—carcinogenic and potentially harmful to human health. So more efforts have been made to seek alternative methods ever since. As summarized in our recent book chapter [78], the methods includes (1) amidoalkylation of crosslinking polystyrene, (2) copolymerization of (chloromethyl)styrene, (3) chloroacetylation of polymers, (4) acetylation of crosslinked polystyrene, and (5) grafting vinyl benzyl trimethyl ammonium chloride onto inert polymers. Some examples will be given below.

2.2.1. Halogenation of Polymers

Considering the economics, operability, and flexibility as in membrane preparation, chloromethylation via the methyl groups in polymers is the best choice for avoiding the use of chloromethyl methyl ether. In our laboratory, a new series of anion exchange membranes were prepared from poly (2,6-dimethyl-1,4-phenylene oxide) (PPO) by bromination and quaternary amination [79]. The procedure was shown in Figure 6a in comparison with the conventional one (Figure 6b). Notably, the crosslinking can be conducted after the formation of a base membrane. It is different from the conventional route in which the base membrane is formed through crosslinking, and thus can mold membranes into the intricate shape as mentioned above. The properties of the membranes can be quantitively controlled by the bromination and amination-crosslinking processes. Table 2 gives the change of membrane properties with benzyl and aryl substitutions, and Table 3 lists the properties of anion exchange membranes prepared by reacting bromomethylated PPO with different amines, such as methyl amine (MA), dimethyl amine (DMA), and trimethyl amines (TMA) [79-81]. The representative schemes for anionic membranes prepared from BPPO with different ammonium groups were shown in Scheme 1 [82].

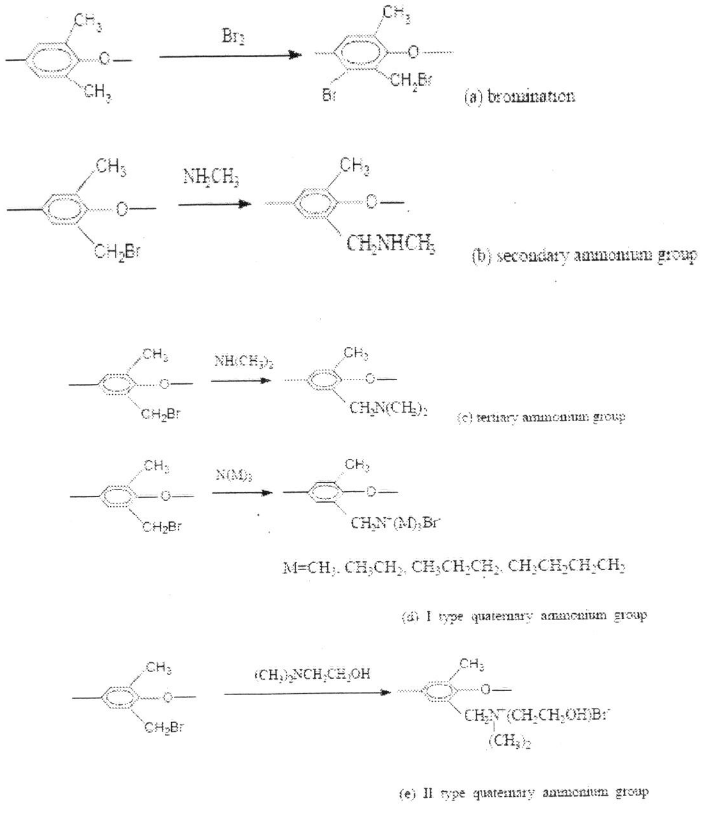

Scheme 1. The bromination and amination reactions and the possible structure of anion exchange layers with different type of functional groups: (a) bromination, (b) secondary amination, (c) tertiary amination, (d) I type quaternary amination and (e) II type quaternary amination.

Table 2. Properties of some membrane series [a]

Membrane Number	ABC[b], %	BBC[b], %	IEC (meq/g dry membrane)	W_R ($g.H_2O$/g dry membrane)	C_R (M)	R_m ($\Omega.cm^2$)	Burst strength, (MPa)
1	0.16	0.116	0.89	0.418	2.13	4.82	
2	0.16	0.173	1.27	0.50	2.52	3.81	
3	0.16	0.285	1.94	0.593	3.26	3.72	>0.8
4	0.16	0.310	2.08	0.74	2.80	1.93	
5	0.16	0.328	2.18	0.77	2.82	1.93	
6	0.10	0.285	1.935	0.673	2.87	1.82	
7	0.28	0.285	1.952	0.478	4.08	3.81	
8	0.38	0.285	1.943	0.442	4.39	4.32	>0.8
9	0.42	0.285	1.933	0.387	4.99	4.93	
10	0.54	0.285	1.920	0.302	6.36	5.25	

[a]. All the measurements were conducted at 25°C.

[b]. ABC: Aryl bromine content.

BBC: benzyl bromine content. Both are assumed Mono-substituted.

Table 3. Main properties of anion exchange layers with different functional groups

Amine	Molecular formula	HLB value*	IEC (meq/g dry membrane)	W_R (g.H_2O/g dry membrane)	R_m ($\Omega.cm^2$)
Methyl amine(MA)	NH_2CH_3	15.93	1.60	0.046	70.3
Dimethyl amine(DMA)	$NH(CH_3)_2$	15.45	1.68	0.081	40.4
Trimethylamine(TMA)	$N(CH_3)_3$	14.98	2.22	0.822	3.01
Triethylamine(TEA)	$N(CH_2CH_3)_3$	13.55	0.53	0.079	74.6
Tripropylamine(TPA)	$N(CH_2CH_2CH_3)_3$	12.13	0.27	0.054	100.1
Tributylamine(TBA)	$N(CH_2CH_2CH_2CH_3)_3$	10.7	0.23	0.033	270.3
Dimethyethanolamine(DMEA)	$(CH_3)_2NCH_2CH_2OH$	16.40	0.78	0.087	52.3

*Calculated from groups' HLB numbers: N (primary, secondary, and tertiary amine) 9.4; $-CH_2-$ 0.475; $-CH_3$ 0.475; -OH 1.9.

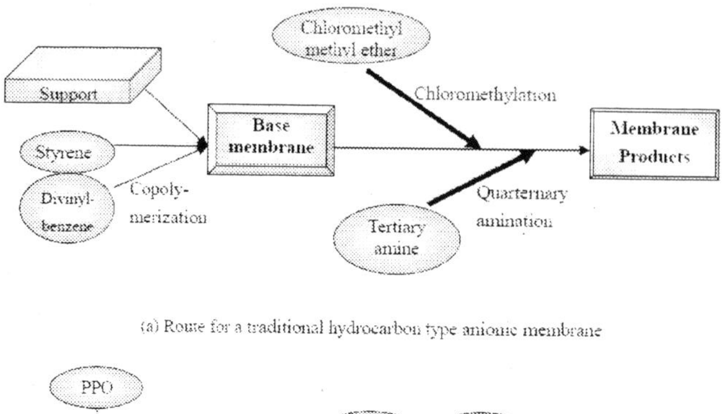

(a) Route for a traditional hydrocarbon type anionic membrane

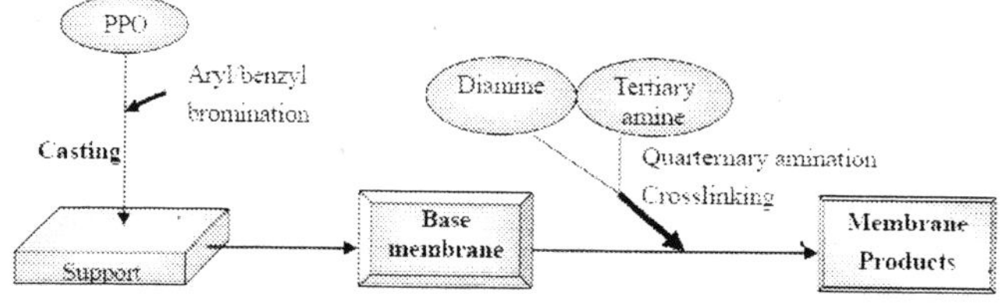

(b) Route for a new anionic membrane prepared from PPO

Figure 6. A comparison of manufacture routes between conventional hydrocarbon type anionic membranes and the anionic membranes prepared from polymer poly (2,6-dimethyl-1,4-phenylene oxide) (PPO) by bromination and quaternary amination.

2.2.2. Chloroacetylation of Polymers- Friedel-Crafts Reaction

The above procedure was limited to the aromatic polymer with alkyl side chains. A more general procedure for aromatic polymer is chloroacelation. It provides an easy way for anion exchangers to avoid the additional crosslinking of the chloromethylated or bromomethylated polymers and not limited by the structure of the aromatic polymers (a side methyl group is not needed). Anionic membranes have been prepared from PPO in this way [83]. As shown in Scheme 2, PPO is treated with 1,3-bis-chloroacetyl in the presence of a Friedel-Crafts catalyst to obtain chloroacetyl groups rather than chloromethyl groups as substituents for aromatic nuclei in the polymer. Then, the chloroacetylated PPO (CPPO) is quaternary aminated. The preliminary experiments demonstrated that the chloroacetylation process was largely influenced by the quantity of the catalyst (anhydrous $AlCl_3$) and reaction temperature. At a given ratio of reactants and catalyst, the substitution degree increases appreciably with reaction temperature. The most suitable degree of chloroacetylation is around 50% because these kinds of substituted polymers are easy to form homogeneous base membranes to the naked eye [83]. Upon the determination of substitution degree, the intrinsic properties of membranes will be largely dependent on the amination process, i.e., amination time, temperature, and TMA concentration. Table 4 lists some properties of the anionic membrane aminated with different concentrations of TMA aqueous solutions (0.7 to 2.5 mol/l) for 48 h at 30 °C. when compared with the anionic membranes prepared from bromomethylated PPO, the conversion ratio for the quaternary amination reaction is relatively low, giving rise to low IEC values and water content, and higher area resistance. Therefore, this comparison demonstrates that the quaternary amination for bromomethylated PPO is quicker than that for CPPO due to the steric hindrance of cloroacetylated groups.

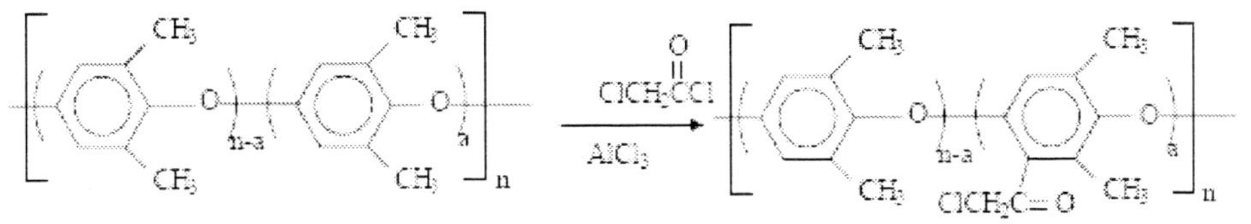

Step 1. Chloroacetylation

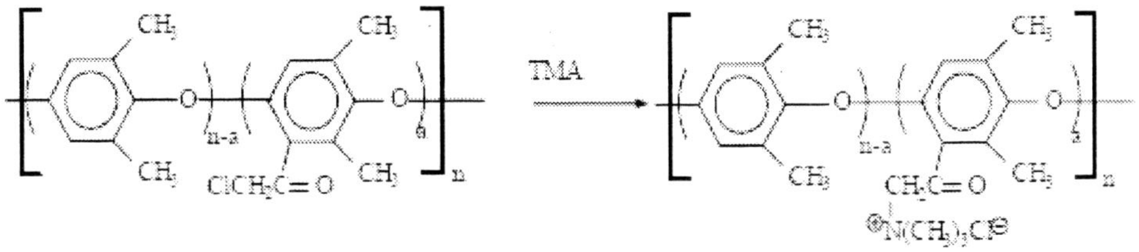

Step 2. Quaternary amination

Scheme 2. The new synthetic route for anionic membranes preparation based on PPO.

It can be expected that an anionic membrane with proper IEC and water content can be obtained by aminating the blend membrane from BPPO and CPPO [84]. Table 5 showed the intrinsic properties of anionic membranes prepared from the blends of CPPO with aryl chloroacetylation degree 50.3% and BPPO with benzyl bromination degree 100%. The results showed that the properties, such as water uptake, IEC, hydroxyl conductivity (electrical resistance), and methanol permeability, decrease as CPPO content increases. By properly balancing these properties, blend membranes with 30-40 wt % CPPO can be expected to apply in direct methanol alkaline fuel cells (DMAFCs) because such membranes showed high hydroxyl ion conductivity (0.021-0.027 $S\ cm^{-1}$), low methanol permeability ($1.35\text{-}1.44\times10^{-7}\ cm^2\ s^{-1}$), and high mechanical strength (33.9 MPa) [84].

Table 4. The intrinsic properties of anionic membranes prepared by chloroacetylation of PPO followed by quaternary amination at various concentrations of TMA. (substitution degree = 50.3 %, amination temperature = 30 °C, amination time = 48 h)

Code number	Concentration of TMA, M	Water uptake (g. water /g wet membrane)	IEC, mmol/g dry membrane	Area resistance, $\Omega.cm^2$
1	0.77	0.619	0.76	188
2	0.91	0.637	0.78	170
3	1.12	0.656	0.8	89.2
4	1.43	0.656	1.03	34.4
5	1.67	0.67	1.17	18.6
6	2.5	0.678	1.16	18.7

Table 5. The intrinsic properties of anionic membranes prepared from the blends of CPPO and BPPO. (benzyl bromination degree was about 100%, chloroacetylation degree = 50.3 %, amination temperature = 25 °C, amination time = 48 h)

CPPO content (wt%)	Water uptake (g. water /g dry membrane)	IEC (mmol/g dry membrane)	Hydroxyl Conductivity (S.cm)	Methanol permeability (cm^2/s)	Tensile strength (MPa)
10	280.3	3.3	(excessively swelling)	-	-
20	179.2	2.9	3.26×10^{-2}	1.46×10^{-7}	28.4
30	137.4	2.2	2.71×10^{-2}	1.44×10^{-7}	29.2
40	79.7	2.0	2.21×10^{-2}	1.35×10^{-7}	33.9
50	44.3	1.5	5.58×10^{-4}	-	25.5

2.2.3. Copolymerization of Epoxy Group-Containing Monomers

Epoxy groups are very active and tend to react with amines to form anion exchange groups. An example given here is a base membrane prepared by copolymerization of divinylbenzene (DVB) with epoxy acrylate monomers (e.g., glycidyl methacrylate, or glycidyl ethacrylate) and sequent amination. Such route is shown as Scheme 3 [85]. The main intrinsic properties of the obtained anionic membrane are as follows: IEC=1.8-2.2 mmol/g dry membrane; water content =23%~33% [85].

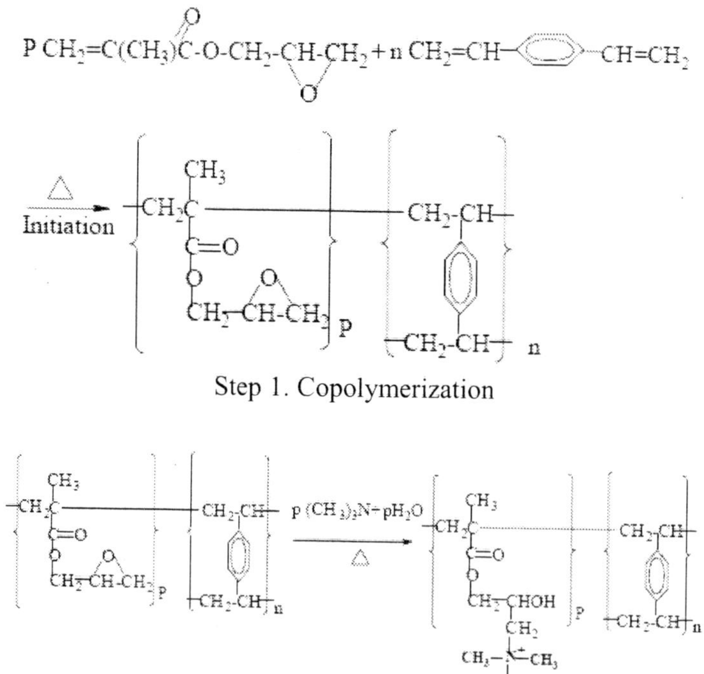

Step 1. Copolymerization

Step 2. Quaternary amination

Scheme 3. Anionic membrane preparation by ring-opening reaction of epoxy groups instead of chloromethylation with chloromethyl methyl ether.

Besides direct copolymerization into base membrane, epoxy acrylate monomers and divinylbenzene (DVB) can also undergo graft polymerization onto an inert polymeric film using the grafting method described above. A porous anionic membrane in flat-sheet form was prepared through co-grafting of an epoxy group-containing monomer, glycidyl methacrylate (GMA), and a cross-linker, divinylbenzene (DVB), onto a porous polyethylene membrane, followed by the introduction of a diethylamino (DEA) group [86]. The remaining epoxy groups in the G/D-AE and G-AE membranes reacted with ethanolamine. The results demonstrated that a monomer mixture of GMA and DVB (containing 1-10 mol % of DVB) provided a DVB-crosslinked graft chain ranging from 15 to 45 mol % of DVB. After the introduction of DEA groups, the membrane swelling in water was reduced, and protein binding capacity increased due to the hydrophobic interaction [86].

An anionic membrane can also be directly prepared from epoxy groups-containing polymers. Such example is the traditional commercial anionic membranes made from polyepichlorohydrin rubber in the 1970s in China. In the presence of triisobutyl aluminium, chloroepoxy propane monomers were polymerized onto a support and then quaternized with TMA to obtain such anionic membrane, whose IEC can attain as high as 1.2-1.6 mmol/g dry membrane.

The membrane structures and their preparation routes described above are just some examples. Some other methods can be employed to avoid using hazard materials: (1) halomethylation by in-situ producing chlorommethylated reagents; (2) (graft) copolymerization with halomethylated vinyl monomers (e.g., vinyl benzene chloride); (3) sol-gel route or using long chain haloalkylethers. The corresponding details were included our recent book chapter [78].

It is noted that, whether anionic or cationic membranes, if the base membrane is prepared through copolymerization, the monomers and crosslinking reagent are not limited to styrene and DVB. Substituted styrenes (e.g., methylstyrene and phenylacetate) and substituted divinylbenzene monomers (e.g., divinylacetylene and butadiene) are often used, which will result in slightly different products. Furthermore, instead of sulfonic acid for cationic membrane preparation, phoposphoric or arsenic acid has been introduced in the cross-linked polystyrene. Most of these membranes, however, have no, or only very little, commercial relevance [2].

$$n(CH_2Cl\text{-}\underset{\diagdown\,O\,\diagup}{CH\text{-}CH}) \xrightarrow{\text{triisobutyl aluminium}} (CH_2\text{-}\underset{\displaystyle CH_2Cl}{\underset{|}{CH}}\text{-}O)_n \xrightarrow{N(CH_3)_3} (CH_2\text{-}\underset{\displaystyle CH_2N^+(CH_2)_3Cl^-}{\underset{|}{CH}}\text{-}O)_n$$

Scheme 4. Anionic membranes preparation from polymerization of chloroepoxy propane monomers.

2.3. Bipolar Membranes

A bipolar membrane (BPM) is a kind of composite membrane that at least consists of a layered ion exchange structure composed of a cation selective layer (with negative fixed charges) and an anion selective layer (with positive fixed charges) [87]. Its typical function is manifested when applied under reverse potential bias (Figure 7). Under this condition, the electrolyte ions will be depleted in the transition region, and the current carriers will stem from the dissociation of solvent molecules. At the present, there are only two kinds of solvents reported to dissociate in bipolar membranes: water (H_2O) and methanol (CH_3OH) [87, 49]. They split into H^+ and OH^-, and H^+ and CH_3O^-, respectively. The process is called electrodialysis with bipolar membranes (EDBM). It can realize some new synthesis processes to achieve the maximal utilization of resources and pollution prevention [21]. It can be flexibly coupled with many other technologies and obtain a better function by means of technological symbiosis. It can realize closing loops when inputting waste materials as the feedstock and carrying out production, resource regeneration, and effluent treatment at the same time [21,30]. In this sense, EDBM plays the same role as "photosynthesizers" in industrial ecosystems and inherently possesses economical and environmental benefits.

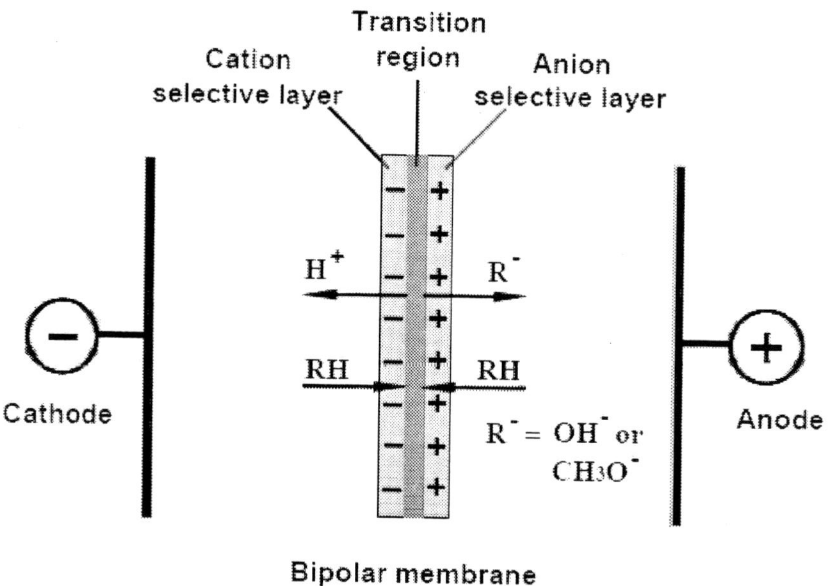

Figure 7. The structure and function of a bipolar membrane. H^+, hydrogen ion; R, OH or CH_3O.

For preparing bipolar membranes, various methods have been initiated. The most important ones are as follows:

- Preparing directly from commercial cationic and anionic membranes by adhering with heat and pressure or with an adhesive paste [88]
- Casting a cation exchange polyelectrolyte solution (or an anion exchange polyelectrolyte solution) on a commercial anion-exchange membrane (or on a cation exchange membrane) respectively [89-90]
- Simultaneous functionalizing the same base membrane at both membrane sides [91-94] or selectively functionalizing on one side to give cation selectivity and on the other side to give anion selectivity. [95].

- Co-extrusion of a cation- and anion-exchange resin, one of which contains the catalytically active component [87].

Among these, the casting method seems to be the most attractive one for preparing such membranes because it is simple, less costly and also allows a bipolar membrane with desired prosperities for commercial use, such as good mechanical strength, ability to operate at high current density, high permselectivity, and low potential drop [96]. Using this method and the basics of the homogeneous membrane preparation as the above, a novel series of bipolar membranes have been recently prepared by casting the sulphonated PPO solution on a series of anionic membranes as shown in Figure 8 [97]. The properties of bipolar membranes are significantly affected by the properties of anion exchange layers, which are dependent on the benzyl substitution and aryl substitution as well as amination time. The results show that the voltage drop across a bipolar membrane at a given current density decreases with benzyl bromine content and amination time, and increase with aryl bromine content. The trends are reasoned out according to the change of anion exchange layer properties with those factors. By properly balancing them, a series of bipolar membranes could be obtained with desired properties for practical applications. For example, a bipolar membrane with water dissociation voltage around 1.0 V can be obtained if the benzyl substitution is high and the amination time is enough. This property is better than that of commercial bipolar membranes, such as BP-1 from Tokuyama, which has a voltage about 1.2 V at the identical conditions. Besides, the properties can be also optimized by changing the charge type of the anion exchange layer [98]. A bipolar membrane with strongly basic groups has larger water dissociation ability than that with weakly basic groups. For membranes with weakly basic groups (secondary or tertiary ammonium groups) or strongly basic groups (quaternary ammonium groups), water dissociation depends on the pK_b value of amines if anion or cation exchange layers processes the close intrinsic properties such as water uptake and conductivity; water dissociation ability increases with an increase in pK_b values [98]

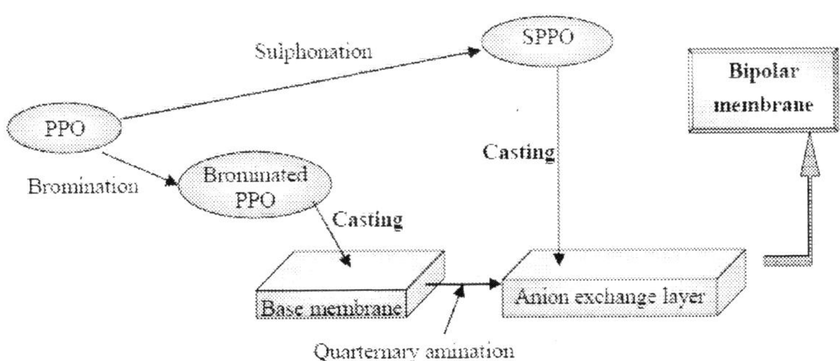

Figure 8. Schematic of novel bipolar membranes prepared from the same base material--poly (2,6-dimethyl-1,4-phenylene oxide) (PPO) [97].

Note that, due to the unique structure and the same swelling degree of both anion and cation exchange layers, the above prepared bipolar membranes possess excellent mechanical stability and chemical stability [97-98].

The two ion exchange layers in a bipolar membrane allow for the selective transport of the water dissociation products—protons and hydroxyl ions and block co-ions in the electrolyte. A bipolar membrane has a contact region, also referred to as the interfacial layer, where the desired water dissociation reaction occurs. The charged groups and structure of this region are of great significance for water dissociation and thus generally are modified to improve the bipolar membrane's performance. Now it is concluded that materials with the best catalytic activity are (1) sufficient amounts of weak acids (and the corresponding bases), such as amino groups, pyridine, carboxylic acid, and phenolic or phosphoric acid group [99-102], and (2) heavy metal ion complexes, such as ruthenium trichloride, chromic nitrate, indium sulphate, and hydrated zirconium oxide [103-106]. The catalytic mechanism is underlined by chemical reaction model of water dissociation, that is, the water splitting could be considered some type of proton-transfer reaction between water molecules and the functional groups or chemicals [101, 107]:

$$B + H_2O \rightleftharpoons BH^+\text{----}OH^- \rightleftharpoons BH^+ + OH^- \quad (1)$$

$$BH^+ + H_2O \rightleftharpoons B\text{----}H_3O^+ \rightleftharpoons B + H_3O^+ \quad (2)$$

or

$$A^- + H_2O \rightleftharpoons AH\text{----}OH^- \rightleftharpoons AH + OH^- \quad (3)$$

$$AH + H_2O \rightleftharpoons A^-\text{----}H_3O^+ \rightleftharpoons A^- + H_3O^+ \quad (4)$$

where BH^+ and A^- refer to the catalytic centers. The catalytic sites provide an alternative path with low effective activation energy for water splitting into hydrogen and hydroxyl ions.

A typical bipolar membrane with this kind of structure was prepared by H. Strathmann et. al.[108] by laminating anion- and cation-exchange layers. In this method, an anion-selective layer with required properties,(IEC = 1.2 mmol/g, thickness = 60 μm, area resistance = 1.05 ohm cm^2, permselectivity = 97.5 % and swelling degree =8%) was obtained by dissolving chloromethylated polysulfone in N-methyl-2- pyrrolidone and reacting with the mono-quaternary salt of 4,4'-diazabicyclo- [2.2.2]-octane (DABCO) (see the following structure).

The crosslinking density of the anion-selective layer can be adjusted by changing the ratio of DABCO to the chloromethylated polysulfone. The cation-selective layer with desired properties (e.g., IEC = 1.0 mmol/g, thickness = 60 μm, area resistance = 1.31 Ohm cm^2, permselectivity = 98.5 %, and swelling degree=12.5%) was prepared by introducing sulfonic acid groups as fixed charges into a polyetheretherketone matrix using chlorosulfonic acid. During the membrane formation, partial cross-linking tends to take place as indicated in the following scheme:

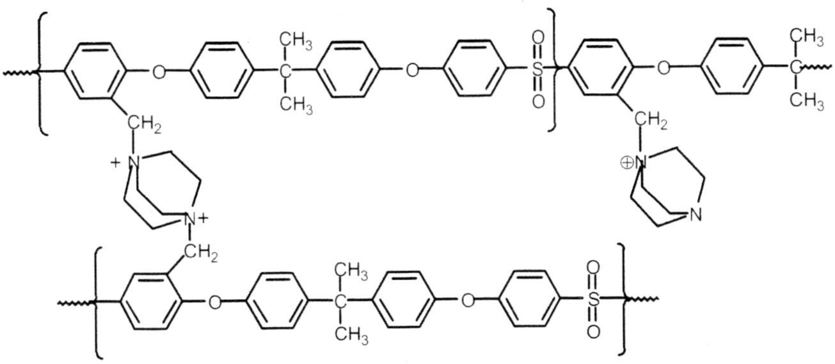

This cross linking will decrease the swelling degree and co-ion transport of the layer. By laminating these two layers, a bipolar membrane with specific structure as follows can be obtained:

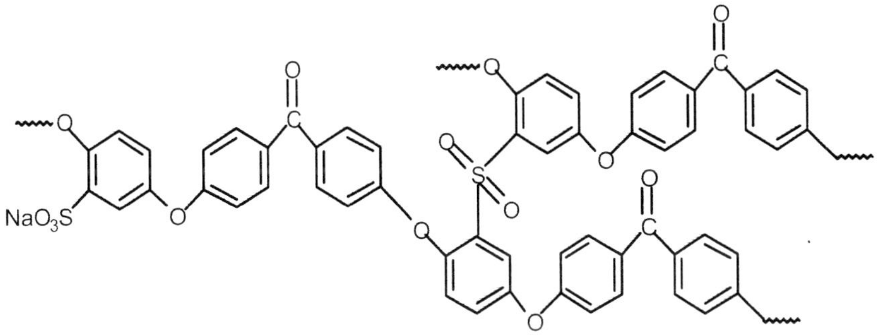

- a highly permselective anion-exchange layer which shows excellent alkaline stability,
- an equally highly permselective cation-exchange layer which exhibits excellent acid stability,
- a transition region with a limited number of catalytically active tertiary ammonium groups.

This bipolar membrane exhibits most of the properties requested for practical applications where the cation-selective layer of the bipolar membrane is generally in contact with an acid solution and thus should be stable in strong acids [108].

A bipolar membrane with satisfactory properties, such as low electrical resistance at high current density and high water dissociation rates, can be also prepared using different macromolecules as the interfacial layers during laminating of anion- and cation-exchange layers. Our current work concentrates on using macromolecules, such as polyethylene glycol (PEG) [109], gelatin containing both carboxylic and amino groups [110], starburst dendrimers [111], and polyvinyl alcohol (PVA) [112] as the interfacial layer to catalyze water dissociation in a bipolar membrane. Here are some surprising findings. PEG has greatly improved the performance of bipolar membranes from the viewpoint of the I-V characteristics curves [109]: the catalytic effect increases with not only the adsorbed concentration but also the molecular weight. But for the similar material PVA, which has more hydroxyl groups than PEG, the function is much different: its catalytic function is available only at low adsorption concentration [112]. Theoretical interpretation relates to the strong hydrophilicity and the interaction of PEG or PVA with water molecules. Although both PEG and PVA are hydrophilic, the hydrogen-bonding and polar interactions between PEG and water molecules are easily formed to increase the solubility of PEG in water, while PVA has a limited solubility above which it is easily crystallized to form a neutral layer [112].

Another interesting approach reported recently is the modification of a bipolar membrane with the starburst dendrimer polyamidoamine (PAMAM) [111], which possesses much higher amino group densities than conventional macromolecules as shown in Figure 9. For example, a third generation PAMAM prepared from ammonia core has 1.24×10^{-4} amine moieties per unit volume (cubic Angstrom units) in contrast to the 1.58×10^{-6} amine moieties per unit volume contained in a conventional star polymer [113]. As expected, these amino groups of high density in PAMAM have appreciable catalytic function when they are incorporated into the interfacial layer of a bipolar membrane [111]. Because of the PAMAM steric effect, the amino groups' catalytic function varies with both PAMAM concentrations and generations [111]: at a given generation, taking G2 as an example, voltage decreases with PAMAM concentration in the low concentration range due to an increase in catalytic sites from amino groups, and increases with PAMAM concentration in the high concentration range due to an increase in steric effect.

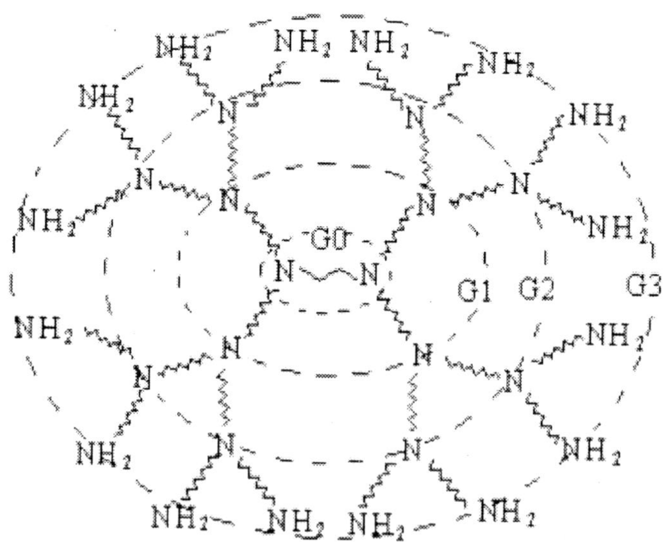

Figure 9. Schematic of PAMAM G3: the initial core is ethylenediamine and the repeating unit is –$CH_2CH_2CONHCH_2CH_2N$–.

The limitation on dentrimer is its high cost and the difficulty in its synthesis. Hyperbranched aliphatic polyesters (Boltorn® series), analogous to dendrimer but cheap and easy to synthesize, can be used as the interface in a bipolar membrane[114]. The Boltorn®

series possess abundant hydroxyl groups. For examples, Boltorn® H20, a second generation dendritic polyester, has an number-average molar mass of 1747 g/mol and 16 hydroxyl end groups per molecule on average. Correspondingly, the third-generation (Boltorn® H30) has 32 hydroxyl end groups per molecule, and the fourth (Boltorn® H40) has 64. The structures of dendritic polyesters of different generation were schematically shown in Figure 10 [115]. These high-density hydroxyl groups have been expected to effectively catalyze the water splitting in a bipolar membrane [114].

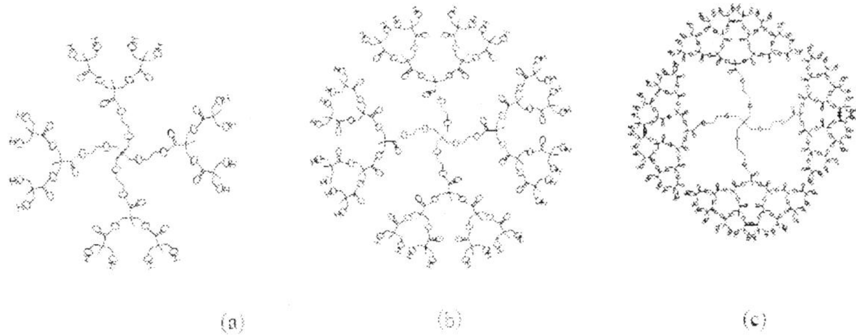

Figure 10. Idealized formula of (a) Boltorn® H20,(b) Boltorn® H30,(c) Boltorn® H40 (Adapted from [115]).

At a proper generation and concentration, dendrimer or dentric polymer not only works as a catalysis itself, but also can be used to coordinate the heavy metal ions, such as Cr^{3+} and Fe^{2+} which have been found to be effective for catalysis of water dissociation [116-117] but easily lost during operation. As shown in Figure 11 [118], the voltage drops cross the bipolar membranes with both PAMAM and Cr(III) show the minimum among the investigated bipolar membranes (one without Cr (III) or PAMAM, ones with only Cr (III) or PAMAM). It indicates a synergetic effect in catalyzing water dissociation. Furthermore, compared with only Cr(III) as an intermediate layer, PAMAM coordinated Cr(III) keeps longer catalytic effect [118].

Bipolar membranes were recently prepared by inserting the cation-exchange fiber (CEF) and anion-exchange fiber (AEF) fabric between a cation exchange layer (CEL, Aciplex K501) and an anion exchange layer (AEL, Aciplex A501) and then compressing them with a pair of poly(tetrafluoroethylene) plates [119]. The main properties of the ion exchange fibers are listed in Table 6. Their current-voltage characteristics under reverse bias are shown in Figure 12 a-b [119]. The voltage drop of water dissociation is remarkably reduced by the CEF fabric (the value of $\Delta I/\Delta V$ decreasing from 1.3 to 0.18), which is attributed to the decrease in the contact area between the CEF fabric and the AEL and the increase in the intermediate layer thickness. Also, the water dissociation is enhanced by the AEF fabric (the value of $\Delta I/\Delta V$ increasing from 1.3 to 1.7), which can be explained by a synergetic effect of the protonation-deprotonation reaction due to the tertiary pyridyl group and the high specific surface area of AEF.

2.4. Miscellaneous

Much effort is made to develop ion-exchange membranes as separators for fuel cells, especially the methanol direct conversion fuel cells. This will be included as a separate chapter of this book. Also, certain electro-organic syntheses require special membranes with satisfactory electrical properties and good chemical stability in the reaction environment, such as anti-fouling anionic membranes, alkaline-stable anionic membranes, anionic membranes of high proton retention, and monovalent ion-permselective membranes. Specific works and literature can be referred to [2-3, 120-122].

Table 6. The physicochemical properties of the CEF and AEF fabrics used as bipolar membrane interface [119]

	CEF fabric	AEF fabric
Ion-exchange capacity (mmol/g)	1.3	0.8
Flow-through porosity (%)	75	80
Average flow-through pore size (μm)	1.9	3.5
BET surface area (m^2/g)	1.7	600
Thickness (mm)	0.05	0.04

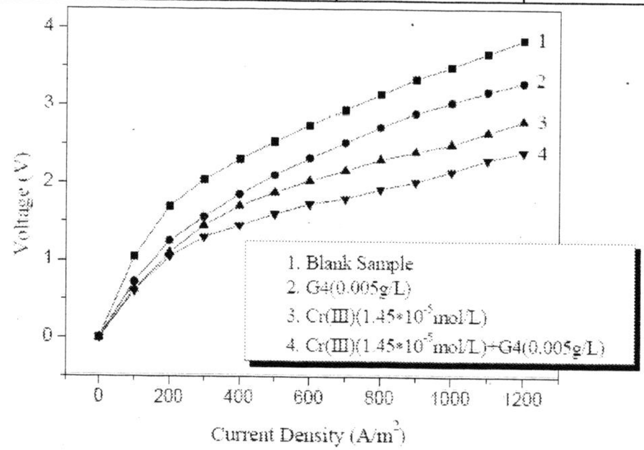

Figure 11. The I-V curves of bipolar membranes with/without Cr(III) or PAMAM G4.

Inorganic-organic composite materials are increasingly important due to their extraordinary properties within a single molecular composite, which arises from the synergism between the properties of the components. These materials have gained much interest due to the remarkable enhancement on mechanical, thermal, electrical, and magnetic properties as compared to pure organic polymers or inorganic materials. For these materials, organic components offer structural flexibility, operability, tunable electronic properties, photoconductivity, efficient luminescence, and the potential for semiconducting and even metallic behavior. Inorganic compounds provide the potential for high carrier mobility, band gap tunability, a range of magnetic and dielectric properties, and thermal and mechanical stability. Correspondingly, the ion exchange membranes prepared from polymer-inoganic hybrids are a new direction in the field and have received great attention in recent years [5, 123]. Ion exchange inorganic-organic hybrid membranes can be made using several routes, including sol-gel process, intercalation, blending, in situ polymerization, and molecular self-assembling. The details can be seen in Chapter 9 of this book.

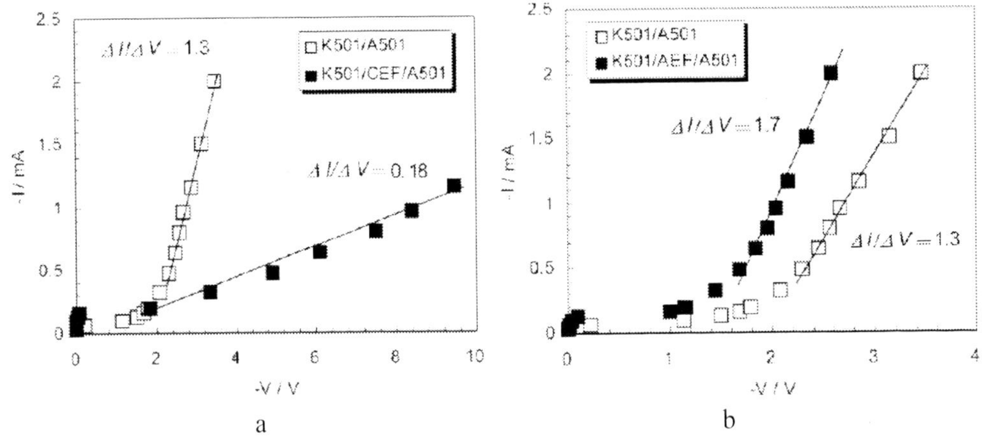

Figure 12 . The current-voltage characteristics of the bipolar membrane with CEF fabric (a) and AEF fabric (b).

Amphoteric ion exchange membranes contain both weak acidic (negative charge) groups and weak basic (positive charge) groups that are randomly distributed within the membrane matrix [124-125]. The sign of the charge groups in these membranes exhibits a pH dependance in an external solution. They are expected to be the next generation of ion exchange membranes for the following features: (1) controllability of the charge property by changing the pH of the external solution, and (2) potential as an anti-fouling material that prevents adsorption of organic molecules and biological macromolecules on the surface. The details for their application and preparation have been reviewed [7].

Different from amphoteric charged membrane (negative and positive fixed ions randomly distributed in a neutral polymer matrix), a charge-mosaic membrane consists of a set of anion and cation exchange elements arranged in parallel, each element providing a continuous pathway from one bathing solution to the other [2, 124]. When a gradient of electrolyte concentration is established across the membrane, anions and cations can flow in parallel through their respective pathways without violating macroscopic electro-neutrality, resulting in a circulation of current between the individual ion exchange elements. As a result of current circulation, the charge-mosaic membrane shows negative osmosis and salt permeability much greater than its permeability to nonelectrolytes. The process is also referred to as piezodialysis. In this process, a hydrostatic pressure is applied to transport salt ions through a mosaic membrane. Since both cations and anions are simultaneously adsorbed by the cation- and anion-exchange domains of the membrane, an electric circuit is formed and allows the salt to pass the membrane. As long as the fixed charge density of the membrane is higher than the ion concentration in the feed, electrolytes will preferentially permeate the membrane under the driving force of hydrostatic pressure and will thus be concentrated in the permeate. In this sense, mosaic membrane can also used to separate salts from water-soluble organic substances, and treat waste streams from dye, food, dairy, fermentation, agriculture, pharmaceuticals and mining industries. [126].

As stated by H. Strathmann [2], mosaic membranes can be prepared by distributing cation- and anion-exchange particles of sub-micron size in a neutral hydrophilic polymer film. Alternatively, mosaic membranes can be prepared (1) by a film casting method using a block polymer composed of two parts into which cation-exchange groups and anion-exchange groups can be introduced, or (2) by some methods describing in our previous review [7]. Unfortunately, mosaic membranes currently seem to be of little technical and commercial relevance [2].

3. Applications of Ion Exchange Membranes

The conventional ion-exchange membrane-based processes include (1) diffusional dialysis (DD), using only anionic or cationic membranes and driven by concentration gradient, (2) electrodialysis (ED) using both anionic and cationic membranes arranged in sereies and driven by potential difference, (3) electrodialysis using bipolar membranes (EDBM) and monopolar membranes and driven by potential difference, and (4) some combinations thereof or integrations with other unit operations, e.g., electrodialysis + ion exchange (EDI), electrolysis + electrodialysis, electrodialysis + membrane reactor (MBR), and electrodialysis + chemical unit operations (extraction, absortion, adsorption). Chapter 7 will give some descriptions of the integration processes. Therefore, in the following sections, the application of ion exchange membranes will be described using single electro-membrane processes, such as DD, ED, and EDBM.

3.1. Diffusional Dialysis

Diffusion dialysis is well established today as a commercial process with a number of applications mainly in the treatment of industrial effluents containing acid and metal impurities from metal processing industries. There are several potential applications in the chemical industry for purification of acids or bases or in the food industry for deacidification of fruit juice. But these applications are presently of minor commercial relevance [2]. Recovery of acids is the main application of diffusion dialysis. It concerns many industries, such as

- Purification of spent sulfuric acid from lead acid batteries;
- Isolation of pure caustic soda from an $NaOH/NaNO_3$ mixture in the Ni/Cd battery manufacture;
- Recovery of excess HCl and NaOH from ion exchange regeneration streams;
- Recovery of spent sulfuric acid from aluminum anodizing baths;
- Mixed acid recovery from stainless steel industries;

- Recovery of purified nitric acid from uranium processing operations;
- Separation and purification of HNO_3 and NH_4NO_3 system;
- Recovery of mineral acids from mining/metal recovery operations;
- Deacidification and purification of rare earth metals
- Deacidification and purification of organic acids from fermentation processes
- Removal of impurities in the acidic waste liquor.

As examples, some industrial tests are summarized below based on the anionic membranes developed from poly(2,6-dimethyl-1,4-phenylene oxide).

3.1.1. Acid Recovery from Aluminum Polishing Waste Liquor

Aluminum surface finishing has been a rapidly growing industry in many countries around the world for the past decade. The products involved in this industry include aluminum cables and wires, structural materials, and electric capacitors. The aluminum finishing industry invariably produces a large amount of waste acid solution. Depending on specific type of aluminum surface finishing, the waste acid solution may contain sulfuric acid, hydrochloric acid, phosphoric acid, additives, and an appreciable amount of dissolved aluminum. A typical waste acid solution is generated in either the aluminum window sash anodizing or the etching of aluminum foil for high-power electric capacitors. Hundreds of thousand tons of such waste acid solution has been generated monthly in China. For example, in a small plant of electric capacitors in middle China, 30 ton per day of such waste containing 15-20% HCl and $AlCl_3$ was produced, equivalent to an exhaustion of 15-20 ton/day 31% commercial HCl solution. Diffusion dialysis with anionic membranes seems desirable for treatment of such waste solution. Since the anionic membranes show extremely high rejection to multivalent cations (Al^{3+} in this case), both the process feasibility and the quality of the recovered acid are quite satisfactory. Now tens and hundreds of industrial set-ups were installed in China. Figure 13 give a demonstration of diffusion dialysis system installed recently in western China with total membranes area 5120 m^2. Whether in sulfuric or HCl systems, the results were quite satisfactory: acid recovery ratio more than 85% and aluminum rejection ratio more than 95%. Note that for this system, anionic membrane with high water uptake was used due to its high acid flux.

3.1.2. Mixed Acid Recovery from Waste Acid in Titanium Materials Processing Industries

Pickling of titanium with an acid mixture, which contains 60 g/l HF, and 300-360 g/l HNO_3, is one of the key steps in titanium materials processing industries. This pickling process is usually accomplished by discharging large quantities of spent liquor. For example, 20-40 tons of such waste liquors will be produced after one ton of titanium material is processed. Generally, this kind of waste is mainly composed of HNO_3 230-260 g/l, HF 3g/l, T_i^{4+} 18-24 g/l, and other metal impurities such as the rare earth metal Zr. The accumulation of metal ions (especially T_i^{4+} ions) in the waste solution will result in a decreased efficiency of the pickling agent and thus after several times' recycling (conventionally, the recycling is within 3 times), the eventual disposal is inevitable. Therefore, recovery of these acids is of great importance. However, different from the steel finishing or aluminum processing waste, in which Fe^{2+} or Al^{3+} exists in cationic form, the mixed acid system (HNO_3+HF) discussed here has a portion of T_i^{4+} ions existing in the form of complex anions, i.e., $[T_iF_6]^{2-}$, due to the following leaching reactions:

Figure 13. Photograph of a diffusion dialysis unit (installed in 2007, 512 m^2 of membrane area) to recover sulfuric acid from aluminum polishing waste liquor using the new PPO-based anionic membrane (by courtesy of Tianwei Membrane Company).

$$Ti + 4\ HNO_3 = H_2TiO_3 + 4\ NO_2 + H_2O \quad (5)$$

$$Ti + 6\ HF = 2\ H^+ + [TiF_6]^{2-} + 2\ H_2 \quad (6)$$

Therefore, a competitive diffusion of complex anions will decrease both the selectivity and recovery ratio. To solve this problem, the membrane should be limited to those which can reject the complex Ti anions more efficiently, so membranes with a mediate water content of 0.59 g H_2O/g dry membrane and IEC of 1.94 mmol/g dry membrane are preferable here. Such apparatus has been running in western China for 5 years. The results showed that the recovery ratio of mixed acid could attain as high as 85% and the total rejection of titanium as high as 85-90% [127]. The estimation of process economics showed that the technology is much competitive [127].

3.1.3. Acid Recovery from Waste Liquor in Hydrometallurgy

On some occasions, the purpose of diffusion dialysis is to decrease the acid concentration of waste to a desired level for the later processes. Hence, the acid recovery ratio is not an important issue. Mostly, the waste contains relatively low concentration of acid and valuable metal ions. Note that in the case of low acid concentration, the water osmosis can not be compensated timely due to the large decrease in flux of hydrated ions; therefore, there will exist a large volumetric expansion at the waste side due to the large flux of water osmosis from the stripping phase to the waste phase [128]. This not only decreases the metal ion's concentration, but also increases the recycling volume if the waste is recycled for recovery of valuable metal. The best way is to control the water osmosis by choosing the membrane with low water content but high IEC. As an example, the membrane crosslinked using ammonium aqueous solution for 8 h (with the water content 0.30 g H_2O/g dry membrane and IEC=2.09 mmol/g dry membrane) is used to treat the waste from a nickel processing plant in western China, which contains H_2SO_4 40-50g/l, Ni^{2+} 70-80g/l, and other metal impurities such as rare earth metals. The diffusion dialysis was conducted at the flow speed of feed 1.2-1.8 and flow ratio of water to feed 1.05-1.1. At these optimum conditions, the volumetric expansion factor (stream ratio of the waste to the recovered) can be controlled within 1.1, nickel leakage is less than 4%, and the acid recover ratio can attain as high as 66-72%. The residual 28-34% acid in the waste can meet the request of acidity in initial leaching process; at the same time, 96% nickel can be recycled by recycling the waste to leaching process as shown in Figure 14.The recovered 66-72% acid can be recycled to the back-extraction stage by adding some fresh acid. Thus, the whole technology discards no waste (zero discarding), and possesses significant environmental benefits apart from the attractive economical benefits. The only equipment is a diffusion dialyser which is very simple in operation and energy saving (the only energy needed is for pumping the waste) and can run automatically without any maintenance.

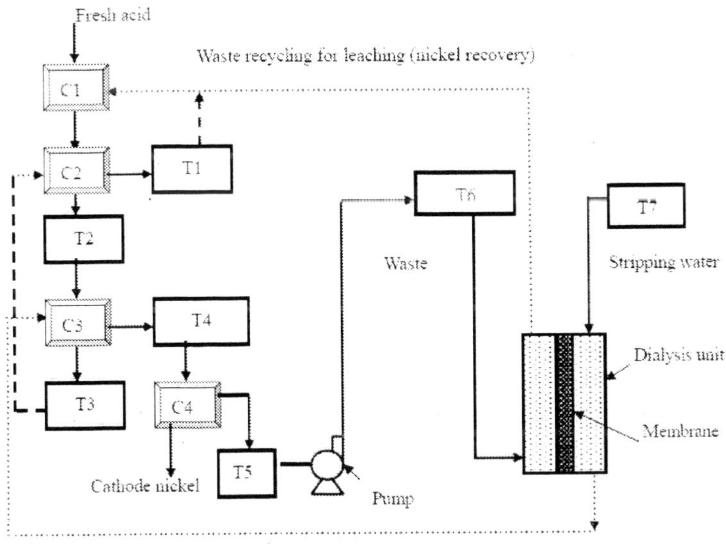

Figure 14. Flow sheet of simultaneous recovery of nickel and acid from nickel hydrometallurgy industries using diffusion dialysis. Tl-raffate aqueous tank; T2-extraction organic phase (nickel-rich phase);T3-regenerated organic phase; T4-nickel aqueous phase; T5-spent electrolysis waste liquor; T6-spent waste liquor; T7-stripping water; C1-leaching cell; C2-extraction cell; C3-back-extraction cell;C4-electrolysis cell.

3.1.4. Others

In addition to the above-mentioned applications, diffusion dialysis with anionic membranes can be also applied in many other fields such as rare earth industries, atomic energy industry (uranium extraction), recovery of an acid from the ion-exchange regeneration process, purification of inorganic acids, or depletion of inorganic acids from fermentation process. The principles are the same as the above and are thus not stated here.

For the same mechanism, if the anionic membrane is replaced by a cationic membrane, bases, such as sodium hydroxide, can be recovered from a mixture with salts by diffusion dialysis. The cationic membranes developed for this process are relatively thin and stable in a strong alkaline environment [2]. The main application of diffusion dialysis is the recovery of sodium hydroxide from etching baths of some metals, such as aluminum, molybdenum, and tungsten. Since the metal anions are relatively bulky, leakage through the membrane is relatively small even in a loosely structured cationic membrane.

3.2. Electrodialysis (Conventional)

Conventional electrodialysis (ED) is a type of technology which arranges cationic and anionic membranes alternately in a direct current field. Owning to its distinguished functions, ED is commercialized the most among all ion-exchange membrane separation processes. Its applications extend from the original water desalination and the pre-concentration of NaCl in sea water to the food, pharmaceutical, and chemical industries (such as the demineralization of whey or deacidification of fruit juice), and from aqueous systems to organic or aqueous-organic systems [2, 7, 27,45,129-134]. Some important examples in biotechnology, chemical synthesis/separation, and non-aqueous systems are described below.

3.2.1. Application of Electrodialysis in Aqueous Systems

Most applications of electrodialysis were used in aqueous medium, which concerns food, pharmaceutical, beverage, fermentation, and chemical industries. Some of them can be considered state-of-the-art processes, e.g., deionization of whey or deacidification of fruit juices. Other applications, such as de-ashing of sugar or removal of salts from protein solutions, have been tested and evaluated on a pilot or laboratory scale [2]. The removal of salts and organic acids (citric acid, glucose acid, lactic acid, and some amino acids) from fermentation broth by electrodialysis has been tested extensively on a laboratory or pilot scale. For the food Industry, ion exchange can be replaced with ED, which provides better economics, simpler operation, and less waste material to deposit. Typical applications of electrodialysis in these areas include [2]

- demineralization of cheese whey or non-fat milk in food industry
- demineralization of molasses, and polysaccharide in sugar industry
- removal of tartrate in wine industry
- deacidification of fruit juice in beverage industry
- demineralization of soy sauce, amino acids, and organic acids; recovery of organic acids in fermentation industry.

An important application of electrodialysis in biotechnology is to demineralize or concentrate organic acids or organic salts. As reviewed in our recent work [129], the electrodialysis stack as shown in Figure 15a has been reported to demineralize lactic acid, glutamine, glycine, and D-α-p-hydroxyphenylglycine; the stack in Figure 15b to concentrate lactate, lysine, glycine, gluconate, propionate, pyruvate, and formic acid, etc. The efficiency of demineralization or concentration mainly depends on counter-ion competence, which increases with the ion quantity and mobility. In the case of demineralization, it is better to acidify the feed so that most of organic anions can exist in the form of acid molecules and stay in the feed, and more inorganic anions can migrate into the adjacent compartment. As for amino acids, a greater efficiency can be achieved if the feed pH is adjusted to the isoelectric point of the amino acid. In the case of concentration of organic acids or organic salts, the opposite measures can be taken to achieve a higher efficiency.

In chemical industries, since ions of different signs move in opposite directions in a direct current field, electrodialysis is often used for chemical reaction and separation. Here we give two examples for electro-metathesis, i.e., electrodialysis for double decomposition [134,135].

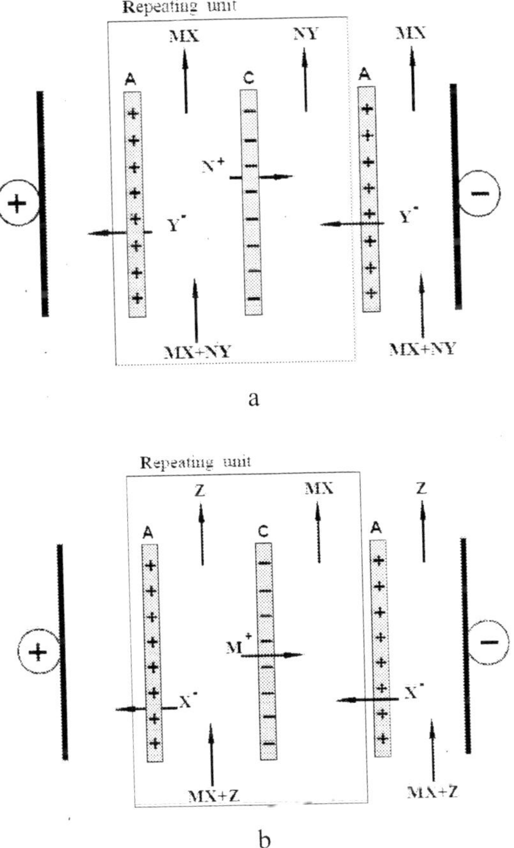

Figure 15. Schematic of the electrodialysis applied to (a) demineralize or (b) concentrate organic acids or organic salts. MX, organic acid or salt; NY, inorganic salt; Z, neutral substances or low concentration of inorganic salts.

Example 1. Production of Tetramethylammonium Hydroxide ($R_4N(OH)$)

The following equation shows the traditional method for $R_4N(OH)$ production, in which a great amount of Ag_2O is consumed though Ag can be recovered.

$$2\ R_4NCl + H_2O + Ag_2O \rightarrow 2\ R_4N(OH) + 2AgCl \downarrow \quad (7)$$

The consumption of Ag_2O can be reduced to a large magnitude if a 4-compartment ED as shown in Figure 16 is employed to produce $R_4N(OH)$ by following the equation below:

$$R_4NCl + NaOH \rightarrow R_4N(OH) + NaCl \quad (8)$$

Moreover, no separation is needed since reactants and products exist, respectively, in 4 compartments and their reservoirs. Naturally, a certain amount of Cl^- will transport into the product compartment due to co-ion leakage or diffusion; however, the required purity can be achieved by using a small amount of Ag_2O.

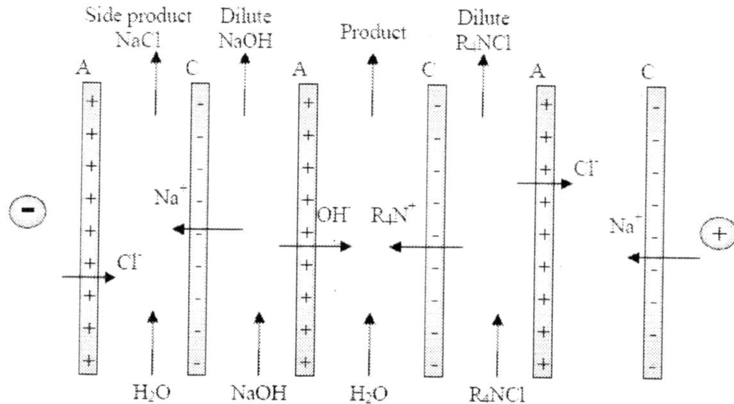

Figure 16. Production of tetramethylammonium hydroxide by conventional electrodialysis. A, anionic membrane; C, cationic membrane.

Example 2. Preparation of Poly Aluminum Chloride from $AlCl_3$

The traditional method for poly aluminum chloride production is to add alkaline in an $AlCl_3$ aqueous solution. Since local extreme pH forms in the process of alkaline addition, much $Al(OH)_3$, instead of poly aluminum chloride, is generated. For improvement, a 3-compartment ED as shown in Figure 17 can be used and the principle is embodied in the following equation:

$$nOH^- + m\ AlCl_3 \rightarrow Al_m(OH)nCl_{3m-n} \quad (9)$$

Since the supply of OH^- is at a constant and slow rate and favors the formation of poly aluminum chloride, the resulted product gains a high quality. Additionally, there is much less Na^+ contamination in the final product. Experimental results showed that the product solution contained $Al_{13}O_4(OH)_{24}{}^{+7}$, and little $Al(OH)_3$ colloid and unreacted Al^{3+}, and that the Al^{3+}

concentration was more than 1 mol/L. Such product can achieve a much better effect of flocculation than conventional commercial poly aluminum chloride.

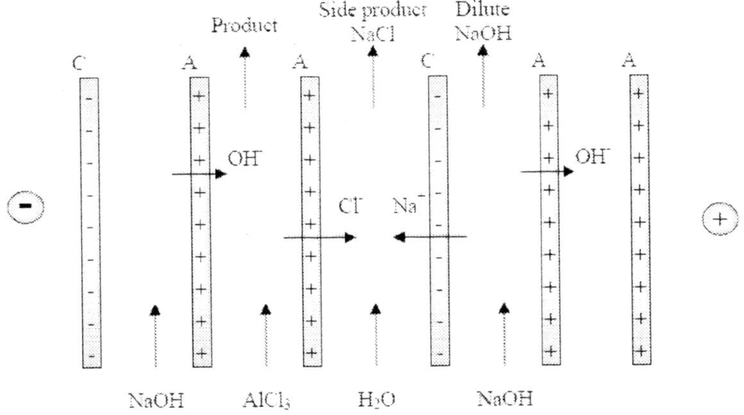

Figure 17. Preparation of poly aluminum chloride by conventional electrodialysis. A, aniónic membrane; C, cationic membrane.

3.2.2. Application of Electrodialysis in Aqueous-Organic Systems

Electrodialysis provides a reliable method for organic production; however, some organic acids and other chemicals can not be produced by following the same procedures since water has a weak solubility for them. These organic acids are, for examples, the alphatic acids with a carbon chain length longer than 8 (caprylic (C8), lauric (C12), myristic(C14), and palmitic (C16)), and salicylic acid. One way out is applying ED in water-alcohol solutions [45, 130,132]. As shown in Figure 18, in this process, the sodium organic salt (NaR) is fed into the middle compartment which is filled with a 50% v/v water-ethanol solution. Naturally, 100% ethanol will solve more NaR, but it does not lead to an optimal result. For one thing, the electrical resistance of ethanol is much higher than that of water, so more energy will be wasted on raising the system temperature if the ED stack is operated in a pure ethanol system. The remedy is to add a considerable amount of electrolyte, but these introduced electrolyte ions will compete with sodium ions, resulting in a low productivity of the organic acid. For another, excessive ethanol will directly lead to a low productivity because the protic solvent will compete for the H^+ ions which are used to produce the organic acid. In general, the ethanol/water ratio has an optimum value. It is noticeable that the solvent in the two electrode compartments can be 100% water or a 50% v/v water-ethanol mixture, but in this circumstance, the former is the better choice. On one hand, there is no need to increase the solubility of organic materials now that inorganic salts act as the rinsing electrolyte in these two compartments. Therefore, water is chosen to dissolve and dissociate these salts and achieve a lower electrical resistance. On the other, the co-ion (organic anions R^-) leakage will decrease because more organic anions are inclined to stay in the aqueous-organic medium instead of the aqueous one.

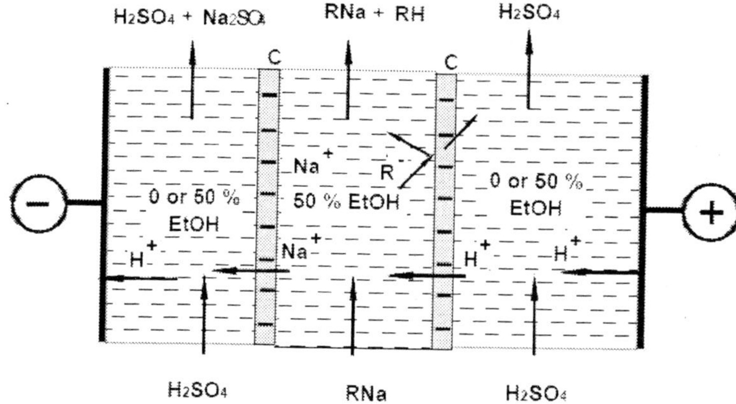

Figure 18. Schematic of the production of low water-soluble organic acids. [45,130] C, cation-selective membrane; EtOH, ethanol; NaR, sodium organic salt; RH, organic acid; R^-, organic anion; ▭, aqueous-organic phase.

Another obstacle for ED to cooperate directly with fermentation is that the broth contains much more sodium organic salts while ED does not possess the specific selectivity for the desired product. Extraction can be used to separate the desired organic acid from other organic components, but it is time-consuming due to the low transport rate and has a weak concentrating capability owing to the dilution by extractant and back-extractant solutions. In order to reinforce ion transport, two-phase electrophoresis (TPE) [133] can be employed. TPE accelerates ion migration by applying a direct current field and selectively collect the organic acid by specific extractants without water electro-osmosis and osmosis. The process was schematically shown in Figure 19.

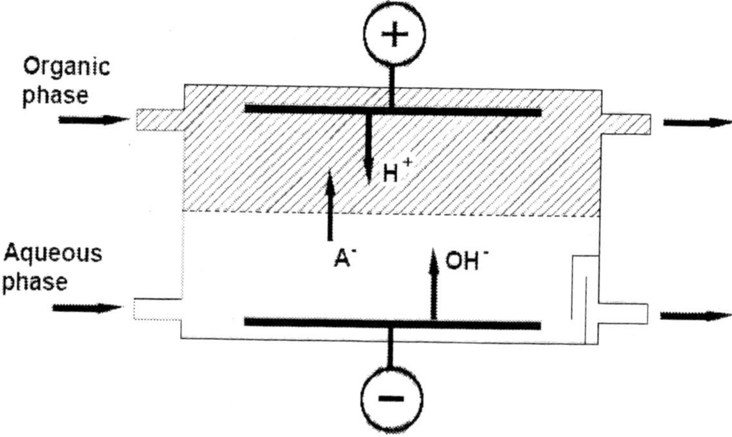

Figure 19. Schematic of organic acid production by using TPE [133].

A^-, organic anion; ▨, organic phase; □, aqueous phase.

3.2.3. Application of Electrodialysis in Organic System

The application of electrodialysis in organic medium is not so common. An example shown here is the preparation of methyl palmitic ester ($PaOOCH_3$) from palmitoyl chloride (PaOCl) [45]. As shown in Figure 20, dichloromethane (CH_2Cl_2) is used to dissolve $PaOOCH_3$ and PaOCl, and tetrapropyl ammonium bromide ($(Pr)_4NBr$) is the supporting electrolyte and used to increase the electrical conductance of this compartment. After dehalogenation, PaOCl reacts with methoxide ions (CH_3O^-) and forms the corresponding ester. The ester can be separated from the mixture by following water solution and evaporation under vacuum. The obtained $PaOOCH_3$ can attain a purity of more than 99 %, which is higher than that of the commercial one.

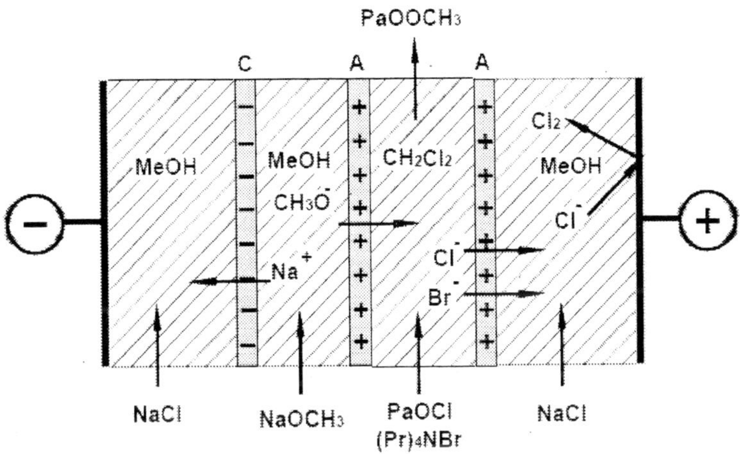

Figure 20 Schematic of the preparation of methyl palmitic ester by conventional electrodialysis.

A, anionic membrane; C, cationic membrane; $PaOCH_3$, methyl palmitic ester; PaOCl, palmitoyl chloride; $(Pr)_4NBr$, tetrapropyl ammonium bromide; CH_2Cl_2, dichloromethane; MeOH, methanol; CH_3ONa, sodium methoxide; CH_3O^-, methoxide ion; ▨, organic phase.

3.3. Electrodialysis with Bipolar Membranes (EDBM)

Electrodialysis with Bipolar Membranes (EDBM), an innovative electrochemical process, has stimulated many applications by its ability to economically reverse neutralization and eliminate salt effluents, and simultaneously recycle acids and bases. In addition, it can be used to acidify or basify streams without adding extra chemicals. Therefore, it has found many applications in the chemical industry, biotechnology, food processing, and especially pollution control (e.g., treating waste water, and recycling resources from industrial effluents) [2,7,20-21, 26-27,30-31]. However, the number of installed EDBM plants worldwide has been relatively small to date. There are many reasons, such as high investment and operating costs, doubts of the technical reliability from the industries, and the limited scope of economical/practical applications. Some relevant applications based on the water or alcohol dissociation will be discussed below in detail.

3.3.1. Application of EDBM Based on Water Dissociation

Till now, many EDBM applications have focused on the water dissociation ability of bipolar membrane. In principle, some examples for conventional ED as mentioned above can be rearranged into EDBM where the needed acid or base are produced in-situ by water dissociation, such as the production of tetramethylammonium hydroxide or poly aluminum chloride (Figures 16 and 17).

A very promising application of electrodialysis with bipolar membranes is the production of organic acids from fermentation processes. Some examples can be found in Chapter 7.

There are many more organic acids that can be produced by using this technique, such as lactic acid, gluconic acid, ascorbic acid, amino acid, and citric acid, etc. Of special interest is the production of lactic acid using EDBM, which was installed by Eurodia Industry in 1997 in France [136]. The plant was initially designed to produce 2.6 T/year of acid (100%) over 8,000 hours of operation. Two EUR20-240 stacks with 81 m^2 of effective cell area were installed in the two-compartment configuration using Neosepta® CMB cation-exchange and BP1 bipolar membranes. The acid conversion rate (purity) was 98 % and the acid concentration was 390 g/l (conductivity, 3 mS/cm). NaOH was produced at a concentration of 6 w% to be reused in the fermentation. The plant operated in a batch mode and consistently met the customer's requirements. After several years of operation, a membrane life of 20,000 hours was reached for the bipolar membranes, while the cation-exchange membranes had been replaced after 18,000 hours. The electrodes were not changed. With the increasing demands, the plant increased the capacity in several steps by adding several EUR20 and EUR40 stacks, and now an additional capacity expansion is being considered. By changing the operation from batch mode to feed & bleed mode, the base concentration could be increased from 6 w% to 8 w% with the same current efficiency (for a new plant, NaOH could be produced at 10 w%). The main operating costs are power at 0.88 kWh/kg of produced acid and membrane (electrodes) replacement at $0.09/kg.

A recent large scale EDBM was installed in China in 2007 for the production of dextrose-δ-lactone (intramolecular condensation of gluconic acid). Unlike most organic acid produced by fermentation technology, the preferable technique for gluconic acid production is catalytic oxidation of glucose into sodium gluconate followed by acidification using ion exchange as shown in Figure 21 (a). The formed sodium gluconate can not be fed direct to ion exchange because there are some impurities in the solution which prevent gluconic acid from crystallization. Consequently, a great amout of coal and electricity is cunsumed to supply the energy for purification of sodium gluconate: condensation under vaccumm, crystallization, centrifugation, and drying. Obviously, this purification process gives rise to air, water, and solid pollution and adds much to the product cost. Besides, ion exchange is not an environmentally sound technique for acidification due to the use of acid and water for resin rehabilitation and the formation of salts during rehabilitation. In contrast, EDBM can be used to achieve a clean production for conversion of gluconate into gluconic acid. In a 3-compartment EDBM as shown in Figure 21 (b), the solution from catalytic oxidation is fed into the salt compartment, and gluconic acid and NaOH are generated in the acid and base compartments, respectively. Under the optimized conditions, the conversion rate can attain as high as 98.6%, and current efficiency as high as 71.5%. The formed gluconic acid has a high purity and can be used directly for crystallization. The formed NaOH can be recycled as the base for catalytic oxidation.

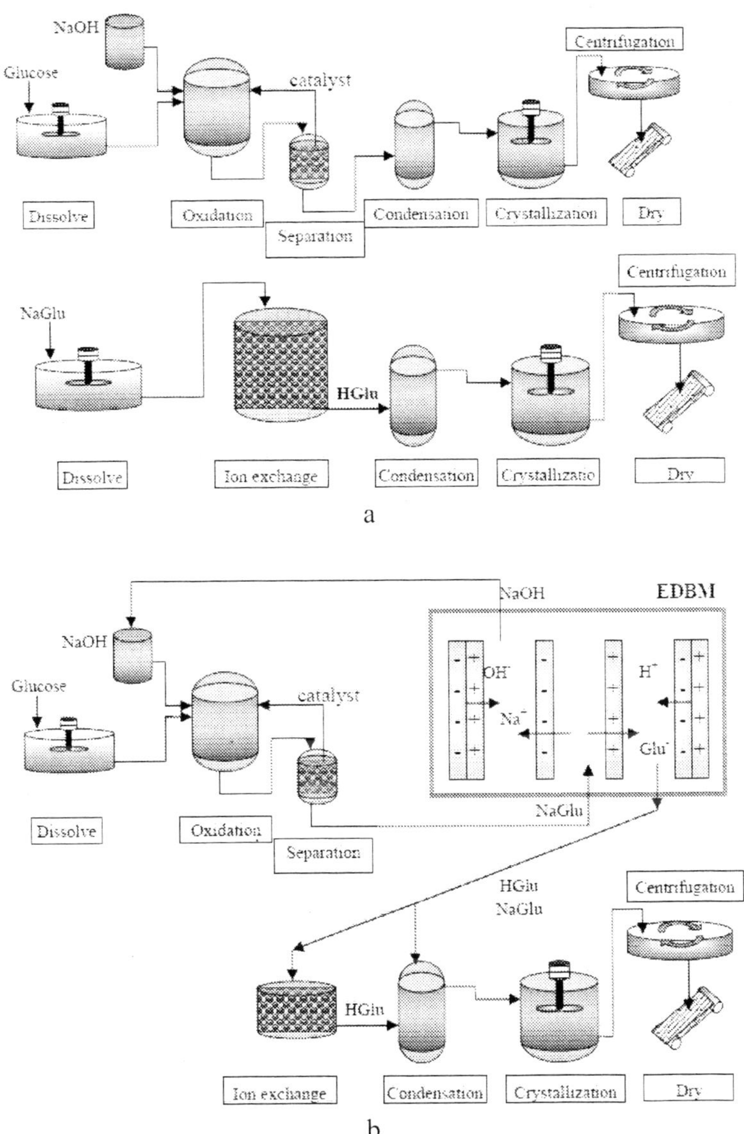

Figure 21. Simplified flow diagram of the production of gluconic acid: (a) traditional method, and (b) continuous production using electrodialysis with bipolar membranes. NaGlu, sodium gluconate; HGlu, gluconic acid.

Another application of EDBM recently developed is the regeneration of desulfurizing agent [137-139]. For example, Piperazine (Pz) is known as a commonly used modifier to remove carbon dioxide from synthesis gas or flue gas because of the fast kinetics of the sorption reaction, and the property for the formed piperazine sulfite ($PzH_2^{2+}SO_3^{2-}$ or $[PzH^+SO_3^{2-}]^-$) to be regenerated by heating. However, more and more piperazine sulfate ($PzH_2^{2+}SO_4^{2-}$ or $[PzH^+SO_4^{2-}]^-$) will be formed in the circulation because of oxidation of sulfur dioxide and/or piperazine sulfite, and it can not be thermally regenerated. The heat stable salt not only decreases the efficiency of desulfurizing operation, but also causes a secondary pollution and waste of resources. One effective measure is to convert such heat stable salts by EDBM. Figure 22 presents a closing loop with desulfurization and recycling of desulfurizing agents, where the thermal regeneration unit is used to convert piperazine sulfites while EDBM is used to convert the heat stable salt--piperazine sulfate. In such way, the desulfurizing process is kept from piperazine loss and runs stably and efficiently. The results of EDBM experiments indicate that the low energy consumption and high current efficiency are achieved when applying electrolyte solutions of middle concentration, piperazine sulfate solution of middle concentration, and EDBM stack of BP-C-C configuration. The results also indicate that when applying a high current density to the EDBM stack, it has a high current efficiency and energy consumption. Besides piperazine, the heat stable salts formed with other alkanolamines, such as monoethanolamine, diethanolamine, and *N,N'*-dimethylethanolamine, can be recovered in the same way [138].

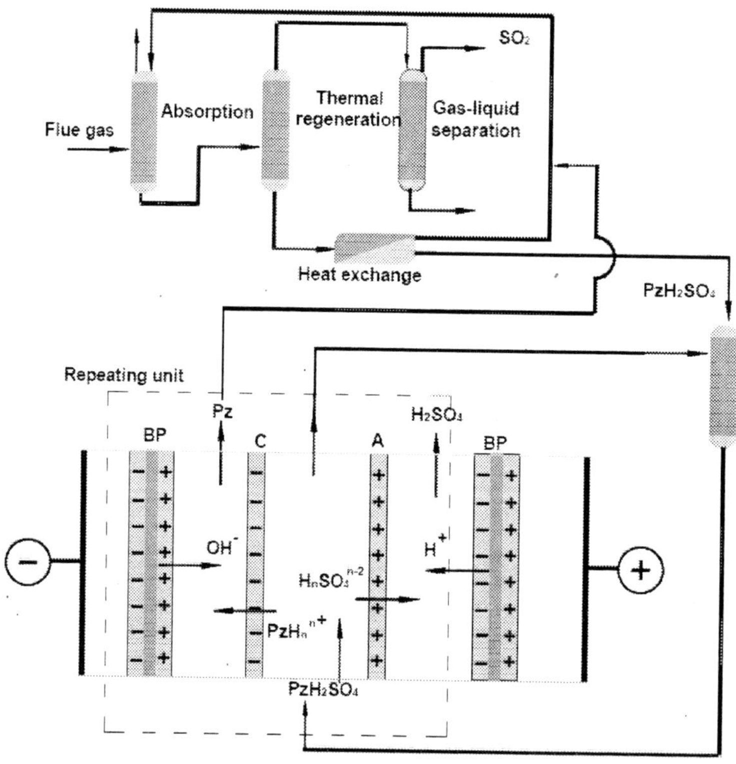

Figure 22. Schematic of the regeneration of heat stable desulfurizing agents by using EDBM [137]. Pz, piperazine; PzH_2SO_4, piperazine sulfate.

3.3.2. Application of EDBM Based on Alcohol Splitting

Bipolar membranes may not only be used for the electrodialytical dissociation of water. They can also be applied for the electrodialytical dissociation of alcohols and thus for the production of alcoholates [49-50, 139], which can realize some novel and "green" paths for organic synthesis. For example, the production of methyl palmitic ester shown in Figure 20 can be directly conducted with EDBM if the middle anionic membrane is replaced by a bipolar membrane. This avoids using an expensive chemical: sodium methoxide.

By means of EDBM, sodium methoxide can be produced from methonol and acetate sodium. In this case, the methanol (CH_3OH) molecules dissociates into methoxyl ions (CH_3O^-) and hydrogen ions (H^+) in the transition region of a bipolar membrane. Then sodium methoxide (CH_3ONa) is formed after the methoxyl ions (CH_3O^-) combine with the Na^+ ions, which migrate from the adjacent compartment filled with a sodium acetate solution. In comparison with the production of CH_3ONa by electro-electrodialysis, this process is free from Kolbe reactions or alcohol oxidation, and thus more cost-effective and environmentally sound. This research is very important not because it decreases the production cost of CH_3ONa or other metal alkoxides but because it initiates the dissociation of non-aqueous solvents by using EDBM, which can realize some novel and "green" paths for organic synthesis. An example shown here is the synthesis of acetoacetic ester. Conventionally, there two main routes: one is from the sodium methoxide as shown in scheme 5(a), and another is via the catalytic reaction between diketene and methanol as shown in scheme 5(b). Both routes demonstrate several disadvantages and thus have been obsoleted. For examples, in the first path, the CH_3ONa employed for enolization is expensive, and adds to the production cost; some by-products are formed, i.e. CH_3OH and Na_2SO_4, which results in a low efficiency of resource utilization. Especially, the formed Na_2SO_4 are not regenerated and thus pollutes the environment after discharge. In the second path, the process has a latent peril caused by diketene, which is susceptible to explosion and needs delicate care in storage.

The new synthesis of acetoacetic ester proposed by Sridhar et al. does not follow the present paths but revives the traditional synthesis by using the EDBM with methanol splitting (Scheme 5(c)). As shown in Figure 23, the technique proposed by Sridhar et al. [50] is realized by a two-compartment EDBM. In the stack, the methoxyl ions generated by methanol splitting react with Na^+ and CH_3COOCH_3 and form the sodium enolate of acetoacetic ester, which is circulated to adjacent compartment to undergo H^+ substitution and then produce acetoacetic ester. Compared with traditional paths, the EDBM show many advantages: (1)the CH_3O^- for enolization and the H^+ for protonation are directly supplied by the CH_3OH splitting in bipolar membranes and thus an extra addition of CH_3ONa and H_2SO_4 is avoided; (b) half the CH_3OH formed at the step of condensation are consumed at the step of protonation; (c) the salt formed is CH_3ONa, which can be reused for the first step. Therefore, Na^+ ions can be recycled in this process and there is no pollutant formed.

$$2CH_3COOCH_3 + CH_3ONa \longrightarrow CH_3\underset{\underset{O^-Na^+}{|}}{C}{=}CHCOOCH_3 + 2CH_3OH$$

$$CH_3\underset{\underset{O^-Na^+}{|}}{C}{=}CHCOOCH_3 + 1/2H_2SO_4 \longrightarrow CH_3COCH_2COOCH_3 + 1/2Na_2SO_4$$

a. The synthesis of acetoacetic ester from the sodium methoxide

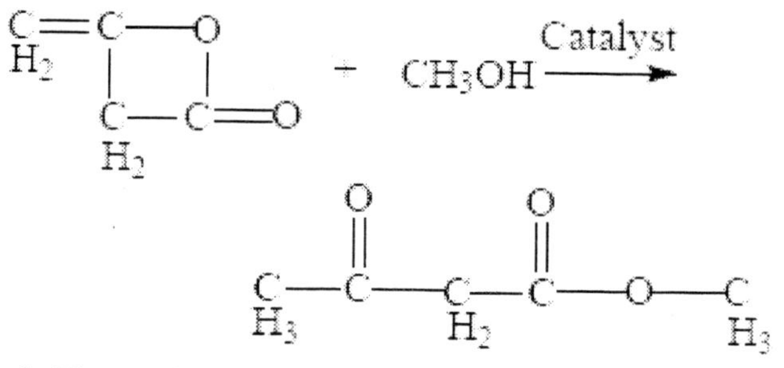

$$\begin{array}{l} \underset{H_2}{C}{=}C{-}O \\ \quad\;\; | \qquad | \\ \quad \underset{H_2}{C}{-}C{=}O \end{array} + CH_3OH \xrightarrow{\text{Catalyst}}$$

$$\underset{H_3}{C}-\overset{\overset{\displaystyle O}{\|}}{C}-\underset{H_2}{C}-\overset{\overset{\displaystyle O}{\|}}{C}-O-\underset{H_3}{C}$$

b. The synthesis of acetoacetic ester via the catalytic reaction between diketene and methanol.

$$2CH_3COOCH_3 + CH_3ONa \longrightarrow$$

$$CH_3\underset{\underset{\displaystyle O^-Na^+}{|}}{C}{=}CHCOOCH_3 + 2CH_3OH$$

$$CH_3\underset{\underset{\displaystyle O^-Na^+}{|}}{C}{=}CHCOOCH_3 + CH_3OH \longrightarrow$$

$$CH_3COCH_2COOCH_3 + CH_3ONa$$

c. The synthesis of acetoacetic ester from sodium methoxide in-situ produced from electrdialysis with bipolar membranes.

Scheme 5. Comparison of the synthesis routes of acetoacetic ester.

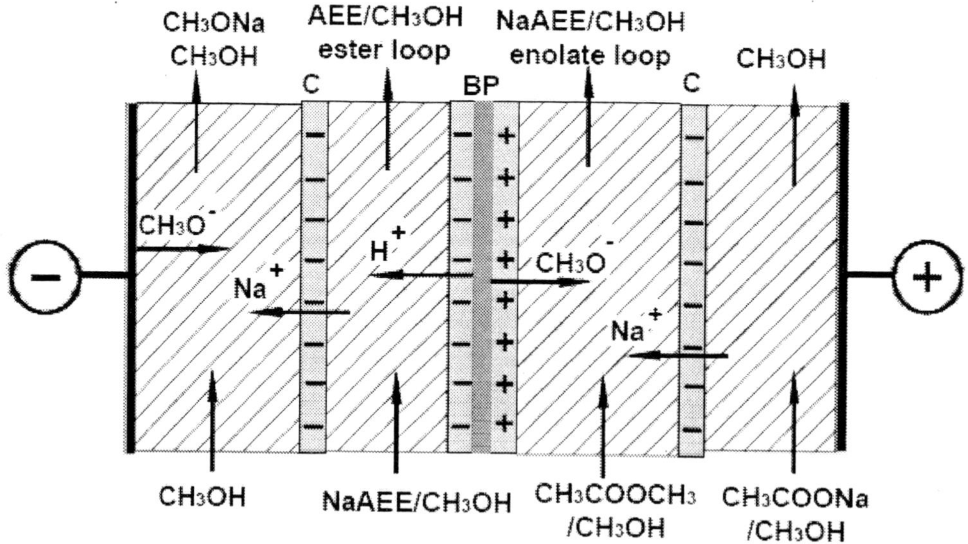

Figure 23. Schematic of the production of acetoacetic ester by EDBM; BP, bipolar membrane; C, cation-selective membrane; AEE, acetoacetic ester; NaAEE, the sodium elonate of acetoacetic ester, ▨, organic phase.

Theoretically, the EDBM with alcohol splitting can be applied to any condensation syntheses using alkoxides to introduce alkoxyl groups to compounds, especially the Claisen condensation and the intramolecular Dickman condensation. As only concerns the sodium methoxide-methanol (MeONa-MeOH), this system can be widely used for organic synthesis, especially for deprotection, methoxydefluorination, debenzoylation, deacetylation base-induced monodehydrohalogenation, and reductive carbonylation. This leaves an enormous space for the further development and application of EDBM. Furthermore, other ED processes also have a great potential to contribute to the cleaner production in organic chemical industries.

It is difficult to list all the developing or to-be developed ion exchange membrane based processes and their applications. As a supplement, Chapter 7 of this book will describe some ion exchange membrane-integrated processes.

4. Conclusive Remarks and the Perspective

Substantial descriptions have been made on the preparation and application of ion exchange membranes. Special interests are given on the clean production and environmental protection using diffusion dialysis, conventional electrodialysis, and elctrodialysis with bipolar membranes. Ion exchange membrane technologies appear to be versatile and capable of solving different industrial problems. We feel that most of the future developments in the area will come from those developers or users who think of these technologies as tools to cope with their specific treatment requirements. But it should be emphasized that, for any purpose, preparation of ion exchange membranes or materials is the most crucial. For this point, a reverse route has been proposed to the conventional one: designing a specific membrane for a specific purpose [7]. It starts from collection of the membrane properties needed for a specific process, determination of membrane properties and membrane structure, and finally design of the membrane at the molecular level, i.e., from molecule to process. The precondition is to make clear the relationship between membrane strucrure and membrane properties.

Apart from the preparation of ion exchange membranes, the technical and commercial relevance of the ion exchange membrane based processes should also be considered. Electrodialysis, a promising technology for cleaner production and environmental protection, has a long way to go into the general industrial stages due to the very limited choice of membranes and the extremely high equipment cost. To date, many applications are limited to aqueous systems. The application in non-aqueous media, even feasible technically, are not available on industrial scale due to the poor stability of membranes and spacers, and the high electrical resistance which results in higher energy consumption and low current efficiency.

Diffusion dialysis, a mature technology, has found many applications in industries. But now all the equipments are in the plate frame and with several shortcomings, such as heavy weight, and low effective membrane area per unit volume. Hence, there is a limitation on processing capacity. A desired configuration is of hollow fiber, which can save much of equipment cost and operation space.

Compared with pressure-driven membrane processes, much fewer applications of ion exchange membrane-based processes can be considered modern technology except conventional desalination or concentrating of electrolyte solutions, production of pure water using continuous electrodeionization, and some specific applications using bipolar membranes (e.g., producing organic acid from fermentation broth, recovering HF and HNO_3 from a waste stream generated by neutralization of a steel pickling bath). Many applications, such as production of acids and bases by electrodialysis with bipolar membranes, are still under the pilot- or laboratory-scale test. The most main reasons are the high cost of membranes (e.g., bipolar membranes), and the stability of electrodes and operation. Much work should be done to reduce the cost of membranes and electrodes and enhance the cell configurations. Fortunately, in some applications, ion exchange membrane based processes provide higher quality products or are more environmentally friendly and thus will be used in spite of the cost disadvantage. The increase of environmental awareness and raw material cost has accerlated the application of ion exchange membrane-based separation especially in highly industrialized and densely populated countries. In the long term, ion exchange membrane-based separation can compete with other mass separation techniques, especially other membrane separation technologies. Their development has benefited from an interdisciplinary approach integrating recent advances in the every field of science and technology; therefore, researchers should give considerations to interdisciplinary knowledge, such as material, inorganic, polymer science and technology, to solve some touchy problems, such as materials design, apparatus design, and operation optimization in electro-membrane processes.

Acknowledgments

Financial supports from the Natural Science Foundation of China (No. 20636050, 20576130, 20774090), National Basic Research Program of China ((No. 2009CB623403)) and Key Foundation of Educational Committee of Anhui Province (KJ2007A016) are greatly appreciated. Gratitude is also given to Mr. Chuanhui Huang for proofreading the manuscript and revising some drawings.

References

[1] Helfferich, F. Ion-exchange. London: McGraw-Hill; 1962.

[2] Strathmann, H. Ion-exchange membrane separation processes. Amsterdam: Elsevier; 2004.

[3] Sata, T. Recent trends in ion-exchange research. *Pure Appl. Chem.* 58 (1986) 1613-1626.

[4] Meares, P. Trends in ion-exchange membrane science and technology. In: Flett, D.S. Ion-Exchange Membranes. Chichester: E. Horwood Ltd.;1983.

[5] Wu, CM; Xu, TW; Liu, JJ. Charged Hybrid Membranes by the Sol-gel Approach: Present States and Future Perspectives. In: Newman, AM. Focus on Solid State Chemistry. New York: Nova Science Publishers Inc.; 2006;1-44.
[6] Wu, YH;Wu, CM; Yu, F; Xu,TW; Fu, YX. Free-standing anion-exchange PEO-SiO_2 hybrid membranes. *J. Membr. Sci.* 307 (2008) 28-36.
[7] Xu, TW. Ion exchange membranes: state of their development and perspective. *J. Membr. Sci.* 263(2005) 1-29.
[8] Ostwald, W. Elektrische Eigenschaften halbdurchlässiger Scheidewände. *Z. Physik. Chemie*. 6 (1890) 71-82.
[9] Donnan, FG. The theory of membrane equilibrium in presence of a non-dialyzable electrolyte, *Z. Electrochem.* 17 (1911) 572-581.
[10] Michaelis, L; Fujita, A. The electric phenomena and ion permeability of membranes. II. Permeability of apple peel. *Biochem. Z.* 158 (1925) 28-37.
[11] Söllner, K. Uber Mosaikmembranen. *Biochem.* Z. 244 (1932) 370-381.
[12] Wassenegger, H; Jaeger, K. Effecting cation-exchange in removing calcium from hard waters. U.S. Patent 2,204,539, 1940.
[13] Meyer, KH; Strauss, H. La perméabilité des membranes VI, Sur le passage du courant électrique a travers des membranes sélective. *Helv. Chim. Acta*, 23 (1940) 795-800.
[14] Juda, M; McRae, WA. Coherent ion-exchange gels and membranes. *J. Am. Chem. Soc.* 72 (1950) 1044-1044.
[15] Winger, AG; Bodamer, GW; Kunin, R. Some electrochemical properties of new synthetic ion-exchange membranes. *J. Electrochem. Soc.* 100 (1953) 178-184.
[16] Nishiwaki, T. Concentration of electrolytes prior to evaporation with an electromembrane process, In: Lacey, RF; Loch S. Industrial process with membranes, New York: Wiley-Interscience; 1972.
[17] Mihara, K; Kato, M. Polarity reversing electrode units and electrical switching means therefore. *U.S. Patent* 3,453,201, 1969.
[18] Grot, WG. Laminates of support material and fluorinated polymer containing pendant side chains containing sulfonyl groups. *U.S. Patent* 3, 770,567, 1973.
[19] Chlanda, FP; Lee, LTC; Liu, KJ. Bipolar membranes and method of making same. *U.S. Patent* 4,116,889, 1976.
[20] Pourcelly, G; Gavach, C. Electrodialysis water splitting-application of electrodialysis with bipolar membranes, In: Kemperman, AJB. Handbook on Bipolar membrane Technology. Enschede: Twente University Press; 2000;17-46.
[21] Huang, CH; Xu, TW. Electrodialysis with bipolar membranes (EDBM) for sustainable development (Critical Review). *Environ.. Sci. Technol.* 40 (2006) 5233-5243.
[22] Daufin, G; Escudier, JP; Carrere, H; Berot, S; Fillaudeau, L; Decloux, M. Recent and emerging applications of membrane processes in the food and dairy industry. *Food Bioprod. Process.* 79 (2001) 89 102.
[23] Tarvainen, T; Svarfvar, B; Akerman, S; Savolainen, J; Karhu, M; Paronen, P; Jarvinen, K. Drug release from a porous ion-exchange membrane in vitro. *Biomaterials*. 20 (1999) 2177-2183.
[24] Kim, YH; Moon, S.H. Lactic acid recovery from fermentation broth using one-stage electrodialysis. *J. Chem. Tech. Biotech.* 76 (2001) 169-178.

[25] Saracco, G. Ionic membrane technologies for the recovery of valuable chemicals from waste waters. *ANN Chim-ROME*. 93 (2003) 817-826.

[26] Bazinet, L; Lamarche, F; Ippersiel, D. Bipolar membrane electrodialysis: applications of electrodialysis in the food industry. *Trends Food Sci. Tech*. 9 (1998) 107-113.

[27] Huang, CH; Xu, TW; Zhang, YP; Xue, YH; Chen, GW. Application of electrodialysis to the production of organic acids: state-of-the-art and recent developments (review). *J. Membr. Sci.* 288 (2007) 1-12.

[28] Matejka, Z. Continuous production of high-purity water by electrodeionization. *J. Appl. Chem. Biotechnol*. 21 (1971) 117-120.

[29] Korngold, E. Electrodialysis processes using ion-exchange resins between membranes. *Desalination* 16 (1975) 225–233.

[30] Xu, TW; Fu, RQ; Huang, CH. Towards the cleaning production using electrodialysis with bipolar membranes-a review (invited review). *Trends in Chem Eng*. 10 (2006) 17-29.

[31] Schlichter, B; Mavrov, V; Erwe, T; Chmiel, H. Regeneration of bonding agents loaded with heavy metals by electrodialysis with bipolar membranes. *J. Membr. Sci.* 32 (2004) 99-105.

[32] Kim, JO; Jeong, JT; Kim, SK; Kim, RH; Lee, YJ. Development of novel wastewater reclamation system using microfiltration with advanced new membrane material and electrodialysis. *Mater. Sci. Forum*. 510-511 (2006) 586-589.

[33] van der Bruggen, B; Milis, R; et al. Electrodialysis and nanofiltration of surface water for subsequent use as infiltration water. *Water Res*. 37 (2003) 3867-3874.

[34] Koprivnjak, JF; Perdue, EM; Pfromm,PH. Coupling reverse osmosis with electrodialysis to isolate natural organic matter from fresh waters. *Water Res*. 40 (2006) 3385-3392.

[35] Shee, FLT; Araya-Farias, M; Arul, J; Mateescu, AM; Brunet, S; Bazinet, L. Chitosan solubilization by bipolar membrane electro-acidification. *Desalination* 200 (2006) 623-623.

[36] Bazinet, L; Lamarche, F; Ippersiel, D; Amiot, J. Bipolar membrane electroacidification to produce bovine milk casein isolate. *J. Agric. Food Chem*. 47 (1999) 5291-5296.

[37] Mondor, M; Ippersiel, D; Lamarche, F; Boye, JI. Production of soy protein concentrates using a combination of electro-acidification and ultrafiltration. *J. Agric. Food Chem*. 52 (2004) 6991–6996.

[38] Mondor, M; Ippersiel, D; Lamarche, F; Boye, JI. Effect of electro-acidification treatment and ionic environment on soy protein extract particle size distribution and ultrafiltration permeate flux. *J. Membr. Sci.* 231 (2004) 169-179.

[39] Alibhai, Z; Mondor, M; Moresoli, C; Ippersiel, D; Lamarche, F. Production of soy protein concentrates/isolates: traditional and membrane technologies. *Desalination* 191 (2006) 351-358.

[40] Poulin, JF; Amiot, J; Bazinet, L. Improved peptide fractionation by electrodialysis with ultrafiltration membrane: influence of ultrafiltration membrane stacking and electrical field strength. *J. Membr. Sci.* 299 (2007) 83-90.

[41] Poulin, JF; Amiot, J; Bazinet, L. Simultaneous separation of acid and basic bioactive peptides by electrodialysis with ultrafiltration membrane. *J. Biotechnol.* 123 (2006) 314-328.

[42] Pellegrino, J; Chapman, M. Desalination: perspectives and state-of-the art. Technical lecture in Gist, Guangju, Korea, 2007, 85-89.

[43] Shee, FLT; Angers, P; Bazinet, L. Precipitation of cheddar cheese whey lipids by electrochemical acidification. *J. Agric. Food Chem.* 53 (2005) 5635-5639.

[44] Wisniewski, C; Persin, F; Cherif, T; Sandeaux, R; Grasmick, A; Gavach, C. Denitrification of drinking water by the association of an electrodialysis process and a membrane bioreactor: feasibility and application. *Desalination* 139 (2001) 199-205.

[45] Xu, FN; Innocent, C; Pourcelly, G. Electrodialysis with ion exchange membranes in organic media. *Sep. Purif. Technol* 43 (2005) 17-24.

[46] Chou, TJ; Tanioka, A. Ionic behavior across charged membranes in methanol–water solutions. 2. Ionic mobility. *J. Phys. Chem. B* 102 (1998) 129-133.

[47] Chou, TJ; Tanioka, A. Membrane potential across charged membranes in organic solutions. *J. Phys. Chem. B* 102 (1998) 7198-7202.

[48] Kameche, M; Xu, FN; Innocent, C; Pourcelly, G. Electrodialysis in water-eathanol solutions: application to the acidification of organic salts. *Desalination* 154 (2003) 9-15.

[49] Sridhar, S. Electrodialysis in non-aqueous medium: production of sodium methoxide. *J. Membr. Sci.* 113 (1996) 73-79.

[50] Sridhar, S; Feldmann, C. Electrodialysis in a non-aqueous medium: a clean process for the production of acetoacetic ester. *J. Membr. Sci.* 124 (1997) 175-179.

[51] Bishop, HK; Bittels, JA; Gutter, GA. Investigation of inorganic ion-exchange membranes for electrodialysis. *Desalination.* 6 (1969) 369-380.

[52] Bregman, JI; Braman, RS. Inorganic ion-exchange membranes. *J. Colloid Sci.* 20 (1965) 913-922.

[53] Shrivastava, K; Jain, AK; Agrawal, S; Singh, RP. Studies with inorganic ion-exchange membranes. *Talanta.* 25 (1978) 157-159.

[54] Ohya, H; Paterson, R; Nomura, T; McFadzean, S; Suzuki, T; Kogure, M. Properties of new inorganic membranes prepared by metal alkoxide methods. Part I. A new permselective cation exchange membrane based on oxides. *J. Membr. Sci.* 105 (1995) 103-112.

[55] Kogure, M; Ohya, H; Paterson, R; Hosaka, M; Kim, J; McFadzean, S. Properties of new inorganic membranes prepared by metal alkoxide methods. Part II. New inorganic-organic anion-exchange membranes prepared by the modified metal alkoxide methods with silane coupling agents. *J. Membr. Sci.* 126 (1997) 161-169.

[56] Ohya, H; Masaoka, K; Aihara, M; Negishi, Y. Properties of new inorganic membranes prepared by metal alkoxide methods. Part III. New inorganic lithium permselective ion exchange membrane. *J. Membr. Sci.* 146 (1998) 9-13.

[57] Wu, CM; Xu, TW; Liu, JS. Charged hybrid membranes by the sol-gel approach: Present states and future perspectives. In: Newman, AM. Focus on Solid State Chemistry. New York: Nova Science Publishers Inc.; 2006; 1-44.

[58] Kariduraganavar, MY; Nagarale, RK; Kittur, AA, Kulkarni SS Ion-exchange membranes: preparative methods for electrodialysis and fuel cell applications. *Desalination* 197 (2006) 225-246

[59] Ginsete, J; Garraud, J; Pourcelly, G. Grafting of acrylic acid with diethyleneglycol dimethacrylate onto radioperoxide polyethylene. *J. Appl. Polym. Sci.* 48 (1993) 2113-2122.

[60] Hegazy, EA; Taher, NH; Kamal, H. Preparation and properties of cationic membranes obtained by radiation grafting of methacrylic acid onto PTFE films. *J. Appl. Polym. Sci.* 38 (1989) 1229-1242.

[61] El-Sawy, NM; Hegazy, EA; Rabie, AM; Hamed, A; Miligy, GA. Radiation-initiated graft copolymerisation of acrylic acid and vinyl acetate onto LDPE films in two individual steps. *Polym. Int.* 32 (1994) 131-135.

[62] Nasefa, MM; Hegazy, ESA. Preparation and applications of ion exchange membranes by radiation-induced graft copolymerization of polar monomers onto non-polar films. *Prog. Polym. Sci.* 29 (2004) 499-561.

[63] Gupta, B; Büchi, F; Scherer, G; Chapiro, A. Crosslinked ion exchange membranes by radiation grafting of styrene/divinylbenzene into FEP films. *J. Membr. Sci.* 118 (1996) 231-238.

[64] Gupta, B; Büchi, F; Staub, M; Grman, D; Scherer, GG. Cation exchange membranes by pre-irradiation grafting of styrene into FEP films. II. Properties of copolymer membranes. *J. Polym. Sci. A.* 34 (1996) 1873-1880.

[65] Nasef, MM; Saidi, H. Preparation of crosslinked cation exchange membranes by radiation grafting of styrene/divinylbenzene mixtures onto PFA films. *J. Membr. Sci.* 216 (2003) 27-38.

[66] Yamaki, T; Asano, M; Maekawa, Y; Morita, Y; Suwa, T; Chen, J; Tsubokawa, N; Kobayashi, K; Kubota, H; Yoshida, M. Radiation grafting of styrene into crosslinked PTFE films and subsequent sulfonation for fuel cell applications. *Radiat Phys. Chem.* 67 (2003) 403-407.

[67] Choi, YJ; Kang, MS; Moon, SH. Characterization of semi-interpenetrating polymer network polystyrene cation-exchange membranes. *J. Appl. Polym. Sci.* 88 (2003) 1488-1496.

[68] Choi, YJ; Kang, MS; Moon, SH. A new preparation method for cation-exchange membrane using monomer sorption into reinforcing materials. *Desalination* 146 (2002) 287-291.

[69] Choi, YJ; Moon, SH; Yamaguchi, T; Nakao, SI. New morphological control for thick, porous membranes with a plasma graft-filling polymerization. *J. Polym. Sci. A* 41 (2003) 1216-1224.

[70] Yamaguchi, T; Miyata, F; Nakao, S. Polymer electrolyte membranes with a pore-filling structure for a direct methanol fuel cell. *Adv. Mater.* 15 (2003) 1198-1201.

[71] Deluca, NW; Elabd, YA. Polymer electrolyte membranes for the direct methanol fuel cell: A review. *J. Polym. Sci. B* 44 (2006) 2201-2225.

[72] B. Smitha, S. Sridhar, A.A. Khan. Solid polymer electrolyte membranes for fuel cell applications - a review. *J. Membr. Sci.*. 259 (2005) 10-26.

[73] Harrison, WL; Hickner, MA; Kim, YS; McGrath, JE. Poly(arylene ether sulfone) copolymers and related systems from disulfonated monomer building blocks: Synthesis, characterization, and performance - A topical review. *Fuel Cells* 5 (2005) 201-212.

[74] Kerres, J; Hein, M; Zhang, W; Graf, S; Nicoloso, N. Development of new blend membranes for polymer electrolyte fuel cell applications. *J. New Mater. Electrochem. Syst.* 6 (2003) 223-230.

[75] Rikukawa, M; Sanui, K. Proton-conducting polymer electrolyte membranes based on hydrocarbon polymers. *Prog. Polym. Sci.* 25 (2000) 1463-1502.

[76] Mizutani, Y; Yamane, R; Ihara, H; Motomura, H. Studies of ion exchange membranes.16. the preparation of ion exchange membranes by the paste method. *Bull. Chem. Soc. Jpn.* 36(1963) 361-366.

[77] Xu,TW; Woo, JJ; Seo, SJ; Moon, SH. In-situ Polymerization: a novel solvent-free route for thermally-stable proton-conductive membranes. to be submitted to *J. Membr. Sci.*

[78] Xu, TW. Alternative routes for anion exchange membranes by avoiding the use of hazardous material chloromethyl methyl ether. In: Hudson RC.Hazardous Materials in the Soil and Atmosphere: Treatment, Removal and Analysis. New York: Nova Science Publishers Inc.; 2006; 55-88.

[79] Xu, TW; Yang, WH. Fundamental study of a new series of anion exchange membranes: Membrane preparation and characterization. *J. Membr. Sci.* 190 (2001) 159-166.

[80] Xu, TW; Zha, FF. Fundamental study of a new series of anion exchange membranes: Effect of simultaneous amination-crosslinking processes on membranes' ion exchange capacity and dimensional stability. *J. Membr. Sci.* 199 (2002) 203-210.

[81] Xu, TW; Yang, WH. Sulfuric acid recovery from titanium white (pigment) waste liquor using diffusion dialysis with a new kind of homogenous anion exchange membrane--Static runs. *J. Membr. Sci.* 183 (2001) 193-200.

[82] Xu, TW; Fu, RQ; Yang, WH; Xue, YH. Fundamental studies on a novel series of bipolar membranes prepared from poly(2,6-dimethyl-1,4-phenylene oxide)(PPO): (II) Effect of functional group type of anion exchange layers on I-V curves of bipolar membranes. *J. Membr. Sci.* 279 (2006) 233-241.

[83] Wu, L; Xu, TW; Yang, WH. Fundamental studies of a new series of anion exchange membranes: membranes prepared through chloroacetylation of poly(2,6-dimethyl-1,4-phenylene oxide)(PPO) followed by quaternary amination. *J. Membr. Sci.* 286 (2006)185-192.

[84] Wu, L; Xu, TW; Wu, D; Zheng, X. Preparation and characterization of CPPO / BPPO blend membranes for potential application in alkaline direct methanol Fuel Cell. *J. Membr. Sci.* 310 (2008) 577-585.

[85] Ge, DC. Preparation of IM-2 homogeneous strongly basic anion exchange membrane and its application in desalination of brackish water. *Chin. J. Membr. Sci. Tech.* (in Chinese). 9 (1989) 26-30.

[86] Sunaga, K; Kim, M; Saito, K; Sugita, K; Sugo, T. Characteristics of porous anion-exchange membranes prepared by cografting of glycidyl methacrylate with divinylbenzene. *Chem. Mater.* 11 (1999) 1986-1989.

[87] Wilhelm, F; Vegt, NVD; Wessling, M; Strathmann, H. Bipolar membrane preparation. In: Kemperman, AJB. Handbook Bipolar Membrane Technology. Enschede: Twente University Press; 2000; 79-108.

[88] Frilette, V. Preparation and characterization of bipolar ion exchange membranes. *J. Phys. Chem.* 60 (1956) 435-439.

[89] Tanaka, Y. Ion Exchange Membrane: Fundamentals and Applications (Membrane Science and Technology Series, 12). Amsterdam: Elsevier; 2007;409-415.

[90] Hao, JH; Chen, CX; Li, L; Yu, LX; Jiang, WJ. Preparation of Bipolar Membranes (I). *J. Appl. Polym. Sci.* 80 (2001) 1658-1663.

[91] Fu, RQ; Xu, TW; Yang, WH; Pan, ZX. Preparation of a mono-sheet bipolar membrane by simultaneous irradiation grafting polymerization of acrylic acid and chloromethylstyrene. *J. Appl. Polym. Sci.* 90 (2003) 572-576.

[92] Jendrychowska-Bonamour, A. Semipermeable membranes synthesized by grafting poly(tetrafluoroethylene) films. Synthesis and study of properties. II. Anionic and cationic mixed membranes. *J. Chim. Phys. Physico. Chim. Biolog.* 70 (1973) 16-22.

[93] Xu, Z; Gao, H; Qian, M; Yu, Y; Wan, G; Chen, B. Preparation of bipolar membranes via radiation peroxidation grafting. *Radiat Phys. Chem.* 42 (1993) 963-966.

[94] Hegazy, EA; Kamal, H; Maziad, N; Dessouki, A. Membranes prepared by radiation grafting of binary monomers for adsorption of heavy metals from industrial wastes. *Nucl. Instrum. Meth. Phys. Res. B.* 151 (1999) 386-392.

[95] El Moussaoui, R; Hurwitz, H. Single-film membrane-process for obtaining it and use thereof. *U.S. Patent* 5,840,192, 1998.

[96] Hao, JH; Yu, LX; Chen, CX; Li, L; Jiang, WJ. Preparation of Bipolar Membranes (II). *J. Appl. Polym. Sci.* 82 (2001) 1733-1738.

[97] Xu, TW; Yang, WH. A novel Series of Bipolar Membranes Prepared from Poly(2,6-dimethyl-1,4-phenylene oxide)(PPO). I. Effect of anion exchange layers on I-V curves of bipolar membranes, *J. Membr. Sci.* 238 (2004) 123-129.

[98] Xu, TW; Fu, RQ; Yang, WH; Xue, YH. Fundamental studies on a novel series of bipolar membranes prepared from poly(2,6-dimethyl-1,4-phenylene oxide)(PPO): (II) Effect of functional group type of anion exchange layers on I-V curves of bipolar membranes. *J. Membr. Sci.* 279 (2006) 233-241.

[99] Sheldeshov, NV; Gnusin, NP; Zabolotskii, VI; Pis'menskaya, ND. Chronopotentiometric study of electrolyte transport in commercial bipolar membranes. *Soviet Electrochem.* 21 (1985) 137-142.

[100] Simons, R. Strong electric field effects on proton transfer between membrane-bound amines and water. *Nature* 280 (1979) 824-826.

[101] Simons, R. Electric field effects on proton transfer between ionizable groups and water in ion exchange membranes. *Electrochim. Acta.* 29 (1984) 151-158.

[102] Simons, R. Water splitting in ion exchange membranes. *Electrochim. Acta.* 30 (1985) 275-282.

[103] Hanada, F; Hirayama, K; Ohmura, N; Tanaka, S. Bipolar membrane and method for its production. *U.S. Patent* 5,221,455, 1993.

[104] Simons, R. A novel method for preparing bipolar membranes. *Electrochim. Acta.* 31 (1986) 1175-1176.

[105] Simons, R. Preparation of a high performance bipolar membrane. *J. Membr. Sci.* 78 (1993) 13-23.

[106] Umemura, K; Naganuma, T; Miyake, H. Bipolar membrane. *U.S. Patent.* 5,401,408, 1995.

[107] Simons, R. The origin and elimination of water splitting in ion exchange membranes during water demineralisation by electrodialysis. *Desaliantion.* 28 (1979) 41-42.

[108] Strathmann, H; Bauer, B; Rapp, HJ. Better bipolar membranes. *CHEMTECH.* 23 (1993) 17-24.

[109] Fu, RQ; Xu, TW; Wang, G; Yang, WH; Pan, ZX. Fundamental studies on the intermediate layer of a bipolar membrane. Part I. PEG–catalytic water splitting in the interface of a bipolar membrane. *J. Colloid Interface Sci.* 263 (2003) 386-390.

[110] Xue, YH; Fu, RQ; Fu, YX; Xu, TW. Fundamental studies on the intermediate layer of a bipolar membrane. part V. Effect of silver halide and its dope in gelatin on water dissociation at the interface of a bipolar membrane. *J. Colloid Interface Sci.* 298 (2006) 313-320.

[111] Fu, RQ; Xu, TW; Cheng, YY; Yang, WH; Pan, ZX. Fundamental studies on the intermediate layer of a bipolar membrane. Part III. Effect of starburst dendrimer PAMAM on water dissociation at the interface of a bipolar membrane. *J. Membr. Sci.* 240 (2004) 141-147.

[112] Fu, RQ; Xu, TW. Fundamental studies on the intermediate layer of a bipolar membrane. Part IV. Effect of polyvinyl alcohol (PVA) on water dissociation at the interface of a bipolar membrane. *J. Colloid Interface Sci.* 285 (2005) 281-287.

[113] Tomalia, DA; Dewald, JR. Dense star polymers having core, core branches, terminal groups. U.S. Patent 4,507,466, 1985; Dense star polymer, U.S. Patent 4,558,120, 1985; Dense star polymers and dendrimers, *U.S. Patent* 4,568,737, 1986.

[114] Xue, YH; Xu, TW; Fu, RQ; Chen, YY; Yang, WH. Catalytic water dissociation using hyperbranched aliphatic polyester (Boltorn® series) as the interface of a bipolar membrane. *J. Colloid Interface Sci.* 316 (2007) 604-611.

[115] Asif, A; Shi, WF; Shenb, XF; Nie, K. Physical and thermal properties of UV curable waterborne polyurethane dispersions incorporating hyperbranched aliphatic polyester of varying generation number. *Polymer* 46 (2005) 11066-11078.

[116] Simons, RG; Bay, R. High performance bipolar membranes, *U.S. Pate*nt 5,227,040, 1993.

[117] Lee, LTC; Dege, GJ; Liu, KJ. High performance, quality controlled bipolar membrane. *U.S. Patent* 4,057,481, 1997.

[118] Fu, RQ; Cheng, YY; Xu, TW; Yang, WH. Fundamental studies on the intermediate layer of a bipolar membrane. Part VI. The effect of the coordinated complex between starburst dendrimer PAMAM and chromium (III) on water dissociation at the interface of a bipolar membrane. *Desalination* 196 (2006) 260-265.

[119] Wakamatsu, Y; Matsumoto, H; Minagawa, M; Tanioka, A. Effect of ion-exchange nanofiber fabrics on water splitting in bipolar membrane. 300 (2006) 442-445.

[120] Hodgdon, RB. Aliphatic anion-exchange membranes having improved resistance to fouling. *US Patent* 5,137,925, 1992.

[121] Gudernatsch, W; Krumbholz, CH; Strathmann, H. Development of an anion-exchange membrane with increased permeability for organic acids of high molecular weight. *Desalination* 79 (1990) 249-260.

[122] Bauer, B; Effenberger, F; Strathmann, H. Anion-exchange membranes with improved alkaline stability. *Desalination* 79 (1990) 125-144.

[123] Liu, JS; Xu, TW; Wu, CM; Shao, GQ. Recent patents on the preparation and application of hybrid materials and membranes (invited review). *Recent Pat. Eng.*, 1 (2007) 214-227.

[124] Söllner, K. Uber Mosaikmembranen. *Biochem. Z.* 244 (1932) 370-381.

[125] Jimbo, T; Tanioka, A; Minoura, N. Characterization of amphoteric-charged layer grafted to pore surface of porous membrane. *Langmuir* 14 (1998) 7112-7118.

[126] Linder, C; Kedem, O. Asymmetric ion exchange mosaic membranes with unique selectivity. *J. Membr. Sci.* 181 (2001) 39-56.

[127] Xu, TW; Yang, WH. Industrial recovery of mixed acid ($HF+HNO_3$).from the titanium spent etching solutions by diffusion dialysis with a new series of anion exchange membranes. *J. Membr. Sci.* 220 (2003) 89-95.

[128] Xu, TW; Yang, WH. Tuning the diffusion dialysis performance by surface crosslinking of PPO anion exchange membranes --simultaneous recovery of sulfuric acid and nickel from electrolysis spent liquor of relatively low acid concentration. *J. Hazard. Mater.* 109 (2004) 157-164.

[129] Kariduraganavar, MY; Nagarale, RK; Kittur, AA; Kulkarni, SS. Ion-exchange membranes: preparative methods for electrodialysis and fuel cell applications. *Desalination* 197 (2006) 225-246.

[130] Huang, CH; Xu, TW.Cleaner production by using electrodialysis in organic or aqueous-organic systems. In: Loeffe, CV. Trends in Conservation and Recycling of Resources. New York: Nova Science Publishers Inc.; 2006; 83-97.

[131] Bazinet, L. Electrodialytic phenomena and their applications in the dairy industry: a review, *Crit. Rev. Food Sci. Nutr.* 45 (2005) 307-326.

[132] Kameche, M; Xu, FN; Innocent, C; Pourcelly, G. Electrodialysis in water-eathanol solutions: application to the acidification of organic salts. *Desalination* 154 (2003) 9-15.

[133] Luo, GS; Liu, JG; Qi, X; Lü,YC; Wang, J.D. Recovery of acetic acid with two-phase electro-electrodialysis. *Chin. J. Membr. Sci. Technol.* 25 (2005) 46-49.

[134] Audinos, R. Ion-exchange membrane processes for clean industrial chemistry. *Chem. Eng. Technol.* 20 (1997) 247-258.

[135] Mo, JX. Application of multi-compartment electrodialysis in chemical industries. Proceeding of 4th national conference on membranes and the the related processes. Oct. 21-24, Nanjing, China, 642-645.

[136] http://www.ameridia.com/html/ebc.html

[137] Huang, CH; Xu, TW. Regenerating flue-gas desulfurizing agents by bipolar membrane electrodialysis. *AIChE J.* 52 (2006) 393-401.

[138] Huang, CH; Xu, TW; Yang, XF. Regenerating fuel-gas desulfurizing agents by using bipolar membrane electrodialysis (BMED): effect of molecular structure of

alkanolamines on the regeneration performance. *Environmenta. Sci. Technol.* 41 (2007) 984-989.

[139] Chou, T; Tanioka, A. Current voltage curves of composite bipolar membrane in alcohol-water solutions. *J. Physic. Chem. B* 102 (1998)7866-7870.

In: Advances in Membrane Science and Technology
Editor: Tongwen Xu
ISBN: 978-1-60692-883-7

Chapter III

Proton Exchange Membranes and Fuel Cells

J. Kerres [*] ***and F. Schönberger***
Inst Chem Engn, Univ Stuttgart,
Boblinger Str 72, D-70199 Stuttgart,
Germany

Abstract

The most important features of fuel cell membranes are their proton conductivity as well as their mechanical and chemical long-term stability under fuel cell operating conditions. The choice of materials depends on the temperature range at which the fuel cells are operated. For fuel cells operated at temperatures below 100°C, cation-exchange materials such as sulfonated polymers are used. They require liquid water for dissociation of the proton from the sulfonic acid group bound to the polymer backbone. Since the partial pressure of water decreases progressively with temperature which leads to a significant drop in proton conductivity at temperatures above 100°C, other amphoteric proton-self conductive materials are necessary for the temperature range between 100°C and 200°C. Phosphoric or phosphonic acid, acidic phosphates or heterocycles which are capable to donate and accept protons without requiring the 'vehicle molecule' water have been used in proton-conducting electrolytes. In this contribution, an overview of the state-of-the-art of the different types of fuel cell electrolytes for H_2 polymer electrolyte fuel cells (PEFC) and direct methanol fuel cells (DMFC) will be given. Beside the illustration of structural features, performances, advantages and disadvantages of each membrane type, the importance of long-term stability and membrane degradation will also be reviewed.

Keywords: Ionomer, polymer electrolyte membranes, cross-linking, low-temperature PEFC, intermediate-temperature PEFC, DMFC, proton conductivity

[*] Correspondence concerning this chapter should be addressed to J. Kerres at jochen.kerres@icvt.uni-stuttgart.de

1. Introduction

Renewable energy sources have been paid much attention in the last decades due to the exhaustable natural resources, the present dependences on imports and the growing global environmental problems.[1] The fuel cell technology is discussed as an innovative and a low-emission energy converter in this context.[2,3,4] Fuel cells are electrochemical systems which convert the chemical energy of the continuously fed fuels (e.g. hydrogen, methanol) directly into electrical energy as long as the integrated materials do not degrade. In contrast to heat engines fuel cells are potentially more effective since they are not subjected to the limitations of Carnot's process.[5] While the basic principle of the fuel cell is known for almost 170 years and goes back to *Christian Friedrich Schönbein's* (1799 – 1868) and *Sir William Grove's* (1811 – 1896)[6,7] investigations the fuel cell technology experienced a renaissance in the early 1950s with the beginning of the aerospace programs. The reasons that such a long period went by between the discovery of the basic principle of the fuel cell and its use as an energy converter might be led back to an insufficient knowledge on the electrochemical processes and especially to material problems as well as to the competition by the steam engine (*Thomas Newcomen*, 1712), by the gas turbine (*John Barber*, 1791), by the electrodynamic generator (*Werner von Siemens*, 1866) and by the combustion engines (*Nikolaus August Otto*, 1863 und *Rudolf Diesel*, 1892). The first fuel cell on the basis of a proton exchange membrane (PEM) dates on the beginning of the 1960's and was firstly used in the *Gemini* aerospace program. A copolymer made up from sulfonated polystyrene and divinyl benzene has been ulitlized as membrane material. However, oxidative degradation was responsible for the short lifetime of such materials making them unattractive for civil utilisation.[3,8] Only with the introduction of poly(perfluoroalkyl)sulfonic acids (PFSAs), originally developed as a permselective separator for the chloralkali electrolysis, by Dupont de Nemours (trade name Nafion®)[9,10] in 1962 a material was provided that could also be used as stable membrane in polymer electrolyte membrane fuel cells (PEMFCs). Although fuel cell technology has made enormous progress (especially in terms of long-term stability) and has gained new impulses by the introduction of this membrane material, there are a few drawbacks up to now that hinder a broad commercial utilization. The most severe issues are correlated with the inherent properties of the membrane material, but there are also unsolved problems in the field of hydrogen storage[11,12,13,14] and of the electrocatalys[15] (e.g. Ostwald ripening of the platinum particles or the corrosion of the carbon support). More details on degradation issues will be given in Section 2.3. Beside the characterisation of Nafion® and the structure-properties relationships, a focus in the present research is the development of alternative membrane materials (for the state-of-the-art: cf. Section 2.2.2.2 to 2.2.2.4)[16,17,18]

This contribution will review both the development of the various alternative membrane materials and the fundamentals of ageing and degradation processes of the membrane material occurring during fuel cell operation with the focal point on radically induced reactions. Although the individual processes that lead to chemical ageing and degradation of the membrane material during the fuel cell operation are only little understood in detail,[19] it is generally assumed in the literature that mainly hydroxyl and hydroperoxyl radicals are responsible for them (for details: cf. Section 2.3).[20,21]

2. State-of-the-Art

2.1. Principle and Types of Fuel Cells

Fuel cells are galvanic couples which produce electrical power as long as the fuel (e.g. hydrogen or methanol) and an oxidant (usually oxygen or air) are fed, as the oxidized products are continuously removed and as the used materials do not degrade severely.[22] The fundamental processes are depicted in Figure 1 at the example of a PEMFC operating with hydrogen as fuel and oxygen as oxidant.

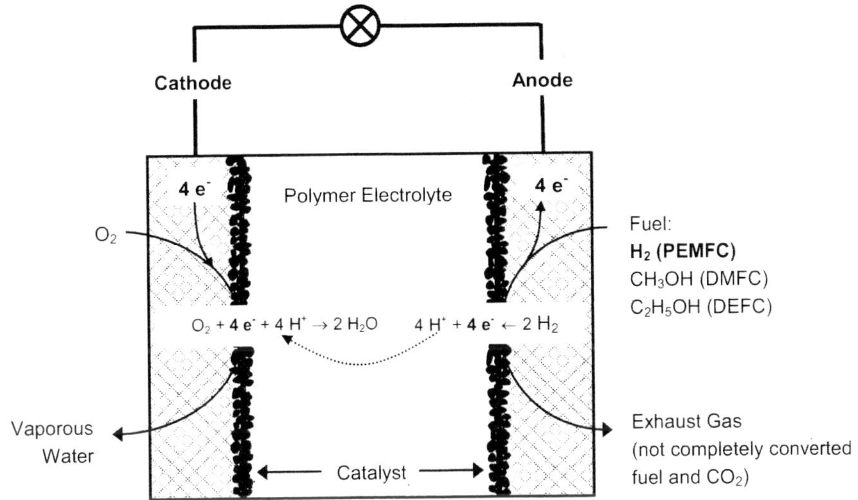

Figure 1. Principle of a PEMFC.

Hydrogen is catalytically oxidized at the anode site.

$$2\ H_2 \rightarrow 4\ H^+ + 4\ e^- \qquad (1)$$

The protons migrate through the membrane driven by the electrical field while the electrons flow through an external circuit and produce electrical power. At the cathode site the oxidant (usually oxygen from air) is reduced to water by electrons and protons according to the following equation:

$$O_2 + 4\ H^+ + 4\ e^- \rightarrow 2\ H_2O \qquad (2)$$

Beside fuel cells on basis of polymer electrolyte membranes (PEMFCs) there are several other types which can be distinguished by the type of the electrolyte and thus be classified into the following types:

d. Alkaline Fuel Cell (AFC)
e. Alkaline Anion-Exchange Membrane Fuel Cell (AAEMFC)[23,24]
f. Phosphoric Acid Fuel Cell (PAFC)

g. Molten Carbonate Fuel Cell (MCFC)
h. Solid Oxide Fuel Cell (SOFC)

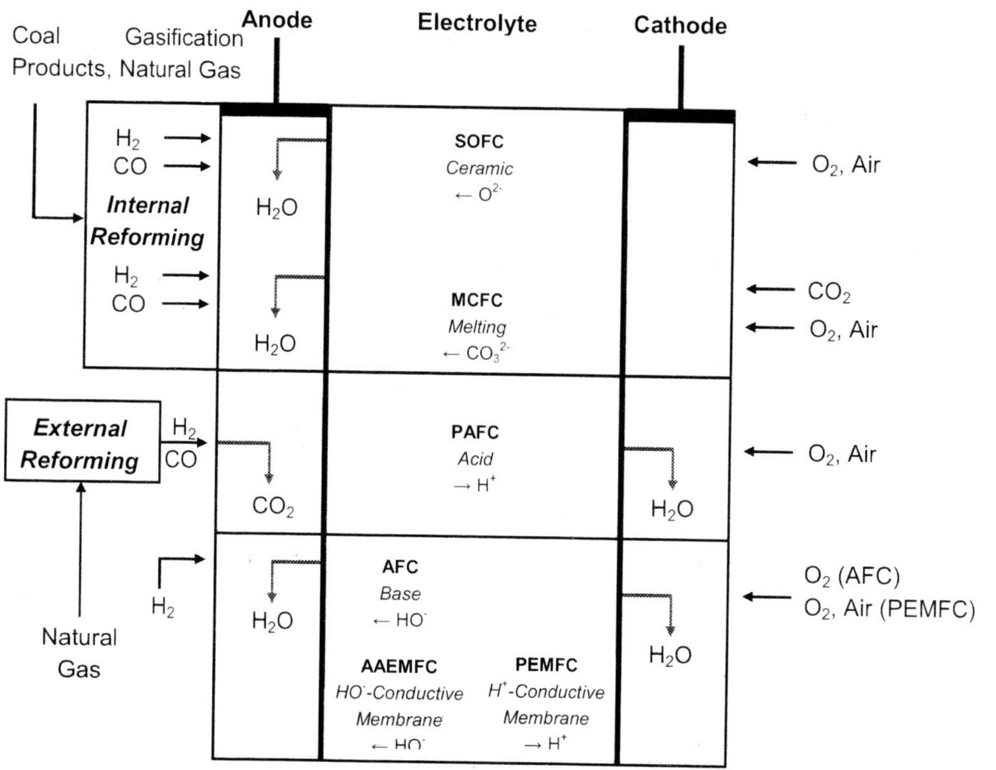

Figure 2. Overview of the different types of fuel cells.

As further categories of the polymer electrolyte membrane fuel cells (PEMFCs) direct methanol (DMFCs) or direct ethanol fuel cells (DEFCs) have to be mentioned here. These types of fuel cells use methanol or ethanol as a fuel which are directly electrochemically converted into CO_2 and H_2O at appropriate catalyst surfaces (e.g. PtRu/C for DMFC[25], PtSn/C for DEFC[26,27]).

2.2. State-of-the-Art of Membrane Development for PEMFCs

2.2.1. Membrane Material Requirements

Beside the function as a proton conductor (and an electrical insulator) the polymer electrolyte membranes have to ensure the spatial separation of the anode from the cathode site. Because of this separation function the membrane is exposed to strongly oxidizing conditions on its one site and to strongly reducing conditions on the other site. The membrane also has to withstand variations in temperature (outside temperature, but also fluctuations in the operating temperature, local inhomogeneities).[28] Additionally, the mechanical stability of the membrane both in the hydrated and in the dried state has to be guaranteed.[8] The permeability for the fuel and the oxidant through the membrane should be as low as possible in order to lower the transport on the opposite sites and thus to minimize any mixed potentials and losses in the cell potential. Especially by operating with liquid fuels (e.g. methanol), this phenomenon which is also termed as methanol crossover is a severe problem.

The cost for the membrane material and its lifetime are key factors from an economical standpoint as they considerably codetermine the overall costs for a PEMFC stack beside the costs for the noble metal catalysts and the bipolar plates.[8]

2.2.2. Membrane Materials

This section will overview the most important research activities on the field of polymer electrolyte membranes and will give a systematization of the various development directions, which are advantageously classified into the following categories and are detailed in a further subsection:[3,29,30,38]

Categories in terms of material type:

- Perfluorinated Polymer Electrolyte Membranes
- Partially Fluorinated Polymer Electrolyte Membranes (except for Poly(aryl) Ionomers)
- Nonfluorinated Polymer Electrolyte Membranes (except for Poly(aryl) Ionomers)
- Partially Fluorinated and Nonfluorinated Poly(aryl) Ionomers
- Hybrid materials

Categories in terms of operation temperature:

1. Low-temperature proton conductors (0 – 100°C)
2. Intermediate-temperature proton conductors (100 – 200°C)

In low-temperature proton conductors the sulfonic acid group is generally the proton-conducting species which requires water for the proton transport both via vehicle and tunnelling mechanism, while in the intermediate temperature operation range of fuel cells phosphoric acid based systems are used. As alternatives for these phosphoric acid systems, imidazole- or phosphonic acid group-containing fuel cell membranes have been proposed. For the fuel cell operation range between the low- and intermediate-temperature proton conductors (i. e. the temperature range between 80 and 130°C) hybrid membrane systems are discussed. These systems typically consist of a sulfonated polymer which provides the protons and a second phase which serves for better retention of water within the membrane, compared to the pure sulfonated polymers, and/or even contributes to proton transport *via* a second proton conduction path. Examples for the second membrane phase are highly porous nanoparticles such as:

1. hygroscopic oxides (e. g. SiO_2, ZrO_2, TiO_2, etc.), which are known for their high water absorption capacity,
2. layered metal phosphates or -phosphonates which also contribute to proton transport by the P–OH groups and
3. heteropolyacids which also significantly contribute to the overall proton transport in the membrane.

In the following, the membrane types will be therefore presented in three subsections:

1. Low-temperature proton conductors (sulfonated polymers)
2. Hybrid proton conductors
3. Intermediate-temperature proton conductors (H_3PO_4 systems, phosphonic acid systems, imidazole systems)

2.2.2.1. Low-Temperature Fuel Cell Membrane Systems (Sulfonated Proton Conductors)

2.2.2.1.1. Poly(Perfluoroalkyl)Sulfonic Acids

Among the ion-exchange membrane materials appropriate for polymer electrolyte membrane fuel cells the poly(perfluoroalkyl)sulfonic acids (PFSA) represent the state-of-the-art.[3,30] These materials are fabricated by copolymerisation of a perfluorinated vinyl ether possessing a sulfonyl fluoride group as a precursor for the sulfonic acid group with tetrafluoroethylene (TFE). They have the chemical structure sketched in Figure 3 (top).

Such PFSA are commercially available in many variants with various numbers of repeating units (x, y, m, n) under different trademarks[30] (e.g. Nafion® from DuPont, Flemion® from Asashi Glass, Aciplex® from Asashi Chemicals, Dow® from Dow Chemicals), from which the first mentioned example might be the most famous and the widestly used and best investigated materials. Since the beginning of the 1980s different models of the morphological structure of Nafion® have been published. The first model goes back to studies made by *Gierke et al* [16] and it is depicted in Figure 3 (bottom). This model describes the structure of hydrated Nafion® with ionical clusters (with about 4 nm in diameter and an average distance of roughly 5 nm) which are approximately spherical in geometry and embedded in an inverse micellar structure within the perfluorinated polymer matrix. The ionical clusters are connected with one another by narrow (ca. 1 nm) channels. Both the swelling by polar solvents and the transport properties (e.g. the excellent proton conductivity and the high permselectivity) can be explained by this model. All other structure models of Nafion® have a similar principal assembling, however they differ – partly significantly – in the geometry and the spatial distribution of the ionical clusters[31,32,33,34,35,36,37] Although such PFSA membranes have excellent longterm performances in fuel cells operating with pure hydrogen[17] at moderate temperature (< 90°C) and under high relative humidity (up to 60,000 operating hours under quasi-stationary conditions),[30,38] some inherent drawbacks of this material get in the way of a commercial application. The high price of these materials (e.g. Nafion®) is mentioned very often at this point. However, it has to be considered that the annual tonnage of Nafion® (with about 65 t/a) is significantly lower than that of other high performance polymers like poly(tetrafluoroethene) (PTFE with 80,000 t/a).[39] This difference in the annual tonnages might also be relevant for the pricing (5000 US $/kg Nafion®, 10 US $/kg PTFE). The limiting factor for the economic application of polymer electrolyte membrane fuel cells are rather the physical properties of the membrane material in combination with the inhibited, strongly temperature-dependent kinetic of the oxygen reduction reaction at the catalyst.[8] However, an increase in the operating temperature which would also improve the tolerance of the catalyst against carbon monoxide (CO) impurities and which would allow the use of hydrogen from reforming gas (e.g. from methanol or natural gas) is restricted by the material properties. The glass transition temperature of Nafion® lies in the range of about 110°C in the dried (and still below that in the hydrated) state limiting the maximum operating temperature to 100°C.[8,40] In fact, a noticeable decrease in the proton conductivity already starts at about 80°C due to accelerated water evaporation which results in significant ohmic losses over the membranes requiring a complex external humidification system.

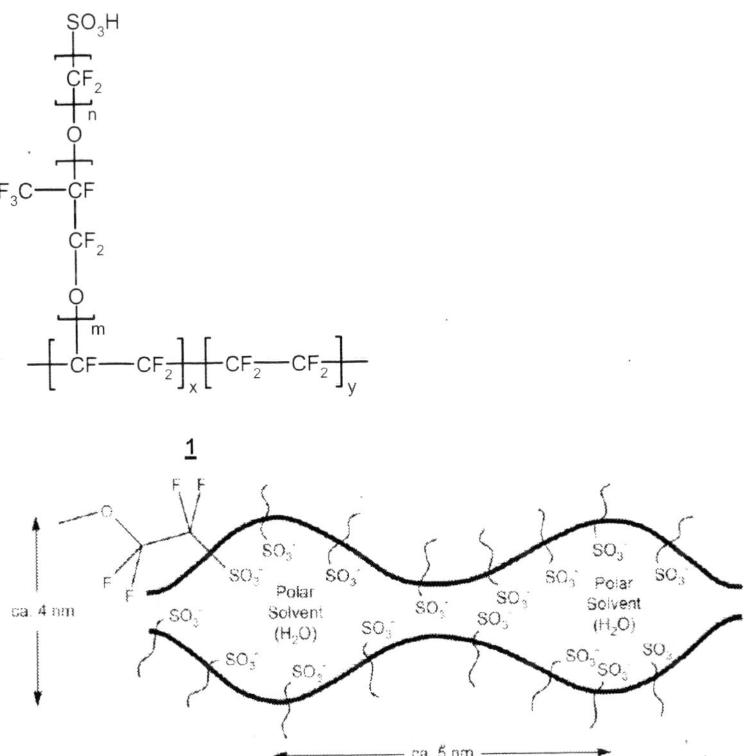

Figure 3. Chemical (top) and morphological (bottom) structure.

A further drawback of these PFSA membranes is their high permeability for water and other polar substances such as methanol ($k_{MeOH} = 1.98 \cdot 10^{-6}\ cm^2 \cdot s^{-1}$)[41] Therefore, their application in direct methanol fuel cells (DMFC) is limited.

The aforementioned disadvantages of these PFSA membrane materials can be minimized to a certain extent by appropriate measures. For example both porous PTFE[42,43] and polypropylene (PP)[44,45] matrices have been used for the preparation of micro-reinforced PFSA membranes. Such membranes having a thickness of only about 25 μm are also commercially avalaible (e.g. from Gore® and Asahi®)[46,47] and have many advantages over the pure PFSA membranes which are significantly thicker (between 57 μm (Nafion® 112) and 183 μm (Nafion® 117)).[30] Therefore the potential drop over the membrane is clearly lower and the water management is simpler since the back diffusion of water from the cathode to the anode is more effective. Moreover, the saved material also reduces the costs of the membrane material.

Other measures for improving the water retention in PFSA membranes at temperature above 80°C and thus the performance of the whole PEMFC or DMFC are the incorporation of hygroscopic oxide (such as SiO_2)[48] or inorganic proton conductors (like zirconium phosphate or heteropoly acids.[49,50,51] All these concepts for the modification of ionomer membranes and especially of PFSA membranes have in common that they effectuate some improvements over the pure membranes but they cannot eliminate the inherent drawbacks of the membrane material itself. For this reason, an active research focus lies on the conceptual and synthetic development of novel membrane materials.[3,29,30,38]

2.2.2.1.2. Partially Fluorinated Polymer Electrolyte Membranes (Except for Poly(aryl)s)

D'Agostino et al. patented a method for a radiation induced grafting polymerisation of α,β,β-trifluorostyrene (TFS) onto different (partially) fluorinated polymer foils (e.g. poly(tetrafluoroethylene-*co*-hexafluoropropylene (FEP), poly(ethylene-*alt*-tetrafluoroethylene) (ETFE, 2 in Figure 4) or poly(vinylidendifluoride) (PVDF)) and the subsequent sulfonation of the phenylene ring in the grafted side chain.[52,53,54]

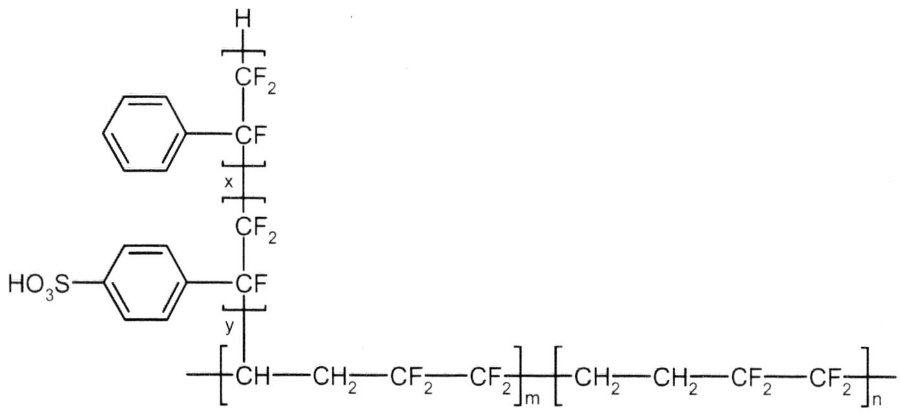

2

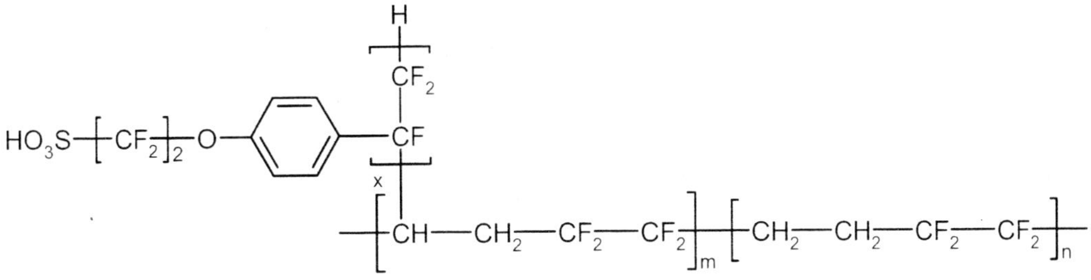

3

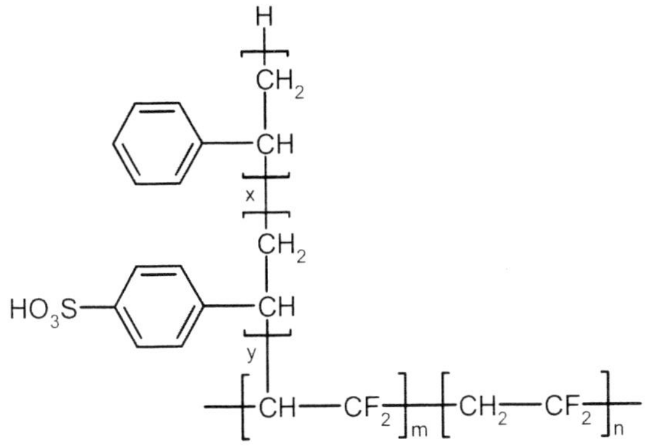

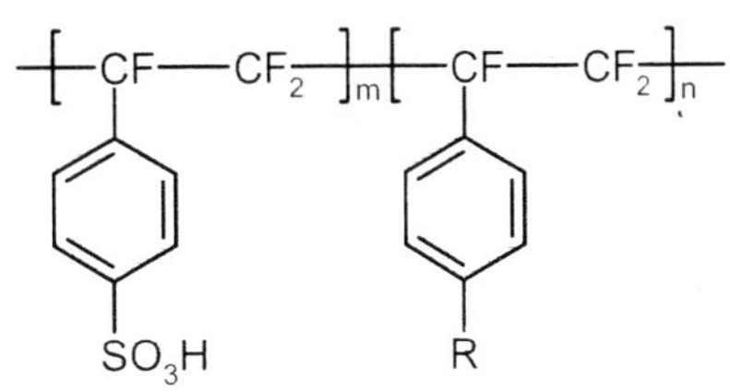

$R = CH_3, OCH_3, OPh, \ldots$

4 **5**

Figure 4. Chemical structure of some partially fluorinated polymer electrolyte membranes: sulfonated ETFE-*g*-PTFS (2), ETFE-*g*-PTFS(*p*-$O(CF_2)_2SO_3H$) (3), sulfonated PVDF-*g*-PSS (4), poly(α,β,β)trifluorostyrene (5).

However, these and similar partially fluorinated materials[55,56] were primarily developed to be applied as membrane in the chlorine alkaline electrolysis. This application requires a high resistance of the membrane material against aggressive chemicals (chlorine, oxygen, hydrogen and alkaline solution) beside sufficient mechanical and thermal stability. Therefore fluorinated materials are often used.[56] In recent years, grafting *co*-polymerisations with styrene and divinyl benzene onto (partially) fluorinated polymer foils were carried out and the resulting membranes have been investigated in terms of possible fuel cell applications.[57,58,59,60] Although this kind of membranes offers a broad leeway for a number of variations and offers the possibility to build up branched or even cross-linked polymer electrolyte systematically by a relatively simple chemistry,[61] the minor chemical stability of the grafted poly(styrenesulfonic acid) (PSSA) side chain gets in the way of a long-term stability. For instance the lifetime of PVDF-*g*-PSSA (**4** in Figure 4) based and hydrogen fed MEA was determined to be about 150 hours when operated potentiostatically at 0.4 V and an operating temperature of 60°C.[59] The successive degradation of the PSSA side chain is mainly held to be responsible for the stepwise decrease of the proton conductivity of the membrane. Especially the benzylic $C_\alpha - H$ bonds were identified as potential points of attack for radicals.[19,62] Recently, *Scherer et al.* reported on the synthesis of a TFS monomer with a flexible $-O(CF_2)_2SO_2F$ substituent, which serves as a precursor for the ionogenic group, and of its grafting polymerisation onto ETFE (5 in Figure 4).[63] A single cell test in the H_2-PEMFC mode at 60°C gave polarisation curves comparable with those of Nafion® 112, even after 500 hours operating time.

Further partially fluorinated polymer electrolyte membranes are sulfonated and differently (e.g. *para*-methyl-, methoxy- and phenoxy-)substituted poly(α,β,β-trifluorostyrene)s developed by Ballard Advanced Materials[64] in which the very sensitive $C_\alpha-H$ bonds are replaced by corresponding $C_\alpha-F$ bonds (**5** in Figure 4). Limiting factors of this type of membranes are not the performance of the material but rather the complicated production process of the monomers, their slow polymerization kinetics,[56] the difficult sulfonation process[38,65] and the lacking knowledge for morphological structuring[3] (e.g. by block formation of differently substituted and sulfonated α,β,β-trifluorostyrenes).

2.2.2.1.3. Non-Fluorinated Polymer Electrolyte Membranes (Except for Poly(aryl)s)

Although non-fluorinated polymer electrolytes with exception of poly(aryl) based systems (cf. Section 2.2.2.1.4) only play an underpart they are listed here since important conclusions for further membrane developing can be drawn from the properties of these membrane types. As already mentioned in Section 1 the first polymer electrolyte membrane fuel cells based on copolymers consisting of sulfonated styrene and divinyl benzene. The short lifetime of such membranes[66] does not allow an economically reasonable use but the corresponding membrane systems are appropriate for the investigation between structure, morphology and proton conductivity. *Holdcroft et al.* published the synthesis of poly(sodiumstyrenesulfonate) (PSSNa) by stable free radical polymerization (SFRP) leading to structures with well-defined chain lengths and an acutely narrow polydispersity. They also reported on the subsequent controlled grafting polymerization of these PSSNa chains onto hydrophobic polystyrene (PS-*g-mac*PSSNa)[67,68] (6 in Figure 5) or polyacrylnitrile (PAN-*g-mac*PSSNa)[69] Structural transitions from isolated ionical domains to networks of ionic channels could be realized by varying the length and the density of the grafted side chains.

Commercially available sulfonated blockcopolymers from Dais Analytic on basis of styrene-ethylene/butylene-styrene[70] (SEBS, 9 in Figure 5) are also worth to be mentioned here. However, their practical use is restricted to portable energy sources with a maximal power of 1 kW and a maximum operating temperature of 60°C due to the relatively low chemical stability.[3]

The high temperature, oxygen and water resistance of silicones[71] due to the high bond dissociation energy of the Si – O bonds might be the reason for the development of sulfonated polysiloxanes (7 in Figure 5). Those membranes show a specific resistance *(ex situ)* comparable to that of Nafion® 117 (0.1 S · cm^{-1}) and high thermal stability (up to 180°C).[72,73,74] Unfortunately no data on fuel cell tests with these membranes are available in the literature which could proof their suitability for this application.[30,75] Other inorganic and highly stable polymer backbones are polyphosphazenes (8 in Figure 5)[76] that can easily be modified by ionogenic groups (e.g. by sulfonic acid,[77] sulfone imides[76] and phosphonic acid[78,79] groups) by cross-linking sites. However, their mechanical stability (especially in the hydrated state) is rather poor which could be counteracted partly by further cross-linking or blending but the very low glass transition temperature of these materials (-28°C to -10°C) [3] might be limiting for a potential application as fuel cell membranes.

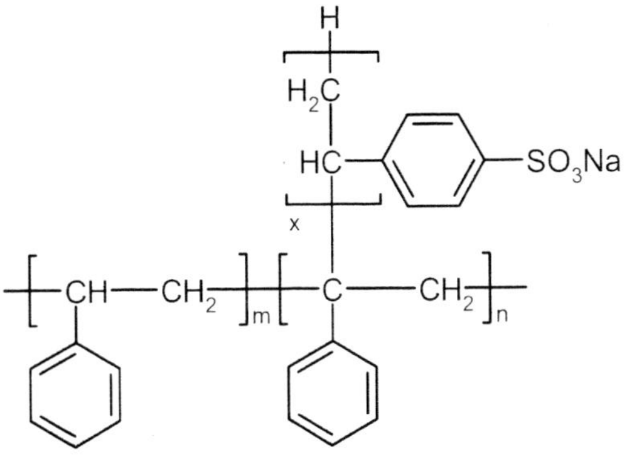

Figure 5. Chemical structure of some nonfluorinated polymer electrolyte membranes: PS-*g*-*mac*PSSNa (6), network of sulfonated polysiloxane (7), sulfonated polyphosphazene (8), sulfonated styrene-ethylene/butylene-styrene blockcopolymer (SEBS).

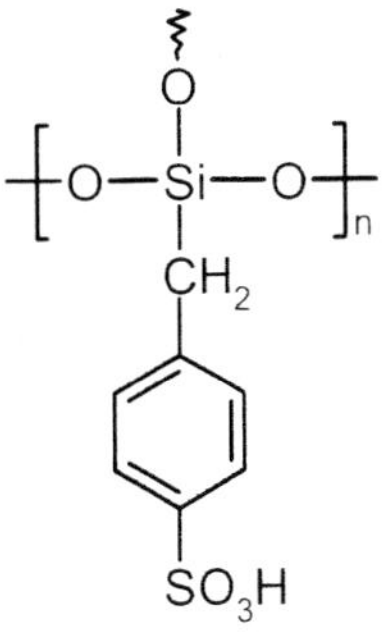

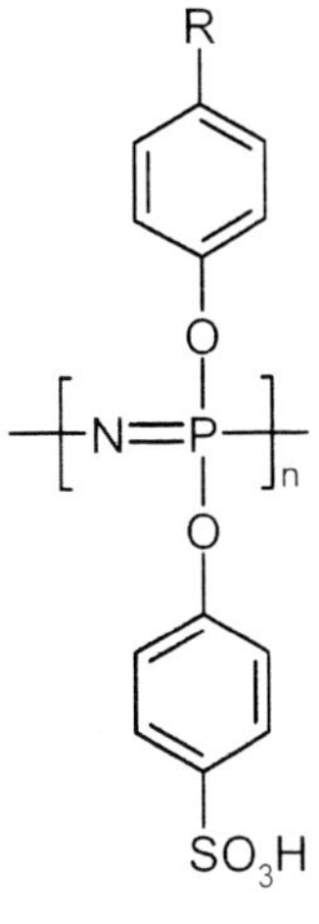

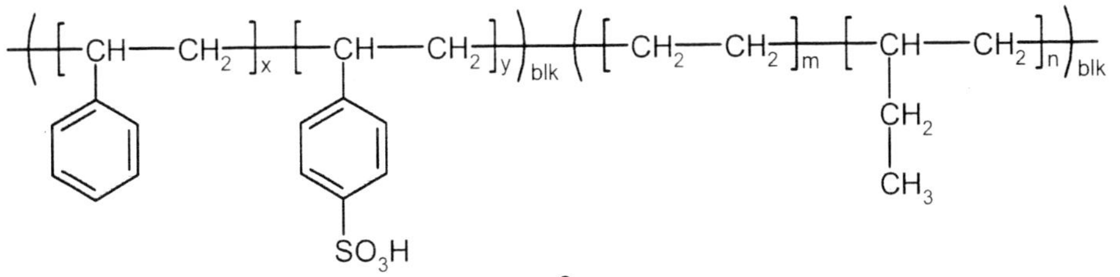

2.2.2.1.4. Partially fluorinated and nonfluorinated Poly(aryl) Ionomers

Poly([hetero]aryl)s are considered as promising candidates for the development of polymer electrolytes due to their general availability, processability, expected stability under fuel cell conditions and due to their simple modifiability. They are thought to overcome the inherent drawbacks of the perfluoroalkylsulfonic acids (e.g. reduction of the electroosmotic drag and of the methanol crossover[17]) soonest among all the alternative membrane materials.[3]

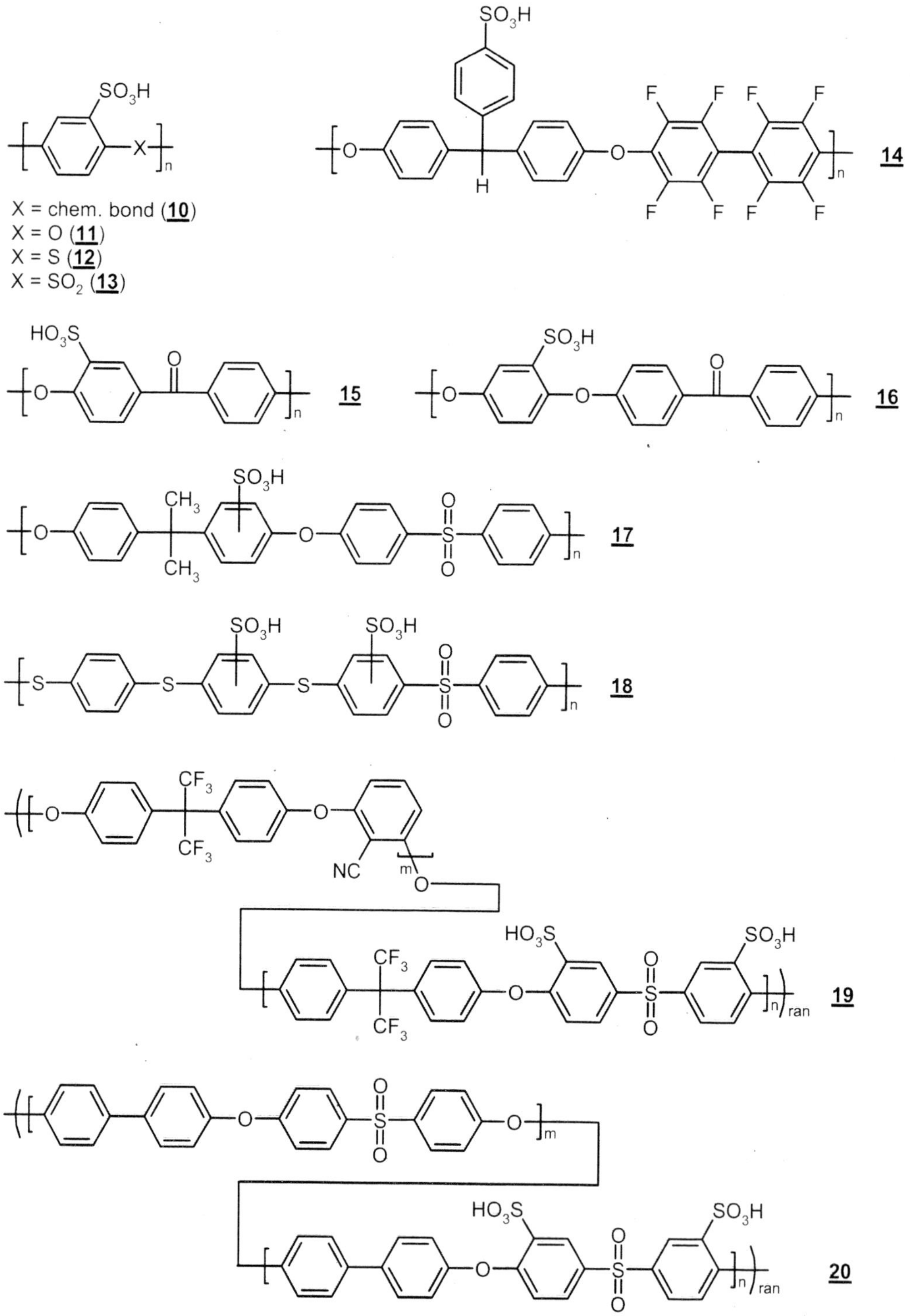

SO_3H
X
n
X = chem. bond (10)
X = O (11)
X = S (12)
X = SO_2 (13)
14
15
16
17
18
19
20

Figure 6. Overview of poly(aryl) ionomers: sulfonated poly-*p*-phenylene (10), polyphenylene oxide (11), poly(phenylene sulphide) (12), poly(arylene sulfone) (13), sulfonated poly(arylene ether) (14), sulfonated poly(ether ketone) (15), poly(arylene ether ether ketone) (16), poly(arylene ether sulfone) (17), poly(arylene sulfide sulfone) (18), poly(arylene ehter sulfone benzonitrile) (19), statistical *co*-poly(arylene ether sulfone) (20).

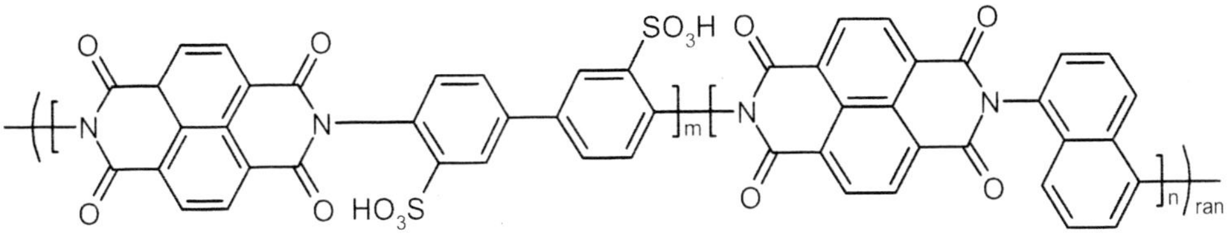

21

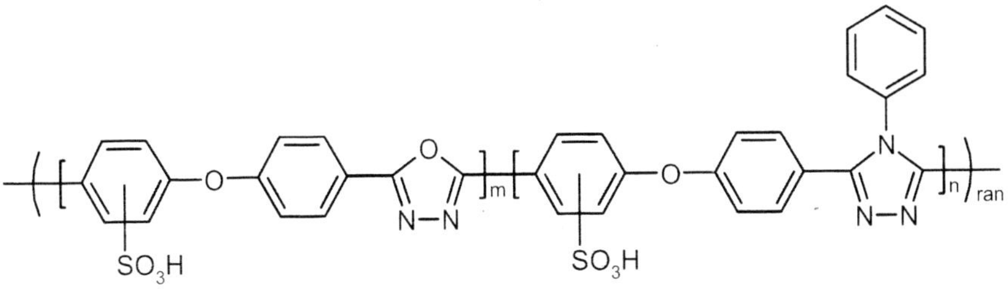

22

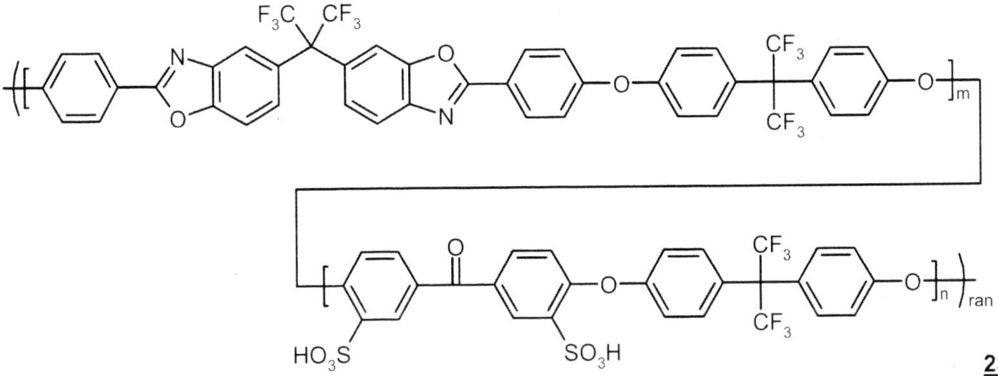

23

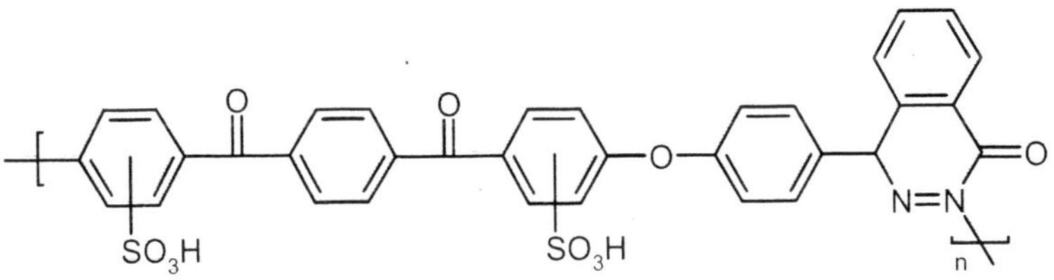

24

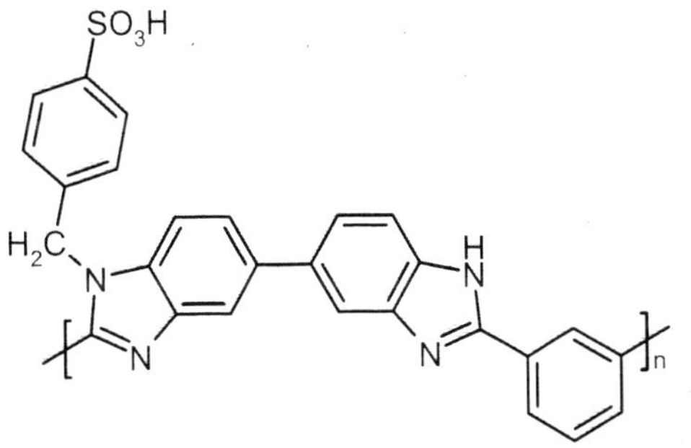

25

Figure 7. Overview of poly(heteroarylene) ionomers: sulfonated poly(arylene-*co*-naphthalimide) (21), poly(arylene ether oxadiazole-*co*-oxatriazole) (22), poly(benzoxadiazole ether ketone) (23), poly(phthalazinone ether ketone) (24), arylsulfonated polybenzimidazole (PBI, 25)

Poly(aryl) ionomers are functionalized (sulfonated, aryl sulfonated, sulfonimidized, carboxylated and/or phosphonated) aromatic condensation polymers (ACPs)[80] whose polymer main chain consists of phenylene rings connected with each other by a chemical bond (poly(arylene, e.g. poly-*p*-phenylene,[81] 10 in Figure 6), a bridging atom (e.g. polyphenylene oxide 11 in Figure 6,[82,83] polyphenylene sulphide, 12 in Figure 6),[84] a bridging group (e.g. polyphenylene sulfone, 13 in Figure 6),[85] and/or annellated (hetero)aromatic ring systems. More common are combinations of those connecting modes, for instance in sulfonated poly(arylene ether)s (14 in Figure 6),[86,87] poly(arylene ether ketone)s (15 in Figure 6),[80,88] poly(arylene ether ether ketone)s (16 in Figure 6),[89,90,91] poly(arylene ether sulfone)s (17 and 20 in Figure 6),[90,92,93] poly(arylene sulfide sulfone)s (18 in Figure 6),[94] poly(arylene ether sulfone benzonitrile)s (19 in Figure 6),[95,96] poly(imide)s (e.g. poly(aryl-*co*-naphthalimide)s: 21 in Figure 7),[97,98,99,100,101,102,103] poly(arylene ether oxadiazole-*co*-oxatriazole)s (22 in Figure 7),[104,105] poly(benzoxazole ether ketone)s (23 in Figure 7),[106] poly(phthalazinone ether ketone ketone)s (24 in Figure 7),[107] and arylsulfonated polybenzimidazoles (PBI: 25 in Figure 7)[108,109,110,111]

The sulfonic acid group normally acts as ionogenic group in these ionomers. Their introduction both onto the arylene main chain and in an aromatic side chain[4,112] can synthetically realized easily by electrophilic aromatic substitution (S_EAr). Beside sulfonated there are also a few examples for carboxylated,[113,114] sulfonimidized[115] and phosphonated (cf. Section 2.2.2.3.3)[116,117,118,119] poly(aryl) ionomers in the literature.

There are two principal methods to introduce the sulfonic acid group onto an aromatic condensation polymer; either by sulfonation of an exisiting poly(aryl) or by polycondensation of sulfonated monomers. For the first method there are a series of various sulfonating reagents, e.g. concentrated[2] or fuming[120,121] sulfuric acid, chlorosulfonic acid,[122,123] complexed sulfur trioxide (e.g. $SO_3 \cdot PO(OEt)_3$),[124] methanesulfonic acid in sulfuric acid,[125] acylsulfate (e.g. acetyl sulfate)[126,127] and trimethylchlorosulfonic acid.[128] Apart from electrophilic sulfonation reactions, a sulfonation method involving the following reaction steps: 1) lithiation with n-buLi; 2) reaction of the lithiated polymer with SO_2; 3) oxidation of the intermediate polymeric sulfinate with an oxidant such as H_2O_2, $KMnO_4$ or NaOCl has been introduced by *Kerres et al.*[92]

A precise control over the position of the sulfonic acid group and the degree of sulfonation is limited in these postsulfonation routes. This method is also fraught with possible side reactions eventually causing molecular weight degradations which strongly depend on the actual substitution pattern of the polymer.[3] *G. Wegner et al.* published a procedure for the polycondensation of already sulfonated monomers, which in principle can also be used for the selective development of statistical and multiblock-*co*-poly(aryl)s.[81] While postsulfonation by S_EAr usually leads to the functionalization of the most electron-rich (most activated) phenylene ring in the repeating unit, the polycondensation of sulfonated monomers can also serve for the introduction of the sulfonic acid group into an electron-deficient part of the polymer structure. *Kerres et al.* reported on a comparative investigation of the stability of the three possible constitution isomers of sulfonated poly(arylene ether sulfone)s (sPES).[167] According to this study the sulfonic acid group is thermally and oxidatively most stable when attached at the most electron-deficient carbon atom i.e. at *meta* position to the sulfone bridge.

Drawbacks of poly(aryl) ionomers are in particular their brittleness in the dried state and their high swellability resulting in mechanical instabilities.[38] Especially the excessive water uptake can cause irreversible changes in the hydrophilic domains that could lead to a significant decohesion and disentanglement of the polymer chain finally making the membrane soft and instable.[129] In order to obtain proton conductivities comparable to that of poly(perfluorosulfonic acid)s, a higher density of sulfonic acid groups (i.e. a higher ion-exchange capacity (IEC)) is necessary in the case of poly(aryl) ionomers due to their lower acidity which, however, entails a stronger interaction with water molecules. Beside the acid strength of the polymer electrolyte its microstructure is of importance for the transport properties and the swelling behaviour of the membrane material. Comparative investigations between sulfonated poly(ether ether ketone ketone)s (sPEEKK with IEC = 1.43 mmol/g) and Nafion® (IEC = 0.91 mmol/g) revealed that the less pronounced separation between hydrophilic and hydrophobic domains in sPEEKK brings about a microstructure consisting of narrower ionic channels with a lower cross-linking density, a longer distance between adjacent sulfonic acid groups and with a higher number of dead-end channels. In the fully hydrated state the electroosmotic drag and the permeation of water (or other solvent) molecules are thus enhanced guaranteeing concomitant high proton conductivity.[17]

It could be shown by atomic force microscopy (AFM) that, in a series of statistical *co*-poly(arylene ether sulfone)s with different ratios between nonsulfonated (m) and disulfonated (n = 1 – m) repeating units (20 in Figure 6), a transition between isolated, hydrophilic domains to a network of ionic channels takes place which is closely correlated with its water uptake behaviour and its proton conductivity.[130,131] The properties of such statistical *co*-poly(aryl) ionomers can be tailored to a certain extent by modifying the chemical structure of the arylene main chain (e.g. of the type of linkage)[86,95,132,133,134,135] or by offsetting the sulfonic acid group onto an aromatic side chain.[112] For instance, the introduction of fluorine containing linkage groups (such as the $C(CF_3)_2$ bridge) results in a stronger microphase separation between hydrophilic and hydrophobic segments which finally leads to an enhanced proton conductivity and a concomitant water uptake.[136] As already discussed in Section 2.2.2.1.1 for PFSA membranes various measures for improving the membrane properties were also applied to poly(aryl) ionomers based membranes, namely the micro-reinforcement by porous matrices,[137,138,139] the incorporation of hygroscopic oxides,[140] phyllo- and tectosilicates,[141,142,143] or of inorganic proton conductors (such as zirconium phosphate[143] or wolframatophosphoric acid[144,145]). The following concepts for an effective minimization of the water uptake of poly(aryl)s while maintaining a sufficiently high proton conductivity are discussed in the literature:

1. Cross-linking (physically, ionically, covalently, ionically-covalently)
2. Molecular design of morphologically controlled (microphase-separated) poly(aryl) ionomers.

The first draft for the cross-linking of polymer electrolyte membranes goes back to *Kerres et al.*[146,147,148] and serves primarily for the improvement of the mechanical properties of poly(aryl) ionomer membranes (reduction of brittleness in the dried state and minimization of excessive swelling in the hydrated state) and lowering of the methanol permeability. More information on the concept of covalent cross-linking can be found in the literature.[38,149] Ionical cross-linking by mixing a polymeric acid and a polymeric base (especially with polybenzimidazole PBI) could be identified to be an effective strategy for the improvement of the mechanical, thermal[150] and oxidative[149] stability in comparison to the pure polymeric acids as well as for the reduction of the methanol permeability in comparison to PFSA based membranes.[151] The different cross-linking concepts will be discussed in the next section. An overview of the area of molecular design of morphologically controlled poly(aryl) ionomers will be given in Section 2.2.2.1.6.

2.2.2.1.5. Ionomer Cross-Linking Concepts

Cross-linked membrane systems can be subdivided into covalently cross-linked membrane systems and physically cross-linked membrane systems. By cross-linking, the swelling and the water uptake of fuel cell membranes can be reduced markedly, or even water-soluble ionomers can be transformed into water-insoluble membranes. The research group of the author of this contribution is active in the development of cross-linked ionomer membranes for fuel cells and other electromembrane applications such as electrodialysis since more than 10 years. This work was recently reviewed;[149] therefore only some highlights of this work will be mentioned here. The covalent cross-linking procedure which was developed by our research group consists of the sulfinate S-alkylation reaction product with α,ω-dihalogenoalkanes (e.g. 1,4-diiodobutane[152]) or with fluoroaromatics (e.g. deca-fluorobiphenyl, decafluorobenzophenone or 4,4'-difluorodiphenylphenyl phosphinoxide). Two types of covalently cross-linked ionomer membranes have been developed. The first type represents blends of sulfinated PSU Udel® with sulfonated PSU Udel®[147] or blends of sulfinated PSU with sulfonated poly(etherketone)s.[153] The second type of ionomer membranes are obtained by cross-linking of the sulfinate groups of PSU-sulfinatesulfonate[154] or by cross-linking of the sulfinate groups of PEEK-sulfinatesulfonate.[155] The resulting membranes show reduced water uptake along with high proton conductivity and therefore good performance in DMFC up to temperatures of 130°C.[155,156] We also developed ionically cross-linked blend membranes by mixing sulfonated ionomers which were in some cases even water-soluble (up to IECs of 4.2 meq/g) with basic polymers, using both self-developed basic polymers from lithiated PSU Udel® or lithiated Radel R® [157,158,159,160,161] and/or highly stable basic polymers such as polybenzimidazole PBI Celazole®.[120,148, 150,162,163,164] Many of these membranes have been tested in PEMFC[150] and DMFC[165,166] and yielded excellent performances in PEMFC at temperatures up to 100°C (maximum power densities of up to 1.2 W/cm^2 were reached at 80°C[150,167,168]) and in DMFC at temperatures up to 130°C (maximum power densities of up to 0.25 W/cm^2 were reached at 110°C,[169] and of up to 0.28 W/cm^2 were reached at 130°C[149,170]). It also turned out that the thermal and radical stability [90,120,164,167] of the arylene main-chain ionomers could be markedly improved by blending with basic polymers. A possible explanation is that the basic polymer blend component in acid-base blend membranes has radical-scavenging properties – it is known from radical chemistry that, for example, imidazole-containing compounds have radical-scavenging properties.[171,172] The concept of blending sulfonated ionomers with basic polymers has also been picked up by other research groups. For example, *Wicisk, Pintauro et al.*[173] and *Ainla and Brandell* prepared Nafion®/PBI blend membranes and obtained a marked increase in DMFC performance, compared to Nafion®.[174] *Wycisk, Lee and Pintauro* prepared acid-base blend membranes from sulfonated poly(phosphazene) and PBI which showed a reduction of the methanol permeability by a factor of 2.6, compared to Nafion®, and a DMFC performance which was slightly worse than that of Nafion® (maximum power density 89 mW/cm^2, vs. 96 mW/cm^2 with Nafion® 117).[175] *Manea* and *Mulder* prepared sulfonated poly(ethersulfone)/PBI blend membranes.[176] The characterization of these membranes revealed that their methanol permeability was significantly lower than that of Nafion®. *Schauer et al.* prepared acid-base blend membranes from sulfonated poly(phenylene oxide) (sPPO) and PBI. They found that the blend membranes were markedly more stable than sPPO alone using Fenton's test procedure.[177] Recently, *Manthiram, Fu, and Guiver* prepared two different types of acid-base blends, from which the first one consists of polysulfone-2-amidobenzimidazole and sulfonated PEEK (sPEEK)[178] and the second one of polysulfone bearing benzimidazole side groups (PSf-BIm) and sPEEK.[179,180] These membranes have been tested in a PEMFC,[179] where better performance at 90 and 100°C was obtained than with pure sPEEK and Nafion® 115 membranes, and in a DMFC,[178,180] in which a remarkably superior long-term performance, compared to Nafion® 112, could be obtained, which is thought to be due to significantly reduced methanol crossover.

Apart from ionically cross-linked blend ionomer membranes, physically cross-linked (*via* hydrogen bridge interactions) ionomer membranes have been prepared. The latter type also includes blends of polyvinylalcohol (PVA) with different sulfonated ionomers, such as Nafion®/PVA,[181] or sulfonated polystyrene/PVA.[182] A significant reduction in the methanol permeability could be obtained using these membranes; however there are concerns about the electrochemical stability of PVA in the fuel cell environment. Other hydrogen-bridge-cross-linked ionomer systems which have been taken into account for fuel cell application are blends between sPEEK and the polyamide Trogamid® and PEI Ultem® [158] – however, the hydrolytical stability of the polyamide or the PEI blend component is questionable.

2.2.2.1.6. Morphologically Controlled Ionomer Systems

Holdcroft et al. reported on highly fluorinated comb-shaped copolymers [183] and ionic graft polymers [184] to mimic the principal architecture of poly(perfluorosulfonic acid)s with the aim to improve the mechanical integrity and to higher the glass transition temperatures in comparison with Nafion®. The backbone of these structures consists of a (partially fluorinated) poly(arylether) structure at which sulfonated poly(α-methylstyrenesulfonic acid) side chain are grafted by the same methods as described in the Sections 2.2.2.1.2 and 2.2.2.1.3. Although the very sensible (benzylic) C_α–H bonds are replaced by the corresponding C_α–CH_3 groups the oxidative stability under fuel cell operating conditions might be questionable. An alternative approach in the area of morphologically controlled ionomer systems is the development of multiblock-*co*-poly(aryl) ionomers consisting of hydrophobic (nonsulfonated) and hydrophobic (sulfonated) blocks. *Scherer et al.* synthesized nonfluorinated sulfonated multiblock-*co*-poly(aryletherketone)s which show a markedly higher proton conductivity despite a considerably reduced water uptake behaviour (and thus increased mechanical stability) in comparison to the corresponding statistical *co*-ionomers.[185] Preliminary single test results of these multiblock-*co*-poly(aryletherketone)s membranes presented in this study show acceptable performance although the not fully optimized electrode structure was rated to be limiting. Sulfonated multiblock-*co*-polyimides show – especially under low relative humidity – an eightfold higher proton conductivity ($\sigma = 10^{-3}$ S · cm^{-1} at 30% RH) at a concomitant lower water uptake than the analogous statistical *co*-polyimides.[186,187] Although single test tests of those multiblock-*co*-polyimides could be successfully carried out up to 5,000 operating hours (at 0.2 A/cm^2, 80°C, 60 and 90% RH) the hydrolytical cleavage of the imide linkages was identified as the main degradation pathway in this kind of polymeric structures.[188] *McGrath et al.* published a novel synthetic method for the preparation of sulfonated multiblock-*co*-poly(arylethersulfone)s consisting of extremely hydrophobic (partially fluorinated) and hydrophilic segments in order to achieve a pronounced microphase-separated domaine structure.[189,190,191,192] However, an exact control over the block lengths and the polydispersity of the molecular weight distributions of such multiblock-*co*-poly(aryl) ionomers might be a challenging future task.[193]

2.2.2.2. Composite Systems

Numerous composite membrane systems have been taken into account for fuel cell membranes. Due to limited capacity in this contribution, only a short selection of these systems can be presented here. The two most important material classes are 1) composite systems of an organopolymer with inorganic particles (Section 2.2.2.2.1), and composite membranes from organopolymer (blends) with heteropolyacids (Section 2.2.2.2.2). In the organopolymer/inorganic particle composite membranes the inorganic particles serve for better retention of water when the membranes are operated in fuel cells at temperatures above the water boiling point, 100°C. As already mentioned, the proton conductivity of sulfonated polymer membranes decreases drastically when they are impoverished in water due to water evaporation.[194] Some of the inorganic particle material classes, such as acidic phosphate or phosphonate salts also contribute to proton transport (see Section 2.2.2.2.2). In the heteropolyacid composite systems, the heteropolyacid blend component strongly contributes to proton conductivity, since the heteropolyacids are superacids containing protons which are delocalized within the heteropolyacid molecules.

2.2.2.2.1. Organopolymer-Inorganic Particle Systems

As mentioned before, the organopolymer/inorganic particle composite membranes can be subdivided into systems, where the inorganic nanoparticle phase of the composite membrane retains water at temperatures above 100°C, and into systems where the inorganic nanophase contributes to proton transport. One example for the first type are Nafion®/silicalite composite membranes which can be either prepared by simply mixing Nafion® suspension with silicalite nanoparticles in an ultrasonic bath, followed by membrane casting,[195,196] or by formation of the SiO_2 nanoparticles inside a Nafion membrane via sol-gel processes by swelling the Nafion® membranes in alkoxysilanes, whereas the sol-gel process inside the membranes is catalyzed by the superacidic Nafion® SO_3H groups.[197,198,199] Another approach for the formation of organopolymer/inorganic nanoparticle composite systems is to perform the sol-gel process in the polymer solution by addition of silanes to the polymer solution, followed by hydrolysis of the silanes to sols and membrane casting.The formation of the nanoparticles starts during membrane formation and solvent evaporation and is completed by acidic membrane posttreatment.[140] Nafion®/SiO_2 membranes have been tested in DMFC at temperatures of 145°C and yielded a good performance at this temperature, approaching a peak power density of 240 mW cm^{-2} at 0.6 A · cm^{-2} and 0.4 V. As mentioned above, also proton-conductive compounds such as layered acidic (hydrogen)phosphates and (hydrogen)phosphonates have been precipitated in organopolymer membranes. In this field, *Alberti*,[200] *Jones and Rozière*[201,202,203,204,205] and *Yang et al.*[206] have performed pioneering work. Nafion®/layered zirconium hydrogen phosphate (Nafion®/ZrP) membranes have been prepared and investigated in DMFC at temperatures up to 140-150°C, wherein maximum power densities of 380 and 260 mW/cm^2 were achieved under oxygen and air feed, respectively.[206] The precipitation of the layered hydrogen phosphates and -phosphonates inside organopolymer membranes can be performed by the following procedure: at first the H^+ ions of sulfonated organopolymer membranes are exchanged against ZrO^{2+} ions by immersion of the membrane in aqueous $ZrOCl_2$ solution, followed by immersion of the membrane in diluted phosphoric acid. Using this method, we have prepared ionically and covalently cross-linked blend ionomer membranes having ZrP contents of up to 50 wt.% which showed improved PEFC performance at temperatures above 100°C, compared to the pure organomembrane (for example, an ionically cross-linked ZrP composite membrane showed a peak power density of 0.6 W/cm^2 at PEFC operation temperature of 115°C, while the corresponding undoped organomembrane showed a peak power density of only 0.35 W/cm^2 at operation temperature of 80°C).[149] Other methods for producing Zr hydrogen phosphate/phosphonate salt-containing hybrid membranes have been reviewed in the literature.[200]

2.2.2.2.2. Low Molecular Acid/High Organopolymer Blend Systems

As mentioned above, heteropolyacids (HPA) such as tungstophosphoric acid (TPA) $H_3PW_{12}O_{40} \cdot x\ H_2O$ are solid inorganic acids. They are strong Bronstedt acids because of proton delocalization within their structure[207,208] resulting in high proton conductivity. Consequently, they are interesting candidates for the application as proton-conducting material in fuel cells. Phosphotungstic acid was investigated as electrolyte in a PEFC. Output electrical power of 700 mW cm^{-2} was obtained at 25 °C and 1 atm, which was stable for more than 300 hours.[209] A problem of the heteropolyacids is their excellent water-solubility which generates the need for immobilization of the heteropolyacid molecules in solid materials such as organopolymers. Therefore, a lot of work has been performed in preparation of composite membranes from sulfonated polymer membranes which were doped with different heteropolyacids. Some examples for such systems will be presented below. Nafion® was impregnated with TPA and investigated in PEFC at temperatures up to 120°C. It was observed that the PEFC performance at 120°C was markedly improved by the added TPA, compared to undoped Nafion®.[210] *McGrath et al.* prepared composite membranes from sulfonated PSU and TPA.[211] Interactions between the sPSU macromolecules and TPA could be observed by FTIR spectroscopy, and are thought to be responsible for the good stability of the TPA in the ionomer matrix. Conductivity measurements showed increased proton conductivity at 130°C, compared to the pure organomembrane. Therefore, these membranes were regarded as promising candidates for PEFC. However, the excellent water solubility of HPAs makes their application in fuel cells dubios since they could be washed out of the organomembrane during fuel cell operation. Therefore much effort has been made to reduce or even exclude the bleeding-out of the HPAs from the membrane matrix. For example, *Nunes et al.* have prepared composite membranes of sPEK containing ZrO_2 or SiO_2 nanoparticles which have been doped with different HPAs such as TPA and molybdophosphoric acid (MoPA).[212] It was found that the bleeding-out of the HPAs could be reduced significantly by adsorbing the HPA particles onto the inorganic nanoparticles. In addition, also the water and meOH permeability of the membranes could be significantly reduced, compared to the pure sPEK membrane. In order to overcome the problem of bleeding-out of HPAs, the same group prepared composite membranes from sPEEK, NH_2-functionalized SiO_2 phase and divacant $[\gamma\text{-}SiW_{10}O_{36}]^{8-}$, containing epoxy groups which reacted with the amino-groups in the oxide phase, leading to covalent tethering of the HPA molecules to the inorganic membrane phase.[213] *Herring et al.* prepared composite membranes using lacunary $H_8SiW_{11}O_{39}$ in polymer matrices based on polyethylene glycol by a sol-gel method.[214] NMR and IR investigations of the membranes indicated that the HPA molecules were covalently bound to the polymer backbone. However, this membrane showed poor performance in a PEFC, compared to Nafion®. Another approach to reduce HPA bleeding-out from the membrane matrix was introduced by *Kerres et al.* who immobilized TPA and MoPA in acid-base blend membranes from sulfonated poly(etherketoneetherketone-ketone) (sPEKEKK) and PBI.[215] Membranes with superior proton-conductivities than Nafion® were obtained. It was found that the bleeding-out of MoPA from the acid-base blend membranes was much higher than the bleeding-out of TPA by aqueous membrane posttreatment. For example, a blend membrane which initially contained 40 wt.% TPA, still contained 34 wt.% TPA after the aqueous posttreatment, while the MoPA was nearly completely washed out from the membrane under the same conditions. *Tan et al.* performed a similar approach: composite polymer membranes based on sulfonated poly(arylene ether sulfone) (SPSU) containing benzimidazole derivatives (BlzD) and TPA have been prepared and characterized.[216] It was reported that washing out of the TPA could be markedly reduced by addition of the BIzD compounds.

2.2.2.3. Intermediate-Temperature Fuel Cell Membranes

The operation temperature of the fuel cell should be as high as possible to allow for fast electrode kinetics and therefore high fuel cell efficiency. However, as already pointed out in Section 2.2.2.2. of this contribution, the SO_3H proton conductors dry out at temperature above 100°C because of evaporation of the SO_3H group hydration shell water, leading to a severe proton conductivity drop and therefore strong decrease of the fuel cell efficiency. Therefore, other types of proton conductors have to be used in ITFCs. Several types of ITFC membranes have been developed within the last decades which can be roughly classified into the following material classes:

1. Basic polymers or basic polymer-rich polymer blends which are doped with (poly)phosphoric acid (Section 2.2.2.3.1.).
2. Heterocyclic amphoteric proton conductors (Section 2.2.2.3.2.).

3. Phosphonic acid proton conductors (Section 2.2.2.3.3.).

2.2.2.3.1. PBI/Phosphoric Acid

Savinell et al. [217,218,219,220,221,222] have introduced a proton conductor which is able to work in the 100 – 200°C operation temperature range. This proton conductor consists of the highly stable basic polymer polybenzimidazole (PBI) [162] which is doped with phosphoric acid, the H_3PO_4 molecules act as proton-conducting species, being bound to the PBI imidazole moieties via acid-base (ionical) and hydrogen bridge interactions. H_3PO_4 is a proton conductor even in the absence of H_2O molecules (which act as proton 'vehicle' in low-temperature fuel cells). This property makes H_3PO_4 suitable for operation temperatures above 100°C – the H_3PO_4 molecule itself is amphoteric.[223] PBI/H_3PO_4 blends can be prepared via two different routes, the first being the immersion of PBI films in highly concentrated H_3PO_4 at elevated temperatures, the second being polymerization of PBI in polyphosphoric acid, followed by the hydrolysis (depolymerisation) of polyphosphoric acid present within the PBI chains.[224,225,226] The PBI/H_3PO_4 membranes show high thermal stability also in fuel cell environment,[220] and an electroosmotic drag coefficient for water in these membranes is nearly 0.[219] Application of these membranes both in PEFC (at 150°C)[218] and in DMFC (at 190°C)[221] showed encouraging results. Long-term stability tests of PBI/H_3PO_4 MEAs, which have been commercialized by the former PEMEAS company (now part of BASF Fuel Cell Corp.) under the trade name Celtec®, showed an excellent performance stability[227] for 14,000 hours at 160°C operation temperature (initial cell voltage at 0.2 A/cm^2: 0.67 V, cell voltage at 0.2 A/cm^2 after 14,000 hours: 0.6 V; operation conditions: H_2/air, ambient pressure, cathode $\lambda = 2.0$, anode $\lambda = 1.2$, active area 45 cm^2, no humidification).

A disadvantage of PBI/H_3PO_4 blend membranes is the detoriation of their mechanical stability particularly at high H_3PO_4 concentrations and high doping temperature, leading even to dissolution. Therefore, apart from pure PBI, also other basic polymers or polymer blends have been doped with H_3PO_4, among them blends of PBI with a copolymer from 2,5-bis(4-hydroxyphenyl)pyridine, bisphenol A, and bis(4-fluorophenyl)-sulfone.[228] The oxidative stability of the blend membranes was tested in Fenton's and found to be stable against oxidative attack. Particularly, the membrane retained its excellent mechanical stability after the Fenton reaction. Doping of the basic blend membranes led to proton conductivities in the range of 10^{-2} S · cm^{-1}, which is a prerequisite for the application in fuel cells above 100°C. *Kallitsis, Bjerrum et al.* also took up the concept of acid-base blend membranes composed of sulfonated arylene main-chain polymers and basic polymers (see Section 2.2.2.1.5) and extended it to ternary blend membranes containing sulfonated PSU Udel®, PBI, and H_3PO_4 where the basic blend component was in molar excess over the sPSU, allowing to bind the H_3PO_4 molecules via acid-base interactions and hydrogen bridges to the PBI blend component.[229,230] They found that the ternary blend membranes showed better H^+-conductivities and better mechanical stability than binary blends of PBI/H_3PO_4. The membranes were investigated in a PEFC up to T = 190°C and showed good performance. Recently, blend membranes composed of 70 wt.% PBI and 30 wt.% of a sulfonated polymer from decafluorobiphenyl and bisphenol AF (M_n =82,400 Da, PDI = 1.65, IEC = 2.1 meq/g) were doped with H_3PO_4 and tested in a PEFC at temperatures up to 200°C, yielding good performance (450 mA/cm^2 at 500 mV). Moreover, the membranes showed better mechanical stability than PBI/H_3PO_4 membranes having comparable proton conductivity.[231,232]

2.2.2.3.2. Heterocyclic Amphoteric Proton-Conducting Systems

It is known that in liquid imidazole (melting point: 89 – 91°C, boiling point: 256°C), correlated proton transfers take place, leading to proton diffusion.[223] Self diffusion coefficients of imidazole-type proton conductors are much higher than that of water. Therefore the idea came up that imidazole could possibly be used as proton solvent or even proton conductor in ITFCs. The research groups of *Kreuer* and of *Meyer* performed pioneering work in the investigation of imidazole as potential proton solvent/conductor for ITFCs. When the use of imidazole as proton conductor in fuel cells is intended, it must be immobilized, because it would otherwise bleed out during fuel cell operation. As strategy for immobilization, the connection of imidazole molecules via a flexible spacer or the tethering of imidazole-containing flexible side-chains to different polymer backbones was chosen.[17,233,234,235] However, one of the problems of these systems was that the intrinsic concentration of protonic charge carriers can only be moderately increased through acid doping. This is particularly the case in fully polymeric systems, which is probably due to a reduced dielectric constant. The combination of such immobilized imidazole systems with low-molecular sulfonic acids like triflic acid leads to conductivities being insufficient for the application to membrane fuel cells ($\sigma < 3 \cdot 10^{-3}$ S $\cdot$ cm^{-1}).[223,233] Another disadvantage of imidazole-containing oligomeric and polymeric systems is their limited thermo-oxidative stability and the very high overpotential for oxygen reduction on platinum.[236]

2.2.2.3.3. Phosphonic Acid Systems

As pointed out in Section 2.2.2.3.1., the use of H_3PO_4, which is bound ionically or via hydrogen bridges to a host polymer, as proton conductor in ITFCs has the disadvantage that the H_3PO_4 molecules can bleed out of the membrane when the fuel cell operation temperature is decreased below 100°C. Therefore, phosphonic acid-containing ionomers has been taken into account for use in ITFCs since the phosphonic acid group is, analogous to H_3PO_4, capable of conducting protons at temperatures above 100°C without the need for water being present inside the membrane, as could be shown with phosphonic acid-containing oligomeric model compounds (conductivity range between 10^{-1} and 10^{-2} S $\cdot$ cm^{-1}).[223,236] Novel perfluorinated polyethers have been developed by *Yamabe et al.*[237] The phosphonated membranes, having an IEC of 2.05 meq/g, had proton conductivities of $\sigma = 7 \cdot 10^{-2}$ S $\cdot$ cm^{-1} (measured at 25°C under full humidification). Conductivity data at temperatures above 100°C and under reduced humidification have not been provided by the authors of this study. *Allcock et al.* have developed phenyl phosphonic acid functionalized poly[aryloxyphosphazenes] and tested their suitability in DMFC.[238,239] The membranes show ion exchange capacities of 1.17 and 1.43 meq/g, proton conductivities between 10^{-2} and 10^{-1} S $\cdot$ cm^{-1} and equilibrium water swelling values between 11 and 32%. Methanol diffusion coefficients for both radiation cross-linked and non-cross-linked membranes were found to be at least 12 times lower than those for Nafion® 117. Phosphonated poly(4-phenoxybenzoyl-1,4-phenylene) (P-PPBP) has been prepared by *Rikukawa et al.* by a three-step reaction.[240] The polymer possesses high thermal and mechanical stability, but P-PPBP films containing

40 mol% phosphonic acid groups showed a proton conductivity of only about 10^{-4} S · cm^{-1} at 90% RH, which is not sufficient for the fuel cell application. Recently, *Steiniger et al.* have prepared and characterized novel phosphonated poly(siloxanes).[241] The phosphonated compounds show conductivities of up to 2 · 10^{-3} S · cm^{-1} at T ≈ 130 °C and RH ≈ 37%, a value not being sufficient for use in fuel cells. *Liu et al.* prepared a fluorinated poly(aryl ether) containing a 4-bromophenyl pendant group and its phosphonated derivative.[119] This polymer was characterized and revealed very good thermal stabilities and very low methanol permeabilities. However, the proton conductivities of a phosphonated ionomer with one pendent PO_3H_2 group per polymer repeat unit are 2.9 · 10^{-6} S · cm^{-1} at 45% relative humidity, and 2.0 · 10^{-6} S · cm^{-1} at 25% relative humidity at 120°C, respectively. The authors of the article conclude that it is required to improve the proton conductivities of the phosphonated polymers by increasing the degree of phosphonation. *Lafitte and Jannasch* prepared new polysulfone ionomers functionalized with benzoyl(difluoromethylenephosphonic acid) side chains.[242] Membranes made from ionomers having 0.90 mmol of phosphonic acid units/g of dry polymer takes up 6 wt.% water when immersed at room temperature, and conductivities up to 5 mS · cm^{-1} at 100°C were recorded. It could be verified by thermogravimetry that the aryl – CF_2 – PO_3H_2 arrangement decomposed at approximately 230°C via cleavage of the C–P bond.

Summarizing the reported effort in preparation of phosphonated ionomer systems, one must state that up to now their proton conductivities at temperature above 100°C and under reduced humidification are too low to be regarded as polymer electrolytes for ITFC. Particularly, this material class might be improved by the following strategies:

- increase in number of PO_3H_2 groups per polymer repeat unit
- increase in acidity strength of the PO_3H_2 groups, since the highestly phosphonated polymers prepared so far have too low acid strength
- avoiding of condensation of PO_3H_2 groups which normally takes place in the temperature region 100 – 150°C if the H_2O partial pressure is too low.

2.3. Ageing and Degradation of Polymer Electrolyte Membranes

A basic understanding of possible degradation routes as well as of their dependences on the operating conditions (such as extreme load, cyclic change of load, impurities, inhomogeneities in the fuel and oxidant supply, in the catalytic reaction, in the proton conduction or in the relative humidity) and on the type of the catalyst are of crucial importance for the development of novel, stable polymer electrolyte membranes.[19,129] Morphological alteration of the membrane electrode assembly (MEA) at the interfaces between the membrane and the electrodes is pointed out in the literature in this context (interfacial degradation). Its origin is mainly led back to an ageing and degradation of the catalyst and catalyst support layer. Especially under load cycling, a potential-dependent dissolution of platinum (Eq. (1)) or alloying component (e.g. Cr_3Pt),[243,244] a chemical dissolution of platinum oxide (Eq. (2)) and Ostwald ripening of the nanocrystalline platinum[245,246,247] can be observed.

$$Pt \rightarrow Pt^{2+} + 2\ e^- \tag{1}$$

$$PtO + 2\ H^+ \rightarrow Pt^{2+} + H_2O \tag{2}$$

Interfacial degradation in MEAs[248] can additionally be caused by a leaching out of the ionomer network from the electrode structure[243] or by corrosion of the carbon catalyst support.[249] *Post mortem* analyses of MEAs showed that soluble oxidized platinum (e.g. Pt^{2+}) species are formed at the cathode site. This soluble platinum species diffuse – driven by a concentration gradient – through the ionomer network in the catalyst layer and the membrane toward the anode. Hydrogen, which permeates through the membrane in the opposite direction, reduces these platinum species resulting in sedimentation of elementary platinum within the membrane and near the interface membrane/cathode.[245] The influence of these processes on the degradation of the membrane material itself is relatively sparsely investigated and only little understood to this day.[129]

As already discussed in the previous sections, the properties and thus the lifetime of a MEA and a membrane respectively are also influenced by their mechanical, thermal and thermohydrolytical (longterm) stability. The complex chemical reactions occurring on a microscopic scale and leading to the degradation of the membrane during the fuel cell operation are sparsely understood. It is generally assumed that mainly hydroxyl (HO·) and hydroperoxyl radicals (HOO·) which are formed during the fuel cell operation act as initiator for membrane degradation.[20,62,129,250,251,252] Hydroxyl radicals can initiate depolymerisation starting at the carboxylic acid endgroups of poly(perfluoroalkyl)sulfonic acids for example according to the following reaction equations:[253]

$$R_f - CF_2COOH + HO\cdot \rightarrow R_f - CF_2\cdot + CO_2 + H_2O \tag{3}$$

$$R_f - CF_2\cdot + \cdot OH \rightarrow R_f - CF_2OH \rightarrow R_f - COF + HF \tag{4}$$

$$R_f - COF + H_2O \rightarrow R_f - COOH + HF \tag{5}$$

Such carboxylic acid endgroups are formed in the presence of water via perfluorocarbinol intermediates from sulphuric acid ester end groups which result from the polymerization initiator ($K_2S_2O_8$).[254]

HO· and HOO· are regular intermediates in the catalytic reduction of oxygen at the cathode[255] and might be responsible for membrane degradation after desorption from the catalyst surface.[252] Another often discussed source for the formation of HO· and HOO· is based on the evolution of small amounts of hydrogen peroxide as side product[250] which especially occurs at low coordinated platinum atoms (e.g. edge and corner atoms):[256,257]

$$O_2 + 4\ H^+ + 4\ e^- \rightarrow 2\ H_2O \qquad E^0 = 1.23\ V \tag{6}$$

$$O_2 + 2\ H^+ + 2\ e^- \rightarrow H_2O_2 \qquad E^0 = 0.67\ V \tag{7}$$

This formation of hydrogen peroxide mainly occurs at the cathode site of the MEA but in principle it can happen at the anode site as well after oxygen is permeated through the membrane.[156] Although a certain percentage of the formed hydrogen peroxide is washed out by the product water and the gas flows, another fraction might diffuse into the membrane and decay into HO· by catalysis of bivalent metal ions at the elevated operation temperature.[250] More recent studies put the involvement of hydrogen peroxide for degradation processes in question, but postulate the simultaneous formation of a highly reactive species (whose chemical nature is not clarified to this day) which is exclusively generated by feeding of hydrogen and oxygen in the presence of the catalyst.[257,258,259] Unfortunately only hydrogen has been used as fuel so far and no information is available in terms of the influence of other fuels (such as methanol). However, it is easily conceivable with this findings that the reactive species might be a radical consisting of two the elements hydrogen and oxygen.

Electron paramagnetic spectroscopic studies of reactions between radicals (especially HO·, HOO·) and model compounds representing structural units of unsaturated sulfonated polymers (such as sPEEK, sPES, sulfonated Polystyrene, FEP-*g*-PSSA) provided some information on possible chemical reactions which might play a role in radical induced degradation processes. In this context the addition of HO· at aromatic rings has been identified as the major reaction route in the degradation of aromatic polymers.[20]

It could be demonstrated at toluene-4-sulfonic acid that acid-catalyzed water elimination yields benzyl radicals (*via* hydroxycyclohexadienyl radicals that are formed after HO· radical addition, Figure 8).[20,62,251] Similar reactions in polymers with benzylic C_α-H bonds (e.g. in FEP-*g*-PSSA) could result in a cleavage of polymer chains.[260]

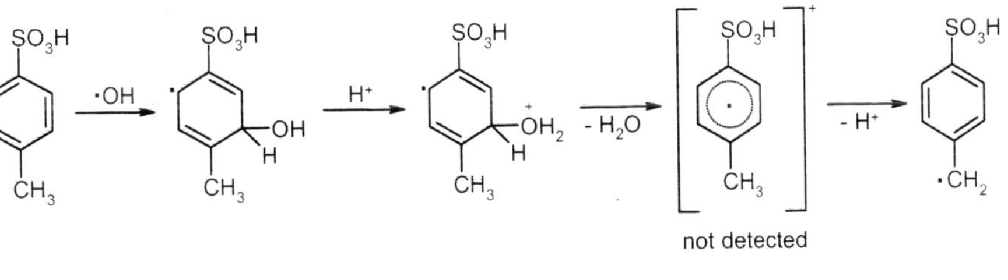

Figure 8. Acid-catalyzed formation of benzyl radicals from hydroxycyclohexadienyl radicals.[20]

Further possible sites of attack for HO· radicals are the *ortho*-positions next to electron-donating groups (e.g. alkyl- or phenoxyl groups) and the *meta*-positions next to the sulfonic acid group. The latter substitution type is usually present in sulfonated poly(aryl)s, for example in sPEK or sPES.[260] ESR spectroscopic investigations of 4-methoxybenzoic acid in the presence of HO· radicals revealed that an *ipso*-substitution of the methoxy group can follow the addition of HO· to the aromatic ring at pH > 5.[20] Since pH inhomogeneities can appear under certain circumstances during fuel cell operation (e.g. under high loads),[19] the authors pointed out the relevance of such reactions in the course of the degradation of ether brigded poly(aryl) ionomers.

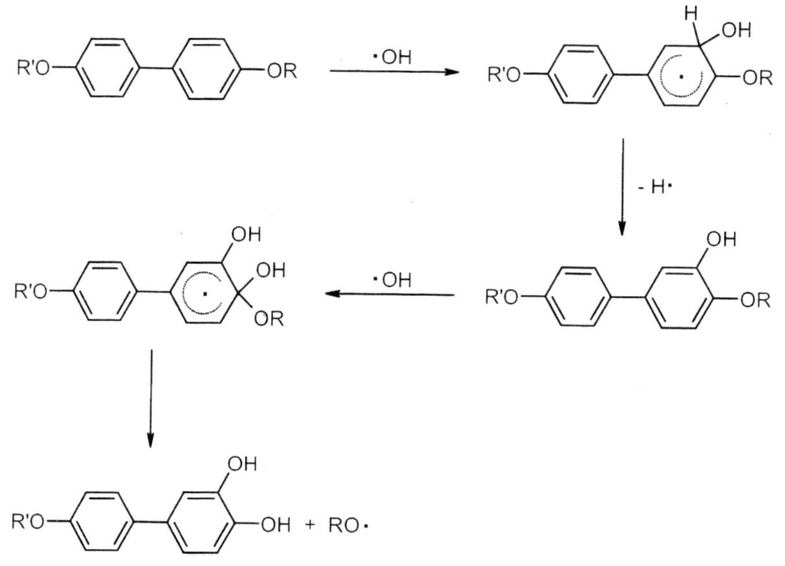

Figure 9. Radical induced cleavage of an ether brigde in poly(aryl)s after addition of ·OH[20,129]

Mukerjee et al. observed significant decrease of the intrinsic viscosimetry and of the IR absorption at 1010 cm^{-1} (absorption band of *para*-substituted arylene ethers) in degradation studies of sulfonated poly(arylethersulfone)s at pH < 5 und postulated a degradation mechanism according to Figure 10:[129]

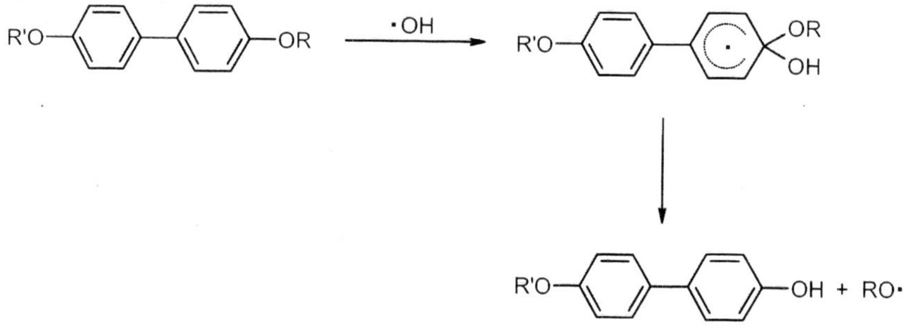

Figure 10. Postulated degradation mechanism of poly(aryl)s by ·OH at pH < 5.[129]

Molecular oxygen ($^3O_2$) which is present in a working fuel cell at the cathode and in the membrane (as permeating species) can also attack the polymer backbone by formation of aromatic peroxyl radicals which could lead to a subsequent cleavage of the aromatic system as well.[20,156,252] *Panchenko* considered reactions between $^3O_2$ and sulfonated hydroxyl-cyclohexadienyl radicals as model compounds for sPEEK and sPES in a theoretical approach based on DFT calculations.[251] Different reaction paths have been taken into account (elimination of $\cdot SO_3H$, formation of bicyclic and epoxide radicals) from which the first one turned out to be the most favourable process in aqueous atmosphere because of energetic reasons. Therefore, this reaction (Figure 11) could also play a role in the degradation process in membrane materials.

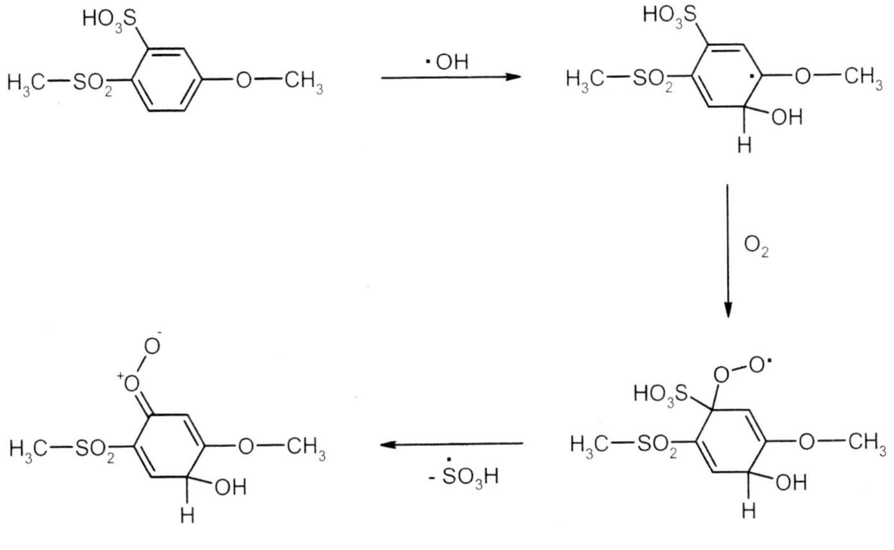

Figure 11. Proposed mechanism for an ·OH induced desulfonation process on the basis of DFT calculations.[251]

References

[1] Entwicklung der erneuerbaren Energien im Jahr 2006 in Deutschland, Bundesministerium für Umwelt, *Naturschutz und Reaktorsicherheit* (21. Februar 2007).

[2] Rikukawa, M.; Sanui, K. Proton-conducting polymer electrolyte membranes based on hydrocarbon polymers. *Prog. Polym. Sci.* 25 (2000) 1463 – 1502.

[3] Hickner, M. A.; Ghassemi, H.; Kim, Y. S.; Einsla, B. R. McGrath, J. E. Alternative polymer systems for proton exchange membranes. *Chem. Rev.* 104 (2004) 4587 – 4612.

[4] Liu, B.; Robertson, G. P.; Kim, D.-S.; Guiver, M. D.; Hu, W.; Jiang, Z. Aromatic poly(ether ketone)s with pendant sulfonic acid phenyl groups prepared by a mild sulfonation method for proton exchange membranes. *Macromolecules.* 40 (2007) 1934 – 1944.

[5] Schönberger, F. Aufbau eines Brennstoffzellenmessstandes und Versuche zur Herstellung BOPP-mikroverstärkter Nafion®-Membranen für Polymerelektrolyt membranbrennstoffzellen, *Diplomarbeit,* Saarbrücken (2004).

[6] Grove, W. R. On voltaic series and the combination of gases by platinum *Phil. Mag. Ser.* III 14 (1839) 127 – 130.

[7] Oertel, D.; Fleischer, T. Brennstoffzellen-Technologie: Hoffnungsträger für den Klimaschutz: Technische, ökonomische und ökologische Aspekte ihres Einsatzes in Verkehr und Energiewirtschaft, *Erich Schmidt Verlag* (2001) 9 – 48.

[8] Dunwoody, D.; Leddy, J. Proton exchange membranes: the view forward and back. *The Electrochemical Society* Interface 14 (3) (2005) 37 – 39.

[9] Connolly, D. J.; Gresham, W. F. Fluorocarbon vinyl ether polymers. US 3282875 (1996).

[10] Heitner-Wirguin, C. Recent advances in perfluorinated ionomer membranes: Structure, porperties and applications. *J. Membr. Sci.* 120 (1996) 1 – 33.

[11] Rowsell, J. L. C.; Yaghi, O. M. Strategies for hydrogen storage in metal-organic frameworks. *Angew. Chem. Int. Ed.* 44 (2005) 4670 – 4679.

[12] Fujii, H.; Ichikawa, T. Recent developments on hydrogen storage materials composed of light elements. *Physica B* 383 (2006) 45 – 48.

[13] Dornheim, M.; Eigen, N.; Barkhordarian, G.; Klassen, T.; Bormann, R. Tailoring hydrogen storage materials towards application. *Adv. Eng. Mater.* 8 (5) (2006) 377 – 385.

[14] Fichtner, M. Nanotechnolgoical aspects in materials for hydrogen storage. *Adv. Eng. Mater.* 7(6) (2007) 443 – 455.

[15] He, C.; Desai, S.; Brown, G.; Bollepalli, S. PEM fuel cell catalysts: Cost, performance, and durability. *The Electrochemical Society Interface.* 14 (3) (2005) 41 – 44.

[16] Hsu, W. Y.; Gierke, T. D. Ion-transport and clustering in Nafion perfluorinated membrane. *J. Membr. Sci.* 13 (1983) 307 – 326.

[17] Kreuer, K. D. On the development of proton conducting polymer membranes for hydrogen and methanol fuel cells. *J. Membr. Sci.* 185 (2001) 29 – 39.

[18] Mauritz, K. A.; Moore, R. B. State of understanding of Nafion, *Chem. Rev.* 104 (2004) 4535 – 4585.

[19] Kerres, J.; Roduner, E. Wechselwirkungen, Die Protonenleitende Membran: Schlüsselkomponente einer Brennstoffzelle, Universität Stuttgart, *Jahrbuch* (2001) 52 – 68.

[20] Hübner, G.; Roduner, E. EPR investigation of HO· radical initiated degradation reactions of sulfonated aromatics as model compounds for fuel cell proton conducting membranes. *J. Mater. Chem.* 9 (1999) 409 – 418.

[21] Kadirov, M. K.; Bosnjakovic, A.; Schlick, S. J.; Membrane-derived fluorinated radicals detected by electron spin resonance in UV-irradiated Nafion and Dow ionomers: Effect of counterions and H_2O_2. *Phys. Chem. B* 109 (2005) 7664 – 7670.

[22] Hamann, C. H.; Vielstich, W. *Elektrochemie,* VCH-Verlag (1998).

[23] Varcoe, J. R.; Slade, R. C. T.; Yee, E. L..H. An alkaline polymer electrochemical interface: A breakthrough in application of alkaline anion-exchange membranes in fuel cells. *Chem. Commun.* (2006) 1428 – 1429.

[24] Varcoe, J. R.; Slade, R. C. T. Prospects for alkaline anion-exchange membranes in low temperature fuel cells. *Fuel Cells*. 5 (2) (2005) 187 – 200.

[25] Li, W.; Wang, X.; Chen, Z.; Waje, M.; Yan, Y. Pt-Ru supported on double-walled carbon nanotubes as high-performance catalysts for direct methanol fuel cells. *J. Phys. Chem.* B 110 (2006) 15353 – 15358.

[26] Zhou, W. J.; Song, S. Q.; Lia, W. Z.; Zhou, Z. H.; Sun, G. Q.; Xin, Q.; Douvartzides, S.; Tsiakaras, P. Direct ethanol fuel cells based on PtSn anodes: the effect of Sn content on the fuel cell performance. *J. Power Sources.* 140 (2005) 140 50 – 58.

[27] Lamy, C. ; Rousseau, S.; Belgsir, E. M.; Coutanceau, C.; Léger, J.-M. Recent progress in the direct ethanol fuel cell: development of new platinum–tin electrocatalysts. *Electrochim. Acta.* 49 (2004) 3901 – 3908

[28] Schönberger, F.; Kerres, J. Sulfonated partially fluorinated ionomers and ionomer blends for PEMFCs: Synthesis, characterization and investigation of the radical stability, 4th International Symposium on High-Tech Polymer Materials. (Beijing, China), 2006, May 14th – 20th, 35 – 36.

[29] Lee, J. S.; Quan, N. D.; Hwang, J. M.; Lee, S. D.; Kim, H.; Lee, H.; Kim, H. S. Polymer electrolyte membranes for fuel cells. *J. Ind. Eng. Chem.* 12(2) (2006) 175 – 183.

[30] Li, Q.; He, R.; Jensen, J. O.; Bjerrum, N. J. Approaches and recent development of polymer electrolyte membranes for fuel cells operating above 100 °C. *Chem. Mater.* 15 (2003) 4896 – 4915.

[31] Fujimura, M.; Hashimoto, T.; Kawai, H. Small-angle X-ray scattering study of perfluorinated ionomer membranes. 1. Origin of two scattering maxima. *Macromolecules.* 14 (1981) 1309 – 1315.

[32] Fujimura, M.; Hashimoto, T.; Kawai, H. Small-angle X-ray scattering study of perfluorinated ionomer membranes. 2. Models for ionic scattering maximum. *Macromolecules.* 15 (1982) 136 – 144.

[33] Gebel, G.; Lambard, J. Small-angle scattering study of water-swollen perfluorinated ionomer membranes. *Macromolecules.* 30 (1997) 7914 – 7920.

[34] Gebel, G.; Moore, R. B. Small-angle scattering study of short pendant chain perfluorosulfonated ionomer membranes. *Macromolecules.* 33 (2000) 4850 – 4855.

[35] Rollet, A. L.; Gebel, G.; Simonin, J.-P.; Turq, P. A SANS determination of the influence of external conditions on the nanostructure of Nafion membrane. *J. Polym. Sci. Pol. Phys.* 39 (2001) 548 – 558.

[36] Rubatat, L.; Rollet, A. L.; Gebel, G.; Diat, O. Evidence of elongated polymeric aggregates in Nafion. *Macromolecules.* 35 (2002) 4050 – 4055.

[37] Dreyfus, B.; Gebel, G. Distribution of the micelles in hydrated perfluorinated ionomer membranes from SANS experiments. *J. Phys.* 51(12) (1990) 1341 – 1354.

[38] Kerres, J. Development of ionomer membranes for fuel cells. *J. Membr. Sci.* 185(1) (2001) 3 – 27.

[39] Mathias, M. F.; Makharia, R.; Gasteiger, H. A.; Conley, J. J.; Fuller, T. J.; Gittleman, C. J.; Kocha, S. S.; Miller, D. P.; Mittelsteadt, C. K.; Xie, T.; Yan, S. G.; Yu, P. T. Two fuel cell cars in every garage? *The Electrochemical Society Interface.* 14 (2005) 24 – 33.

[40] Miyatake, K.; Wanatabe, M. Emerging membrane materials for high temperature polymer electrolyte fuel cells: Durable hydrocarbon ionomers. *J. Mater. Chem.* 16(46) (2006) 4465 – 4467.

[41] Deluca, N.; Elabd, Y. A. Polymer electrolyte membranes for direct methanol fuel cell: A review. *J. Polym. Sci. Pol. Phys.* 44 (2006) 2201 – 2225.

[42] Liu, F.; Yi, B.; Xing, D.; Yu, J.; Zhang, H. Nafion/PTFE composite membranes for fuel cell applications. *J. Membr. Sci.* 212 (2003) 213 – 223.

[43] Shim, J.; Ha, H. Y.; Hong, S.-A.; Oh, I.-H. Characteristics of the Nafion ionomer-impregnated composite membrane for polymer electrolyte fuel cells. *J. Power Sources.* 109 (2002) 412 – 417.

[44] Nouel, K. M.; Fedkiw, P. S. Nafion-based composite polymer electrolyte membranes. *Electrochim. Acta* 43(16-17) (1998) 2381 – 2387.
[45] Bae, B.; Chun, B.-H.; Ha, H.-Y.; Oh, I.-H.; Kim, D. Preparation and characterization of plasma treated PP composite electrolyte membranes. *J. Membr. Sci.* 202 (2002) 245 – 252.
[46] Bahar, B.; Hobson, A. R., Kolde, J. A., Zuckerbrod, D. Ultra-thin integral composite membrane. US 5547551, 1996.
[47] Higuchi, Y.; Terada, N.; Shimoda, H.; Hommura, S. Electrolyte membrane for solid polymer type fuel cell and producing method thereof. EP 1139472 B2, 2001.
[48] Antonucci, P. L., Aricò, A. S.; Cretì, P.; Ramunni, E.; Antonucci, V. Investigation of a direct methanol fuel cell based on a composite Nafion-silica electrolyte for high temperature operation. *Solid State Ionics.* 125 (1999) 431 – 437.
[49] Alberti, G.; Casciola, M.; Costantino, U.; Vivani, R. Layered and pillared metal(IV) phosphates and phosphonates. *Adv. Mater.* 8(4) (1996) 291 – 303.
[50] Alberti, G.; Casciola, M.; Layered metal(IV) phosphonates, a large class of inorgano-organo proton conductors. *Solid State Ionics.* 97 (1997) 177 – 186.
[51] Yang, C.; Srinivasan, S., Bocarsly, A. B.; Tulyani, S.; Benziger, J. B. A comparison of physical properties and fuel cell performance of Nafion and zirconium phosphate/Nafion composite membranes. *J. Membr. Sci.* 237 (2004) 145 – 161.
[52] D'Agostino, V. F., Lee, J. Y.; Cook, E. H. Trifluorostyrene sulfonic acid membranes. US 4012303, 1977.
[53] D'Agostino, V. F., Lee, J. Y.; Cook, E. H. Process for electrolysing sodium chloride or hydrochloric acid and an electrolytic cell, employing trifluorostyrene sulfonic acid membrane. US 4107005, 1978.
[54] D'Agostino, V. F., Lee, J. Y.; Cook, E. H. Trifluorostyrene sulfonic acid membranes. US 4113922, 1978.
[55]. Momose, T.; Tmoiie, K.; Ishigaki, I., Okamoto, J. Radiation grafting ofα,β,β-trifluorostyrene onto various polymer films by preirradiation method. *J. Appl. Polym. Sci.* 37 (1989) 2165 – 2168.
[56] Momose, T.; Yoshioka, H.; Ishigaki, I., Okamoto, J. Radiation grafting of α,β,β-trifluorostyrene onto poly(ethylene-tetrafluoroethylene) film by preirradiation method. II. Properties of cation-exchange membrane obtained by sulfonation and hydrolysis of the grafted film. *J. Appl. Polym. Sci.* 38 (1989) 2091 – 2101.
[57] Gupta, B.; Scherer, G. G. Proton exchange membranes by radiation-induced graft copolymerization of monomers into Teflon-FEP films. *Chimia.* 48 (1994) 127 – 137.
[58] Gubler, L.; Beck, N.; Gürsel, S. A.; Hajbolouri, F.; Kramer, D.; Reiner, A.; Steiger, B.; Scherer, G. G.; Wokaun, A.; Rajesh, B.; Thampi, K. R. Materials for polymer electrolyte fuel cells. *Chimia.* 58 (2004) 826 – 836.
[59] Kallio, T.; Jokela, K.; Ericson, H.; Serimaa, R.; Sundholm, G.; Jacobsson, P.; Sundholm, F. Effects of a fuel cell test on the structure of irradiation grafted ion exchange membranes based on different fluoropolymers. *J. Appl. Electrochem.* 33 (2003) 505 – 514.

[60] Kallio, T.; Lundström, M.; Sundholm, G.; Walsby, N.; Sundholm, F. Electrochemical characterization of radiation-grafted ion-exchange membranes based on different matrix polymers. *J. Appl. Electrochem. 32* (2002) 11 – 18.

[61] Holmberg, S.; Holmlund, P.; Nicolas, R.; Wilén, C.-E.; Kallio, T.; Sundholm, G.; Sundholm, F.; Versatile synthetic route to tailor-made proton exchange membranes for fuel cell applications by combination of radiation chemistry of polymers with nitroxide-mediated living free radical graft polymerization. *Macromolecules.* 37 (2004) 9909 – 9915.

[62] Panchenko, A.; Dilger, H.; Möller, E.; Sixt, T.; Roduner, E. In situ EPR investigation of polymer electrolyte membrane degradation in fuel cell applications. *J. Power Sources.* 127 (2004) 325 – 330.

[63] Gürsel, S. A.; Yang, Z.; Choudhury, B.; Roelofs, M. G.; Scherer, G. G. Radiation-grafted membranes using a trifluorostyrene derivative. *J. Electrochem. Soc.* 153(10) (2006) A1964 – A1970.

[64] Burnaby, J. W.; Vancouver, C. S.; Steck, A. E.; Vancouver, W. Trifluorostyrene and substituted trifluorostyrene copolymeric compositions and ion-exchange membranes formed therefrom. US 5422411, 1995.

[65] Hodgdon, R. B.; Enos, J. F.; Aiken, E. J. Sulfonated polymers of α,β,β-trifluorostyrene, with applications to structures and cells. US 3341366, 1967.

[66] Yu, J.; Yi. B.; Xing, D.; Liu, F.; Shao, Z.; Fu, Y.; Zhang, H. Degradation mechanism of polystyrene sulfonic acid membrane and application of its composite membranes in fuel cells. *Phys. Chem. Chem. Phys.* 5 (2003) 611 – 615.

[67] Ding, J.; Chuy, C.; Holdcroft, S. A self-organized network of nanochannels enhances ion conductivity through polymer films. *Chem. Mater.* 13(7) (2001) 2231 – 2233.

[68] Ding, J.; Chuy, C.; Holdcroft, S. Solid polymer electrolyte based on ionic graft polymers: Effect of graft chain length on nano-structured, ionic networks. *Adv. Funct. Mater.* 12 (2006) 389 – 394.

[69] Chuy, C.; Ding, J.; Swanson, E.; Holdcroft, S.; Horsfall, J.; Lovell, K. V. Conductivity and electrochemical ORR mass transport properties of solid polymer electrolytes containing poly(styrene sulfonic acid) graft chains. *J. Electrochem. Soc.* 150(5) (2003) E271 – E279.

[70] Ehrenberg, S. G.; Troy, J. S.; Wnek, G. E.; Rider, J. N. Fuel cell incorporating novel ion-conducting membrane. US 5468574, 1995.

[71] Hollemann, A. F.; Wiberg, E. Lehrbuch der Anorganischen Chemie, Walter de Gruyter, 1995, 950 – 953. (nachprüfen, ob Zitation stimmt (vor allem Seitenzahl))

[72] Gautier-Luneau, I. ; Denoyelle, A. ; Sanchez, J. Y. ; Poinsigon, C. Organic-inorganic protonic polymer electrolytes as membrane for low-temperature fuel cell, *Electrochim. Acta.* 37(9) (1992) 1615 – 1618.

[73] Popall, M.; Du, X.-M. Inorganic-organic copolymers as solid state ionic conductors with grafted anions, *Electrochim. Acta.* 40(13-14) (1995) 2305 – 2308.

[74] Depre, L.; Kappel, J.; Popall, M. Inorganic-organic proton conductors based on alkylsulfone functionalities and their patterning by photoinduced methods, *Electrochim. Acta.* 43(10-11) (1998) 1301 – 1306.

[75] Jeske, M.; Soltmann, C.; Ellenberg, C.; Wilhelm, M.; Koch, D.; Grathwohl, G. Proton conducting membranes for the high temperature polymer electrolyte membrane fuel cell (HT-PEMFC) based on functionalized polysiloxanes, *Fuel Cells*. 7(1) (2007) 40 – 46.

[76] Hofmann, M. A.; Ambler,C. M.; Maher, A. E.; Chalkova, E.; Zhou, X. Y.; Lvov, S. N.; Allcock, H. R. Synthesis of Polyphosphazenes with Sulfonimide Side Groups, *Macromolecules* 2002, *35*, 6490-6493. (b) Allcock, H. R.; Wood, R. M. Design and synthesis of ion-conductive polyphosphazenes for fuel cell applications: Review, *J. Polym. Sci. Pol. Phys*. 44 (2006) 2358 – 2368.

[77] Wycisk, R.; Lee, J. K.; Pintauro, P. N. Sulfonated polyphosphazene-polybenzimidazole membranes for DMFCs, *J. Electrochem. Soc*. 152(5) (2005) A892 – A898.

[78] Allcock, H. R.; Hofmann, M. A.; Wood, R. M. Phosphonation of aryloxyphosphazenes, *Macromolecules*. 34 (2001) 6915 – 6921.

[79] Allcock, H. R.; Hofmann, M. A.; Ambler, C. M.; Morford, R. V. Phenylphosphonic acid functionalized poly[aryloxyphosphazenes], *Macromolecules*. 35 (2002) 3484 – 3489.

[80] Rusanov, A. L.; Likhatchev, D.; Kostoglodov, P. V.; Müllen, K.; Klapper M. Proton-exchanging electrolyte membranes based on aromatic condensation polymers, *Adv. Polym. Sci.* 179 (2005) 83 – 134.

[81] Rulkens, R.; Schulze, M.; Wegner, G. Rigid-rod polyelectrolytes: Synthesis of sulfonated poly(p-phenylene)s, *Macromol. Rapid Comm*. 15 (1994) 669 – 676.

[82] Schauer, J.; Albrecht, W.; Weigel, T.; Kudela, V.; Pientka, Z. Microporous membranes prepared from blends of polysulfone and sulfonated poly(2,6-dimethyl-1,4-phenylene oxide), *J. Appl. Polym. Sci.* 81 (2001) 134 – 142.

[83] Nakabayashi, K.; Shibasaki, Y.; Ueda, M. Synthesis and properties of sulfonated poly(phenylene), *Polym. Prepr.* 48(2) (2007) 332 – 333.

[84] Miyatake, K.; Iyotani, H.; Yamamoto, K.; Tsuchida, E. Synthesis of poly(phenylene sulfide sulfonic acid) via poly(sulfonium cation) as a thermostable proton-conducting polymer, *Macromolecules*. 29 (1996) 6969 – 6971.

[85] Schuster, M.; Kreuer, K.-D.; Andersen, H. T.; Maier, J. Sulfonated poly(phenylene sulfone) polymers as hydrolytically and thermooxidatively stable proton conducting ionomers, *Macromolecules*. 40 (2007) 598 – 607.

[86] Harrison, W. L.; Wang, F.; Mecham, J. B.; Bhanu, V. A.; Hill, M.; Kim, Y. S.; McGrath, J. E. Influence of the bisphenol structure on the direct synthesis of sulfonated poly(arylene ether) copolymers, *J. Polym. Sci. Pol. Chem*. 41 (2003) 2264 – 2276.

[87] Wang, L.; Meng, Y. Z.; Wang, S. J.; Shang, X. Y.; Li, L.; Hay, A. S. Synthesis and sulfonation of poly(arylethers) containing triphenyl methane and tetraphenyl methane moieties from isocyanate-masked bisphenols *Macromolecules*. 37 (2004) 3151 – 3158.

[88] Ulrich, H.-H.; Rafler, G. Sulfonated poly(aryl ether ketone)s, Angew. *Makromol. Chem*. 263 (1998) 71 – 78.

[89] Shibuya, N.; Porter, R. S. Kinetics of PEEK sulfonation in concentrated sulfuric acid, *Macromolecules*. 25 (1992) 6495 – 6499.

[90] Kerres, J. Covalent-ionically cross-linked poly(etheretherketone)-basic polysulfone blend ionomer membranes, *Fuel Cells*. 6(3-4) (2006) 251 – 260.

[91] Xing, P.; Robertson, G. P.; Guiver, M. D.; Mikhailenko, S. D.; Kaliaguine, S. Sulfonated poly(aryl ether ketone)s containing naphthalene moieties obtained by direct copolymerization as novel polymers for proton exchange membranes, *J. Polym. Sci. Pol. Chem.* 42 (2004) 2866 – 2876.

[92] Kerres, J.; Cui, W.; Reichle, S. New sulfonated engineering polymers via the metalation route. 1. Sulfonated poly(ethersulfone) PSU Udel via metalation-sulfination-oxidation *J. Polym. Sci. Polym. Chem.* 34 (1996) 2421 – 2438.

[93] Deimede, V.; Voyiatzis, G. A.; Kallitsis, J. K.; Li, Q.; Bjerrum, N. J. Miscibility behavior of polybenzimidazol/sulfonated polysulfone blends for use in fuel cell applications, *Macromoelcules.* 33 (2000) 7609 – 7617.

[94] Wiles, K. B.; Wang, F.; McGrath, J. E. Directly copolymerized poly(arylene sulfide sulfone) disulfonated copolymers for PEM-based fuel cell systems. I. Synthesis and characterisation, *J. Polym. Sci. Pol. Chem.*43 (2005) 2964 – 2976.

[95] Kim, Y. S.; Sumner, M. J.; Harrison, W. L.; Riffle, J. S.; McGrath, J. E.; Pivovar, B. S. Direct methanol fuel cell performance of disulfonated poly(arylene ether benzonitrile) copolymers, *J. Electrochem. Soc.*151(12) (2004) A2150 – A2156.

[96] Gao, Y.; Robertson, G. P.; Kim, D.-S.; Guiver, M. D.; Mikhailenko, S. D.; Li, X.; Kaliaguine, S. Comparison of PEM properties of copoly(aryl ether ether nitrile)s containing sulfonic acid bonded to naphthalene in structurally different ways, *Macromolecules.* 40 (2007) 1512 – 1520.

[97] Miyatake, K.; Asano, N.; Wanatabe, M. Synthesis and properties of novel sulfonated polyimides containing 1,5-naphthylene moieties *J. Polym. Sci. Pol. Chem.* 41 (2003) 3901 – 3907.

[98] Lee, C.; Sundar, S.; Kwon, J.; Han, H. Structure-property correlations of sulfonated polyimides. II. Effect of substituent groups on membrane properties *J. Polym. Sci. Pol. Chem.* 42 (2004) 3621 – 3630.

[99] Lee, C.; Sundar, S.; Kwon, J.; Han, H. Structure-property correlations of sulfonated polyimides. I. Effect of bridging groups on membrane properties, *J. Polym. Sci. Pol. Chem.*, 42 (2004) 3612 – 3620.

[100] Sundar, S.; Jang, W.; Lee, C.; Shul, Y.; Han, H. Crosslinked sulfonated polyimide networks as polymer electrolyte membranes in fuel cells, *J. Polym. Sci. Pol. Phys.* 43 (2005) 2370 – 2379.

[101] Yin, Y.; Yamada, O.; Hayashi, S.; Tanaka, K.; Kita, H.; Okamoto, K.-I. Chemically modified proton-conducting membranes based on sulfonated polyimides: Improved water stability and fuel cell performance, *J. Polym. Sci. Pol. Chem.* 44 (2006) 3751 – 3762.

[102] Qiu, Z.; Wu, S.; Li, Z.; Zhang, S.; Xing, W.; Liu, C. Sulfonated poly(arylene-co-naphthalimide)s synthesised by copolymerisation of primarily sulfonated monomer and fluorinated naphthalimide dichlorides as novel polymers for proton exchange membranes, *Macromolecules.* 39 (2006) 6425 – 6432.

[103] Okazaki, Y.; Nagaoka, S.; Kawakami, H. Proton conductive membranes based on blends of polyimides, *J. Polym. Sci. Pol. Phys* .45 (2007) 1325 – 1332.

[104] Shang, X. Y.; Shu, D.; Wang, S. J.; Xiao, M.; Meng, Y. Z. Fluorene-containing sulfonated poly(arylene ether 1,3,4-oxadiazole) as proton-exchange membrane for PEM fuel cell application, *J. Membr. Sci.* 291 (2007) 140 – 147.

[105] Gomes, D.; Roeder, J.; Ponce, M. L.; Nunes, S. P. Characterization of partially sulfonated polyoxadiazoles and oxadiazole-triazole copolymers, *J. Membr. Sci.* 295 (2007) 121 – 129.

[106] Li, J.; Yu, H. Synthesis and characterization of sulfonated poly(benzoxazole ether ketone)s by direct copolymerization as novel polymers for proton exchange membranes, *J. Polym. Sci. Pol. Chem.* 45 (2007) 2273 – 2286.

[107] Shen, L.; Xiao, G.; Yan, D.; Sun, G.; Zhu, P.; Na, Z. Comparison of the water uptake and swelling between sulfonated poly(phthalazinone ether keton ketone)s and sulfonated poly(arylene ether ketone ketone)s, *Polym. Bull.* 54 (2005) 57 – 64.

[108] Glipa, X.; Haddad, M. E.; Jones, D. J.; Rozière, J. Synthesis and characterisation of sulfonated polybenzimidazole: a highly conductiong proton exchange polymer, *Solid State Ionics*. 97 (1997) 323 – 331.

[109] Asensio, J. A.; Borrós, S.; Gómez-Romero, P. Proton-conducting polymers based on benzimidazoles and sulfonated benzimidazoles, *J. Polym. Sci. Pol. Chem.* 40 (2002) 3703 – 3710.

[110] Pu, H. ; Liu, Q. Methanol permeability and proton conductivity of polybenzimidazole and sulfonated polybenzimidazole, *Polym. Int.* 53(10) (2004) 1512 – 1516.

[111] Jouanneau, J. ; Mercier, R. ; Gonon, L.; Gebel, G. Synthesis of sulfonated polybenzimidazoles from functionalized monomers: Preparation of ionic conducting membranes, *Macromolecules*. 40 (2007) 983 – 990.

[112] Li, Z. ; Ding, J. ; Robertson, G. P. ; Guiver, M. D. A novel bisphenol monomer with grafting capability and the resulting poly(arylene ether sulfone)s, *Macromolecules*. 39 (2006) 6990 – 6996.

[113] Poppe, D.; Frey, H.; Kreuer, K. D.; Heinzel, A.; Mülhaupt, R. Carboxylated and sulfonated poly(arylene-co-arylene sulfone)s: Thermostable polyelectrolytes for fuel cell applications, *Macromolecules*. 35 (2002) 7936 – 7941.

[114] Kim, D. S. ; Shin, K. H. ; Park, H. B. ; Chung, Y. S. ; Nam, S. Y. ; Lee, Y. M. Synthesis and characterization of sulfonated poly(arylene ether sulfone) copolymers containing carboxyl groups for direct methanol fuel cells, *J. Membr. Sci.* 278 (2006) 428 – 436.

[115] Rahman, M. K. ; Aiba, G. ; Susan, M. A. B. H. ; Sasaya, Y. ; Ota, K.-I. ; Watanabe, M. Synthesis, characterization and copolymerisation of a series of novel acid monomers based on sulfonimides for proton conducting membranes, *Macromolecules*. 37 (2004) 5572 – 5577

[116] Miyatake, K.; Hay, A. S. New poly(arylene ether)s with pendant phosphonic acid groups, *J. Polym. Sci. Pol. Chem.* 39 (2001) 3770 – 3779.

[117] Jakoby, K.; Peinemann, K. V.; Nunes, S. P. Palladium-catalyzed phosphonation of polyphenylsulfone, *Macromol. Chem. Physic.* 204 (2003) 61 – 67.

[118] Lafitte, B., Jannasch, P. Phosphonation of polysulfones via lithiation and reaction with chlorophosphonic acid esters, *J. Polym. Sci. Pol. Chem.* 43 (2005) 273 – 286.

[119] Liu, B.; Robertson, G. P.; Guiver, M. D.; Shi, Z.; Navessin, T.; Holdcroft, S. Fluorinated poly(aryl ether) containing a 4-bromophenyl pendant group and its phosphonated derivative, *Macromol. Rapid Comm.* 27 (2006) 1411 – 1417.

[120] Kerres, J. A.; Xing, D.; Schönberger, F. Comparative investigation of novel PBI blend ionomer membranes from nonfluorinated and partially fluorinated poly arylene ethers, *J. Polym. Sci. Pol. Phys.* 44 (2006) 2311 – 2326.

[121] Schönberger, F.; Hein, M., Kerres, J. Preparation and characterisation of sulfonated partially fluorinated statistical poly(arylene ether sulfone)s and their blends with PBI, *Solid State Ionics.* 178 (2007) 547 – 554.

[122] Lu, D.; Zou, H.; Guan, R.; Dai, H.; Lu, L. Sulfonation of polyethersulfone by chlorosulfonic acid, *Polym. Bull.* 54 (2005) 21 – 28.

[123] Shang, X., Tian, S.; Kong, L. Meng, Y. Synthesis and characterization of sulfonated fluorene-containing poly(arylene ether ketone) for proton exchange membrane, *J. Membr. Sci.* 266 (2005) 94 – 101.

[124] Noshay, A.; Robeson, L. M. Sulfonated polysulfone, *J. Appl. Polym. Sci.* 20 (1976) 1885 – 1903.

[125] Kido, M.; Hu, Z.; Ogo, T.; Okamoto, Y. S. K.-I.; Fang, J. Novel preparation method of cross-linked sulfonated polyimide membranes for fuel cell application, *Chem. Lett.* 36(2) (2007) 272 – 273.

[126] Thaler, W. A. Hydrocarbon-soluble sulfonating reagents. Sulfonation of aromatic polymers in hydrocarbon solution using soluble acyl sulfates, *Macromolecules.* 16 (1983) 623 – 628.

[127] Orler, E. B.; Yontz, D. J.; Moore, R. B. Sulfonation of syndiotactic polystyrene for model semicrystalline ionomer investigations, *Macromolecules.* 26 (1993) 5157 – 5160.

[128] Chao, H. S.; Kesley, D. R. Process for preparing sulfonated poly(aryl ether) resins, 1996, US 6425000.

[129] Zhang, L.; Mukerjee, S. Investigation of durability issues for selected nonfluorinated proton exchange membranes for fuel cell application, *J. Electrochem. Soc.* 153(6) (2006) A1062 – A1072.

[130] Wang, F.; Hickner, M.; Ji, Q.; Harrison, W., Mecham, J.; Zawodzinski, T. A.; McGrath, J. E. Synthesis of highly sulfonated poly(arylene ether sulfone) random (statistical) copolymers via direct polymerization, *Macromol. Symp.* 175 (2001) 387 – 395.

[131] Wang, F.; Hickner, M.; Kim, Y. S.; Zawodzinski, T. A.; McGrath, J. E. Direct polymerization of sulfonated poly(arylene ether sulfone) random (statistical) copolymers: candidates for new proton exchange membranes, *J. Membr. Sci.* 197 (2002) 231 – 242.

[132] Gao, Y.; Robertson, G. P.; Guiver, M. D.; Mikhailenko, S. D.; Li, X.; Kaliaguine, S. Synthesis of copoly(aryl ether ether nitrile)s containing sulfonic acid groups for PEM application, *Macromolecules.* 38 (2005) 3237 – 3245.

[133] Li, Y.; Wang, F.; Yang, J.; Liu, D.; Roy, A.; Case, S.; Lesko, J.; McGrath, J. E. Synthesis and characterization of controlled molecular weight disulfonated

poly(arylene ether suflone) copolymers and their applications to proton exchange membranes, *Polymer.* 47 (2006) 4210 – 4217.

[134] Liu, B.; Robertson, G. P.; Guiver, M. D.; Sun, Y.-M.; Liu, Y.-L.; Lai, J.-Y.; Mikhailenko, S., Kaliaguine, S. Sulfonated poly(aryl ether ether ketone ketone)s containing fluorinated moieties as proton exchange membrane materials *J. Polym. Sci. Pol. Phys.* 44 (2006) 2299 – 2310.

[135] Wiles, K. B.; de Diego, C. M.; de Abajo, J.; McGrath, J. E. Directly copolymerized partially fluorinated disulfonated poly(arylene ether sulfone) random copolymers for PEM fuel cell systems: Synthesis, fabrication and characterization of membranes and membrane-electrode assemblies for fuel cell applications, *J. Membr. Sci.* 294 (2007) 22 – 29.

[136] Kim, Y. S.; Einsla, B.; Sankir, M.; Harrison, W.; Pivovar, B. S. Structure-property-performance relationships of sulfonated poly(arylene ether sulfone)s as a polymer electrolyte for fuel cell applications, *Polymer.* 47 (2006) 4026 – 4035.

[137] Zhu, X.-B.; Zhang, H.-M.; Liang, Y.-M.; Zhang, Y.; Yi, B.-L. A novel PTFE-reinforced multilayer self-humidifying composite membrane for PEM fuel cells, Electrochem. *Solid St.* 9(2) (2006) A49 – A52.

[138] Zhu, X.; Zhang, H.; Liang, Y.; Zhang, Y.; Luo, Q.; Bi, C.; Yi, B. Challeging reinforced composite polymer electrolyte membranes based on disulfonated poly(arylene ether sulfone)-impregnated expanded PTFE for fuel cell applications, *J. Mater. Chem.* 17(4) (2007) 386 – 397.

[139] Zhang, Y.; Zhang, H.-M.; Zhun, X.-B.; Liang, Y.-M. A low-cost PTFE-reinforced integral multilayerd self-humidifying membranes for PEM fuel cells, *Electrochem. Solid St.* 9(7) (2006) A332 – A335.

[140] Nunes, S. P.; Ruffmann, B.; Rikowski, E.; Vetter, S.; Richau, K. Inorganic modification of proton conductive polymer membranes for direct methanol fuel cells, *J. Membr. Sci.* 203 (2002) 215 – 225.

[141] Gaowen, Z.; Zhentao, Z. Organic/inorganic composite membranes for application in DMFC, *J. Membr. Sci.* 261 (2005) 107 – 113.

[142] Mathuraiveeran, T.; Roelofs, K.; Senftleben, D.; Schiestel, T. Proton conducting composite membranes with low ethanol crossover for DEFC, *Desalination.* 200 (2006) 662 – 663.

[143] Kim, J.-D.; Mori, T. Honma, I. Organic-inorganic hybrid membranes for a PEMFC operation at intermediate temperatures, *J. Electrochem. Soc.* 153 (3) (2006) A508 – A514

[144] Shan, J.; Vaivars, G.; Luo, H.; Mohamed, R.; Linkov, V. Sulfonated polyether ether ketone (PEEK-WC)/phosphotungstic acid composite: Preparation and characterization of the fuel cell membranes, *Pure Appl. Chem.* 78(9) (2006) 1781 – 1791.

[145] Mecheri, B.; D'Epifanio, A.; Di Vona, M. L.; Traversa, E.; Licocccia, S.; Miyayama, M. Sulfonated polyether ether keton-based composite membranes doped with a tungsten-based inorganic proton conductor for fuel cell applications, *J. Electrochem. Soc.* 153(3) (2006) A463 – A467.

[146] Kerres, J.; Cui, W.; Disson, R.; Neubrand, W. Development and characterization of crosslinked ionomer membranes based upon sulfinated and sulfonated PSU:

Crosslinked PSU blend membranes by disproportionation of sulfinic acid groups, *J. Membr. Sci.* 139 (1998) 211 – 225.

[147] Kerres, J.; Cui, W.; Junginger, M. Development and characterization of crosslinked ionomer membranes based upon sulfinated and sulfonated PSU: Crosslinked PSU blend membranes by alkylation of sulfinate groups with dihalogenoalkanes, *J. Membr. Sci.* 139 (1998) 227 – 241.

[148] Kerres, J.; Ullrich, A.; Meier, F.; Häring, T. Synthesis and characterization of novel acid-base polymer blends for application in membrane fuel cells, *Solid State Ionics.* 125 (1999) 243 – 249.

[149] Kerres, J. A. Blended and cross-linked ionomer membranes for application in membrane fuel cells, *Fuel Cells.* 5 (2005) 230 – 247.

[150] Kerres, J.; Hein, M.; Zhang, W.; Nicoloso, N., Graf, S. Development of new blend membranes for polymer electrolyte fuel cell applications, *J. New Mat. Electrochem. Syst.* 6 (2003) 223 – 229.

[151] Walker, M.; Baumgärtner, K.-M.; Kaiser, M.; Kerres, J.; Ullrich, A.; Räuchle, E. Proton-conducting polymers with reduced methanol permeation, *J. Appl. Polym. Sci.* 74 (1999) 67 – 73.

[152] Kerres, J.; Cui, W.; Schnurnberger, W. Vernetzung von modifizierten Engineering Thermoplasten, 1997, DE 19622337C1; Cross-linking of modified engineering thermoplastics, 2001, US 6221923; 2003, US 6552135.

[153] Kerres, J.; Zhang, W.; Häring, T. *Covalently cross-linked ionomer (blend) membranes for fuel cells*, *J. New Mat. Electrochem. Syst.* 7 (2004) 299 – 309.

[154] Kerres, J.; Zhang, W.; Cui, W. New sulfonated engineering polymers via the metalation route. II. Sulfinated/sulfonated poly (ether sulfone) PSU Udel and its crosslinking, *J. Polym. Sci. Pol. Chem.* 36 (1998) 1441 – 1448.

[155] Zhang, W.; Gogel, V.; Friedrich, K. A.; Kerres, J. Novel covalently cross-linked poly(etheretherketone) ionomer membranes, *J. Power Sources.* 155 (2006) 3 – 12.

[156] Mitov, S.; Vogel, B.; Roduner, E.; Zhang, H.; Zhu, X.; Gogel, V.; Jörissen, L.; Hein, M.; Xing, D.; Schönberger, F.; Kerres, J. Preparation and characterization of stable ionomers and ionomer membranes for fuel cells, *Fuel Cells.* 6 (2006) 413 – 424.

[157] Kerres, J.; Cui, W. Acid-base polymer blends and their application in membrane processes, 2001, US 6194474B1; 2001, US 6300381B1; Säure-Base-Polymerblends und ihre Verwendung in Membranprozessen, 2004, EP 1073690.

[158] Cui, W.; Kerres, J.; Eigenberger, G. Development and characterization of ion-exchange polymer blend membranes, *Sep. Purif. Technol.* 14 (1998) 145 – 154.

[159] Kerres, J.; Ullrich, A. Synthesis of novel engineering polymers containing basic side groups and their application in acid–base polymer blend membranes, *Sep. Purif. Technol.* 22 (2001) 1 – 15.

[160] Kerres, J.; Ullrich, A.; Hein, M. Preparation and characterization of novel basic PSU polymers, *J. Polym. Sci. Pol. Chem.* 39 (2001) 2874 – 2888.

[161] Kerres, J.; Ullrich, A.; Häring, T. Modification of polymers with basic N-groups and ion exchange groups in the lateral chain, 2003, US 6590067; 2004, EP 1105433 B.

[162] Coffin, D. R., Serad, G. A., Hicks, H. L.; Montgomery, R. T. Properties and applications of Celanese PBI fiber, *Text. Res. J.* 52 (1982) 466 – 472.

[163] Kerres, J.; Ullrich, A.; Häring, T. Engineering ionomeric blends and engineering ionomeric blend membranes, 2004, EP 1076676; 2004, US 6723757.

[164] Lee, J.; Kerres, J. Synthesis and characterization of sulfonated poly(arylene thioether)s and their blends with polybenzimidazole for proton exchange membranes, *J. Membr. Sci.* 294 (2007) 75 – 83.

[165] Kerres, J., Zhang, W.; Jörissen, L.; Gogel, V. Application of different types of polyaryl blend membranes in DMFC, *J. New Mat. Electrochem. Syst.* 5 (2002) 97 – 107.

[166] Meier, F.; Kerres, J.; Eigenberger, G. *Characterization of polyaryl blend membranes for DMFC applications*, J. *New Mat. Electrochem. Syst.* 5 (2002) 91 – 96.

[167] Xing, D.; Kerres, J. Improved performance of sulfonated polyarylene ethers for proton exchange membrane fuel cells, *Polym. Advan. Technol.* 17 (2006) 591 – 597.

[168] Xing, D.; Kerres, J. *Improvement of synthesis procedure and characterization of sulfonated poly(arylene ether sulfone) for proton exchange membranes*, J. New Mat. Electrochem. Syst. 9 (2006) 51 – 60.

[169] Jörissen, L.; Gogel, V.; Kerres, J.; Garche, J. New membranes for direct methanol fuel cells, *J. Power Sources.* 105 (2002) 267 – 273.

[170] Kerres, J.; Ullrich, A.; Hein, M.; Gogel, V.; Friedrich, K. A.; Jörissen, L. Cross-linked polyaryl blend membranes polymer electrolyte fuel cells, *Fuel Cells.* 4 (2004) 105 – 112.

[171] *Agon,* V. V.; *Bubb,* W. A.; *Wright,* A.; *Hawkins,* C. L.; *Davies,* M. J. *Sensitizer-mediated photooxidation of histidine residues: Evidence for the formation of reactive side-chain peroxides, Free Radical Bio. Med.* 40 (2006) 698 – 710.

[172] *Reybier,* K.; *Boyer,* J.; *Farines,* V.; *Camus,* F. ; *Souchard,* J.-P.; *Monje,* M.-C.; *Bernardes-Genisson,* V.; *Goldstein,* S.; *Nepveu,* F. *Radical trapping properties of imidazolyl nitrones, Free Radical Res.* 40 (2006) 11 – 20.

[173] *Wycisk,* R.; *Chisholm,* J.; *Lee,* J.; *Lin,* J.; *Pintauro,* P. N. *Direct methanol fuel cell membranes from Nafion-polybenzimidazole blends, J. Power Sources.* 163 (2006) 9 – 17.

[174] *Ainla,* A.; *Brandell,* D. *Nafion-polybenzimidazole (PBI) composite membranes for DMFC applications, Solid State Ionics.* 178 (2007) 581 – 585.

[175] Wycisk, R.; Lee, J. K.; Pintauro, P. N. Sulfonated polyphosphazene-polybenzimidazole membranes for DMFCs, *J. Electrochem. Soc.* 152(5) (2005) A892 – A898.

[176] *Manea,* C.; *Mulder,* M. *New polymeric electrolyte membranes based on proton donor-proton acceptor properties for direct methanol fuel cells, Desalination.* 147 (2002) 179 – 182.

[177] Kosmala, B.; Schauer, J. J. *Ion-exchange membranes prepared by blending sulfonated poly(2,6-dimethyl-1,4-phenylene oxide) with polybenzimidazole, J. Appl. Polym. Sci.* 85 (2002) 1118 – 1127.

[178] Fu, Y.; *Manthiram,* A.; *Guiver,* M. D. *Acid-base blend membranes based on 2-amino-benzimidazole and sulfonated poly(ether ether ketone) for direct methanol fuel cells, Electrochem. Commun.* 9 (2007) 905 – 910.

[179] *Fu,* Y.; *Manthiram,* A.; *Guiver,* M. D. *Blend membranes based on sulfonated poly(ether ether ketone) and polysulfone bearing benzimidazole side groups for proton exchange membrane fuel cells, Electrochem. Commun.* 8 (2006) 1386 – 1390.

[180] *Fu,* Y.-Z.; *Manthiram,* A.; *Guiver,* M. D. *Blend membranes based on sulfonated poly(ether ether ketone) and polysulfone bearing benzimidazole side groups for DMFCs, Electrochem. Solid St.* 10 (2007) B70 – B73.

[181] *DeLuca,* N. W.; *Elabd,* Y. A. *Nafion/poly(vinyl alcohol) blends: Effect of composition and annealing temperature on transport properties, J. Membr. Sci.* 282 (2006) 217 – 224.

[182] *Shao,* Z.-G., *Wang,* X.; *Hsing,* I.-M. *Composite Nafion/polyvinyl alcohol membranes for the direct methanol fuel cell, J. Membr. Sci.* 210 (2002) 147 – 153.

[183] Norsten, T. L.; Guiver, M. D.; Murphy, J.; Astill, T.; Navessin, T.; Holdcroft, S.; Frankamp, B. L.; Rotello, V. M.; Ding, J. Highly fluorinated comb-shaped copolymers as proton exchange membranes (PEMs): Improving PEM properties through rational design, *Adv. Funct. Mater.* 16 (2006) 1814 – 1822.

[184] Ding, J.; Chuy, C.; Holdcroft, S. Solid polymer electrolyte based on ionic graft polymers: Effect of graft chain length on nano-structured, ionic networks, *Adv. Funct. Mater.* 12 (2006) 389 – 394.

[185] Shin, C. K.; Maier, G.; Andreaus, B.; Scherer, G. G. Block copolymer ionomers for ion conductive membranes, *J. Membr. Sci.* 245 (2004) 147 – 161.

[186] Nakano, T.; Nagaoka, S.; Kawakami, H. Preparation of novel sulfonated block copolyimides for proton conductivity membranes, *Polym. Adv. Technol.* 16 (2005) 753 – 757.

[187] Asano, N.; Miyatake, K.; Watanabe, M. Sulfonated block polyimide copolymeres as proton-conductive membrane, *J. Polym. Sci. Pol. Chem.* 44 (2006) 2744 – 2748.

[188] Aoki, M.; Asano, N.; Miyatake, K.; Uchida, H.; Watanabe, M. Durability of sulfonated polyimide membrane evaluated by long-term polymer electrolyte fuel cell operation, *J. Electrochem. Soc.* 153(6) (2006) A1151 – A1158.

[189] Ghassemi, H.; Harrison, W.; Zawodzinski, T. A.; McGrath, J. E. New multiblock copolymers containing hydrophilic-hydrophobic segments for proton exchange membranes *Polym. Prepr.* 4581) (2004) 68 – 69.

[190] Roy, A.; Hickner, M. A.; Yu, X.; Li, Y.; Glass, T. E.; McGrath, J. E. Influence of chemical composition and sequence length on the transport porperties of proton exchange membranes *J. Polym. Sci. Pol. Phys.* 44 (2006) 2226 – 2239.

[191] Ghassemi, H., McGrath, J. E.; Zawodzinski Jr, T. A. Multiblock sulfonated-fluorinated poly(arylene ether)s for a proton exchange membrane fuel cell, *Polymer.* 47 (2006) 4132 – 4139.

[192] McGrath, J. E.; Roy, A.; Yu, X.; Lee, H.-S.; Badami, A.; Wang, H.; Hill, M. Li, Y. Multiblock hydrophilic-hydrophobic proton exchange membranes for direct methanol fuel cells, *Meet. Abstr. – Electrochem. Soc.* 601 (2006) 1137.

[193] Schönberger, F.; Kerres, J. *Novel multiblock-co-ionomers as potential polymer electrolyte membrane materials*, J. Polym. Sci. Pol. Chem. 45(22) (2007) 5237 – 5255.

[194] Adjemian, K. T., Srinivasan, S., Benziger, J.; Bocarsly, A. B. *Investigation of PEMFC operation above 100 degrees C employing perfluorosulfonic acid silicon oxide composite membranes*, J. Power Sources. 109(2) (2002) 356 – 364.

[195] Antonucci, V.; Aricò, A. S.; Modica, E.; Cretì, P.; Staiti, P.; Antonucci, P. L. Polymer-silica composite membranes for direct methanol fuel cells, *Stud. Surf. Sci. Catal.* 140 (2001) 37 - 45

[196] *Aricò*, A. S.; *Cretì*, P.; *Antonucci*, P. L.; *Antonucci*, V. *Comparison of ethanol and methanol oxidation in a liquid-feed solid polymer electrolyte fuel cell at high temperature*, *Electrochem. Solid St.* 1 (1998) 66 – 68

[197] Siuzdak, D. A.; Mauritz, K. A. Surlyn®/[Silicon Oxide] Hybrid Materials via Polymer In Situ Sol-Gel *Chemistry Polym. Prepr.* 38 (1997) 245 – 246

[198] Payne, J. T.; Reuschle, D. A.; Mauritz, K. A. Inorganic Modification of Elastomeric Poly(styrene-b-Isobutylene-b-styrene) Block Copolymer Ionomers via in Situ Sol-Gel Reactions of Tetraethoxysilane Polym. *Prepr.* 38 (1997) 247 – 248

[199] Deng, Q.; Moore, R. B.; Mauritz, K. A. *Nafion (SiO_2, ORMOSIL, and dimethylsiloxane) hybrids via in situ sol-gel reactions: Characterization of fundamental properties*, J. Appl. Polym. Sci. 68 (1998) 747 – 763.

[200] *Alberti*, G.; *Casciola*, M. *Composite membranes for medium-temperature PEM fuel cells*, *Ann. Rev. Mater. Res.* 33 (2003) 129 – 154.

[201] Jones, D. J.; Rozière, J.; Marrony, M.; Lamy, C.; Coutanceau, C.; Léger, J. M.; Hutchinson, H.; Dupont, M. High-temperature DMFC stack operating with non-fluorinated membranes, *Fuel Cells Bull.* 10 (2005) 12 - 15

[202] Rozière, J.; Jones, D. J. Handbook of Fuel Cell Technology, Ed. W. Vielstich, A. Lamm, H. Gasteiger, John Wiley and Sons, 2003, 3, 447 – 455.

[203] Bauer, B.; Jones, D. J.; Rozière, J.; Tchicaya, L.; Alberti, G.; Massinelli, L.; Casciola, M.; Peraio, A. , Ramunni, E. Hybrid organic-inorganic membranes for a medium temperature fuel cell, *J. New Mat. Electrochem. Syst.* 3 (2000) 87 – 92.

[204] Jones, D. J.; Roziere, J. *Recent advances in the functionalisation of polybenzimidazole and polyetherketone for fuel cell applications*, J. Membr. Sci. 185 (2001) 41 – 58.

[205] Tchicaya-Bouckary, L.; Jones, D. J.; Rozière, J. Hybrid Polyaryletherketone Membranes for Fuel Cell Applications, *Fuel Cells.* 2 (2002) 40 – 45.

[206] Yang, C., Srinivasan, S.; Aricò, A. S.; Cretì, P.; Baglio, V.; Antonucci, V. *Composition Nafion/zirconium phosphate membranes for direct methanol fuel cell operation at high temperature*, *Electrochem. Solid St.* 4 (2001) A31 – A34

[207] Wu, Q.; Lin, H.; Meng, G. Studies on preparation and conductivity of solid high-proton conductor molybdotungstovanadogermanic heteropoly acid, *Mater. Lett.* 39 (1999) 129–132.

[208] Kreuer, K. D. *Proton conductivity: Materials and applications*, *Chem. Mater.* 8(3) (1996) 610 – 641.

[209] Giordano, N.; Staiti, P.; Hocevar, S.; Aricò, A. S. *High performance fuel cell based on phosphotungstic acid as proton conducting electrolyte*, *Electrochim. Acta* 41 (1996) 397 – 403.

[210] *Malhotra*, S.; *Datta*, R. J. *Membrane-supported nonvolatile acidic electrolytes allow higher temperature operation of proton-exchange membrane fuel cells*, *J. Electrochem. Soc.* 144 (1997) L23 – L26.

[211] *Kim*, Y. S.; Wang, F.; *Hickner*, M.; *Zawodzinski*, T. A.; *McGrath*, J. E. *Fabrication and characterization of heteropolyacid ($H_3PW_{12}O_{40}$)/directly polymerized sulfonated*

poly(arylene ether sulfone) copolymer composite membranes for higher temperature fuel cell applications, *J. Membr. Sci.* 212 (2003) 263 – 282.

[212] *Ponce*, M. L., *Prado*, L.; *Ruffmann*, B., *Richau*, K., *Mohr*, R.; *Nunes*, S. P. *Reduction of methanol permeability in polyetherketone-heteropolyacid membranes*, *J. Membr. Sci.* 217 (2003) 5 – 15.

[213] *Ponce*, M. L. ; *Prado*, L. ; *Silva*, V.; *Nunes*, S. P. *Membranes for direct methanol fuel cell based on modified heteropolyacids*, *Desalination*. 162 (2004) 383 – 391.

[214] *Vernon*, D. R.; *Meng*, F.; *Dec*, S. F.; *Williamson*, D. L.; *Turner*, J. A.; *Herring*, A. M. *Synthesis, characterization, and conductivity measurements of hybrid membranes containing a mono-lacunary heteropolyacid for PEM fuel cell applications*, *J. Power Sources*. 139 (2005) 141 – 151.

[215] Kerres, J.; Tang, C.-M.; Graf, C. *Improvement of properties of poly(ether ketone) ionomer membranes by blending and cross-linking*, Ind. Eng. Chem. Res. 43 (2004) 4571 – 4579.

[216] Tan, A. R.; de Carvalho, L. M.; de Souza Gomes, A. *Nanostructured proton-conducting membranes for fuel cell applications*, *Macromol. Symp.* 229 (2005) 168 – 178.

[217] Wainright, J. S.; Wang, J.-T. Weng, D.; Savinell, R. F.; Litt, M. Acid-doped polyimidazoles – A new polymer electrolyte, *J. Electrochem. Soc.* 142 (1995) L121 – L123.

[218] Wang, J.-T.; Savinell, R. F., Wainright, J.; Litt, M.; Yu, H. *A H_2/O_2 fuel cell using acid doped polybenzimidazole as polymer electrolyte*, *Electrochim. Acta* 41 (1996) 193 – 197.

[219] Weng, D.; Wainright, J. S.; Landau, U.; Savinell, R. F. *Electro-osmotic drag coefficient of water and methanol in polymer electrolytes at elevated temperatures*, *J. Electrochem. Soc.* 143 (1996) 1260 – 1263.

[220] Samms, S. R.; Wasmus, S.; Savinell, R. F. *Thermal stability of proton conducting acid doped polybenzimidazole in simulated fuel cell environments*, J. Electrochem. Soc. 143 (1996) 1225 – 1232.

[221] Wang, J.-T.; Wainright, J. S.; Savinell, R. F.; Litt, M. *A direct methanol fuel cell using acid-doped polybenzimidazole as polymer electrolyte*, *J. Appl. Electrochem.* 26 (1996) 751 – 756.

[222] Savinell; R. F., Litt; M. H., Proton conducting polymers used as membranes, 1996, US 5525436

[223] Kreuer, K. D.; Paddison, S. J.; Spohr, E.; Schuster, M. Transport in proton conductors for fuel cell applications: Simulations, elementary reactions and phenomenology, *Chem. Rev.* 104 (2004) 4637 – 4678

[224] Perry, K. A.; Eisman, G. A.; Benicewicz, B. C. Electrochemical hydrogen pumping using a high-temperature polybenzimidazole (PBI) membrane, *J. Power Sources*. 177 (2008) 478–484

[225] *Xiao*, L. ; *Zhang*, H. ; *Scanlon*, *E.* ; *Ramanathan*, L. S. ; *Choe*, E.-W. ; *Rogers*, D. ; *Apple*, T.; *Benicewicz*, B. C. *High-temperature polybenzimidazole fuel cell membranes via a sol-gel process*, *Chem. Mater.* 17 (2005) 5328 – 5333.

[226] *Jayakody*, J. R. P.; *Chung*, S. H.; *Durantino*, L.; *Zhang*, H.; *Xiao*, L.; *Benicewicz*, B. C.; *Greenbaum*, S. G. *NMR studies of mass transport in high-acid-content fuel cell membranes based on phosphoric acid and polybenzimidazole*, *J. Electrochem. Soc.* 154 (2007) B242 – B246.

[227] http://www.basf-fb.com/de.html

[228] Daletou, M. K.; Gourdoupi, N.; Kallitsis, J. K. *Proton conducting membranes based on blends of PBI with aromatic polyethers containing pyridine units*, *J. Membr. Sci.* 252 (2005) 115 – 122.

[229] Hasiotis, C.; Li, Q.; Deimede, V. Kallitsis, J. K.; Kontoyannis, C. G.; Bjerrum, N. J. *Development and characterization of acid-doped polybenzimidazole/sulfonated polysulfone blend polymer electrolytes for fuel cells*, J. Electrochem. Soc. 148 (2001) A513 – A519.

[230] Deimede, V.; Voyiatzis, G. A.; Kallitsis, J. K.; Li, Q.; Bjerrum, N. J. Miscibility behavior of polybenzimidazol/sulfonated polysulfone blends for use in fuel cell applications, *Macromoelcules*. 33 (2000) 7609 – 7617.

[231] Kerres, J.; Schönberger, F.; Chromik, A.; Häring, T.; Li, Q.; Jensen, J. O.; Pan C.; Noyé P.; Bjerrum, N.J. Partially fluorinated arylene polyethers and their ternary blend membranes with PBI and H_3PO_4. Part I. Synthesis and Characterization of polymers and binary blend membranes, *Fuel Cells* 2008, 3 – 4, 175 – 187.

[232] Li, Q.; Jensen, J. O.; Pan, C.; Bandur, V.; Nilsson, M. S.; Schönberger, F.; Chromik, A.; Hein, M.; Häring, T.; Kerres, J.; Bjerrum N.J. Partially fluorinated arylene polyethers and their ternary blend membranes with PBI and H_3PO_4. Part II. Characterization and fuel cell tests of ternary membranes, *Fuel Cells* 2008, 3 – 4, 188 – 199.

[233] Schuster, M. F. H.; Meyer, W. H.; Wegner, G.; Herz, H. G.; Ise, M.; Kreuer, K. D. J. Maier, *Proton mobility in oligomer-bound proton solvents: imidazole immobilization via flexible spacers*, *Solid State Ionics*. 145 (2001) 85 – 92.

[234] Schuster, M. E.; Meyer, W. H. *Anhydrous proton-conducting polymers*, *Ann. Rev. Mater. Res.* 33 (2003) 233 – 261.

[235] Schuster, M. F. H.; Meyer, W. H.; Schuster, M.; Kreuer, K. D. *Toward a new type of anhydrous organic proton conductor based on immobilized imidazole*, *Chem. Mater.* 16 (2004) 329 – 337.

[236] *Schuster*, M.; *Rager*, T.; *Noda*, A.; *Kreuer*, K. D. *About the choice of the protogenic group in PEM separator materials for intermediate temperature, low humidity operation: A critical comparison of sulfonic acid, phosphonic acid and imidazole functionalized model compounds*, *J.* Maier, *Fuel Cells*. 5 (2005) 355 – 365.

[237] Yamabe, M.; Akiyama, K.; Akatsuka, Y.; Kato, M. *Novel phosphonated perfluorocarbon polymers*, *Eur. Polym. J.* 36 (2000) 1035 – 1041.

[238] *Zhou*, X.; Weston, J.; *Chalkova*, E.; *Hofmann*, M. A.; *Ambler*, C. M.; *Allcock*, H. R.; *Lvov*, S. N. High temperature transport properties of polyphosphazene membranes for direct methanol fuel cells, *Electrochim. Acta*. 48 (2003) 2173 – 2180.

[239] *Allcock*, H. R.; *Hofmann*, M. A.; *Ambler*, C. M.; *Lvov*, S. N.; *Zhou*, X. Y.; *Chalkova*, E.; *Weston*, J. *Phenyl phosphonic acid functionalized poly[aryloxyphosphazenes] as*

proton-conducting membranes for direct methanol fuel cells, *J. Membr. Sci.* 201 (2002) 47 – 54.

[240] *Yanagimachi*, S.; *Kaneko*, K.; *Takeoka*, Y.; *Rikukawa*, M. *Synthesis and evaluation of phosphonated poly(4-phenoxybenzoyl-1,4-phenylene)*, *Synthetic Met.* 135 (2003) 69 – 70.

[241] *Steininger*, H.; *Schuster*, M.; *Kreuer*, K. D.; *Maier*, J. *Intermediate temperature proton conductors based on phosphonic acid functionalized oligosiloxanes*, *Solid State Ionics.* 177 (2006) 2457 – 2462.

[242] *Lafitte*, B.; *Jannasch*, P. *Polysulfone ionomers functionalized with benzoyl(difluoromethylenephosphonic acid) side chains for proton-conducting fuel-cell membranes*, *J. Polym. Sci. Polym. Chem.* 45 (2007) 269 – 283.

[243] Xie, J. Wood, D. L.; Wayne, D. M.; Zawodzinksi, T. A.; Atanassov, P.; Borup, R. L. *Durability of PEFCs at high humidity conditions*, *J. Electrochem. Soc.* 152(1) (2005) A104 – A113.

[244] Xie, J.; Wood, D. L.; More, K. L.; Atanassov, P.; Borup, R. L. *Microstructural changes of membrane electrode assemblies during PEFC durability testing at high humidity conditions*, J. Electrochem. Soc. 152 (2005) A1011 – A1020.

[245] Ferreira, P. J.; la O', G. J.; Shao-Horn, Y.; Morgan, D.; Makharia, R. ; Kocha, S.; Gasteiger, H. A. Instability of Pt/C electrocatalysts in proton exchange membrane fuel cells: A mechanistic investigation *J. Electrochem. Soc.* 152(11) (2005) A2256 – A2271

[246] Darling, R. M.; Meyers, J. P. Kinetic model of platinum dissolution in PEMFCs, *J. Electrochem. Soc.* 150(11) (2003) A1523 – A1527.

[247] Darling, R. M.; Meyers, J. P. Mathematical model of platnium movement in PEM fuel cells, *J. Electrochem. Soc.* 152(1) (2005) A242 – A247.

[248] Franco, A. A.; Tembely, M. Transient multiscale modeling of aging mechanisms in a PEFC cathode, *J. Electrochem. Soc.* 154(7) (2007) B712 – B723.

[249] Meyers, J. P.; Darling, R. M. Model of carbon corrosion in PEM fuel cells, *J. Electrochem. Soc.* 153(8) (2006) A1432 – A1442.

[250] Panchenko, A.; Dilger, H.; Kerres, J.; Hein, M.; Ullrich, A.; Kaz, T.; Roduner, E. In-situ spin trap electron paramagnetic resonance study of fuel cell processes, *Phys. Chem. Chem. Phys.* 6 (2004) 2891 – 2894.

[251] Panchenko, A. DFT investigation of the polymer electrolyte membrane degradation caused by OH radicals in fuel cells, *J. Membr. Sci.* 278(1-2) (2006) 269 – 278.

[252] Mitov, S.; Delmer, O.; Kerres, J.; Roduner, E. Oxidative and photochemically stability of ionomers for fuel cell membranes, *Helv. Chim. Acta.* 89 (2006) 2354 – 2370.

[253] Curtin, D. E.; Lousenberg, R. D.; Henry, T. J.; Tangeman, P. C.; Tisack, M. E. Advanced materials for improved PEMFC performance and life, *J. Power Sources.* 131 (2004) 41 – 48.

[254] Bro, M. I.; Sperati, C. A. Endgroups in tetrafluoroethylene polymers *J. Polym. Sci.* 38(134) (1959) 289 – 295.

[255] Anderson, A. B.; Albu, T. V. Catalytic effect of platinum on oxygen reduction: An ab initio model influencing electrode potential dependence, *J. Electrochem. Soc.* 147(1) (2000) 4229 – 4238.

[256] Paulus, U. A.; Schmidt, T. J.; Gasteiger, H. A.; Behm, R. J. Oxygen reduction on a high-surface area Pt/Vulcan carbon catalyst: a thin-film rotating ring-disk electrode study, *J. Electroanal. Chem.* 492 (2001) 134 – 145.

[257] Mittal, V. O.; Kunz, H. R.; Fenton, J. M. Is H_2O_2 involved in the membrane degradation mechanism in PEMFC? *Electrochem. Solid St.* 9(6) (2006) A299 – A302.

[258] Mittal, V. O.; Kunz, H. R.; Fenton, J. M. Effect of catalyst properties on membrane degradation rate and the underlying degradation mechanism in PEMFCs, *J. Electrochem. Soc.* 153(9) (2006) A1755 – A1759.

[259] Mittal, V. O.; Kunz, H. R.; Fenton, J. M. Membrane degradation mechanism in PEMFCs, *J. Electrochem. Soc.* 154(7) (2007) B652 – B656.

[260] Mitov, S. In-situ radiation grafting of polymer films and degradation studies of monomers for application in fuel cell membranes, PhD thesis, Stuttgart, 2007.

In: Advances in Membrane Science and Technology
Editor: Tongwen Xu
ISBN: 978-1-60692-883-7

Chapter IV

Organic/Inorganic Hybrid Membranes: Overview and Perspective

Cuiming Wu[a] ***and Tongwen Xu***[*b]

[a] School of Chemical Engineering, Hefei University of Technology, Hefei 230009, Anhui, P. R. China

[b]Laboratory of Functional Membranes, School of Chemistry and Material Science, University of Science and Technology of China, Hefei 230026, P. R. China

Abstract

During the last two decades, hybrid membranes (organic-inorganic and inorganic-organic) have rapidly become a fascinating new field of research and have found important applications in various fields. The most attractive point of hybrid membranes lies in their excellent potential and opportunity in keeping and enhancing the best properties of both organic and inorganic membranes. Here in this chapter, the general concept, preparation methods, important properties and applications of hybrid membranes will be briefly overviewed. Sol-gel process is the most important means by which hybrid membranes are prepared, so the membranes from sol-gel process will be specifically concentrated on. Though not intended to be a comprehensive review of the enormous research reports of hybrid membranes, this chapter will try to make the readers appreciate the high level of diversity in the properties, the applications and the production of this kind of membranes.

This chapter includes 6 sections. Section 1 is an introduction of the general background of membranes and membrane materials. Then, in sections 2 and 3, classification and brief history of hybrid membranes are introduced. Section 4 focuses on the different preparation methods of hybrid membranes, among which the sol-gel process is introduced in more detail. Chapter 4 documents the potential applications of the hybrid membranes, especially in gas separation, pervaporation, UF, NF, RO, chemical analysis and fuel cells. Chapter 6 is the conclusions and remarks.

[*] Correspondence concerning this chapter should be addressed to Tongwen Xu at twxu@ustc.edu.cn

Keywords: Hybrid membranes, organic-inorganic, inorganic-organic, classification, history, preparation, application, sol-gel, separation, pervaporation, fuel cells.

1. Introduction: Membranes and Membrane Materials

Membrane Science and technology have indeed "come of age" [1]. "When one inserts the word *membrane* in a search of Chemical Abstracts for the past 10 years, the result is over 100,000 hits." This quote taken from a 1995 paper by Humphrey [2] demonstrates the high rate at which this technological area has expanded over the last few decades. Development during the recent 10 years is even more dramatic and imposing and the membrane technologies touch almost all aspects of human life nowadays [3]. In addition to the earliest-developed desalination process, applications have reached areas such as energy, biochemistry, microbiology, industrial separations and environmental protection, to mention just a few, with noticeable success.

The "heart" of a membrane process is the membrane itself [4]. Effort in identification and search for suitable membrane materials has as long a history as the membrane science and technology themselves. The earliest developed membrane materials were based on organic polymers, inartificial or man-made. Though inherent characteristics decide that organic polymeric materials can not be as outstanding as the inorganic materials (such as metal and ceramic) in the respect of strength and stability (chemical or thermal), merits in other ways such as ease of processing, low density, high flexibility, functionalization and selectivity, make the organic polymeric materials good enough for most of the conventional membrane application fields. Natural polymers, such as cellulose were taken for use in the early stage. Then, different industrial polymers, such as hydrocarbon polymers, polysulfone, polyamide, polyimide, polyketone, polyether, and etc, were applied to membrane preparation. Also, new kinds of high-performance polymer materials are continually being invented over the years, which triggers a number of the breakthroughs in membrane applications, as summarized in some previous reviews and books [3, 5-7].

On the other hand, beginning from 1940s, the inorganic membranes immerged and gradually came into commercialization. Strong points of inorganic membranes are excellent strength and stabilities. Also, the permeability to gases or liquids can far exceed the permeability of organic membranes. In 1940s mesoporous inorganic membranes were firstly successfully used in gas diffusion process for uranium isotope separation. Even at present, much information about the mesoporous inorganic membranes used in the gas diffusion plants is still classified. Besides, dense inorganic membranes were also fabricated. The first group of dense inorganic membranes studied extensively was metallic membranes, primarily palladium alloy membranes for hydrogen separation. Different types of inorganic membranes and new application fields were found and developed following these early studies. In this regard the interested reader is referred to some other books and reviews [8-10].

With the parallel development of organic and inorganic membranes, many applicational needs have been satisfied and much success obtained. However, as the membrane technology continues to develop, more application fields are being found and higher requirements brought forward. Especially, strong needs exist for the combination of the merits of both inorganic and organic membrane materials. Membrane gas separation can be taken for a good example [4]. For polymeric membranes, there is an upper limit in their performance for gas separation, as predicted by Robeson in early 1990s [11]. A rather general trade-off exists between permeability and selectivity, with an ''upper-bound''. So the increasing of permeability is generally accompanied with undesirable decreasing of selectivity, and vice versa, when membrane materials with separation properties near this limit are modified based on the traditional structure-property relation. On the other hand, permeability and selectivity of some inorganic materials can far exceed the upper-bound limit for the organic polymers [12-13]. However, there are some obstacles that hinder the wide application of inorganic membranes, such as the lack of technology to form continuous and defect-free membranes, the extremely high cost for the membrane production, and handling issues [14-15]. In view of this situation, the mixed matrix membranes (MMMs) consisting of organic polymer and inorganic particle phases, came into being. Permeability and selectivity of MMMs can break through the upper-bound limit for the organic polymers, while the flexibility can be maintained. The MMMs belong to a new kind of membranes, that is, the organic-inorganic hybrid membranes, which are the focus of this chapter.

2. Classification and Nomenclature of Hybrid Membranes

For classification of hybrid membranes, the corresponding classification of hybrid materials should be checked first. Hybrid materials are the materials consisting of both organic and inorganic components. Generally no macro-phase separation exists between the two components since dispersion at the micro-scale or chemical bonding at molecular level is commonly utilized. According to the type of chemical interaction between the components, hybrid materials can be classified into two major groups [16-17]. Class I materials are those in which organic and inorganic components exchange rather weak bonds, mainly through van der Waals, hydrogen, or ionic interactions; and class II corresponds to hybrid compounds where organic and inorganic components are bonded through stronger covalent or iono-covalent chemical bonds. Some typical chemical structures of the class I and II of hybrid materials are shown in figure 1 [18]. According to the nature of the host and guest phases, the hybrid materials may also be classified into two major groups (and a category in between them) [19] i. e., the organic-inorganic materials where the organic phase is host to an inorganic guest, and the inorganic-organic materials with inorganic hosts and organic guests. It should be noted that the above classifications are just for general guides. In a number of hybrid materials, different kinds of interactions co-exist between the organic and inorganic components. Also, the distinguishing between "host" and "guest" phases may be hard for some hybrid materials. So the title "organic-inorganic" or "inorganic-organic" might be ambiguously used in some research reports.

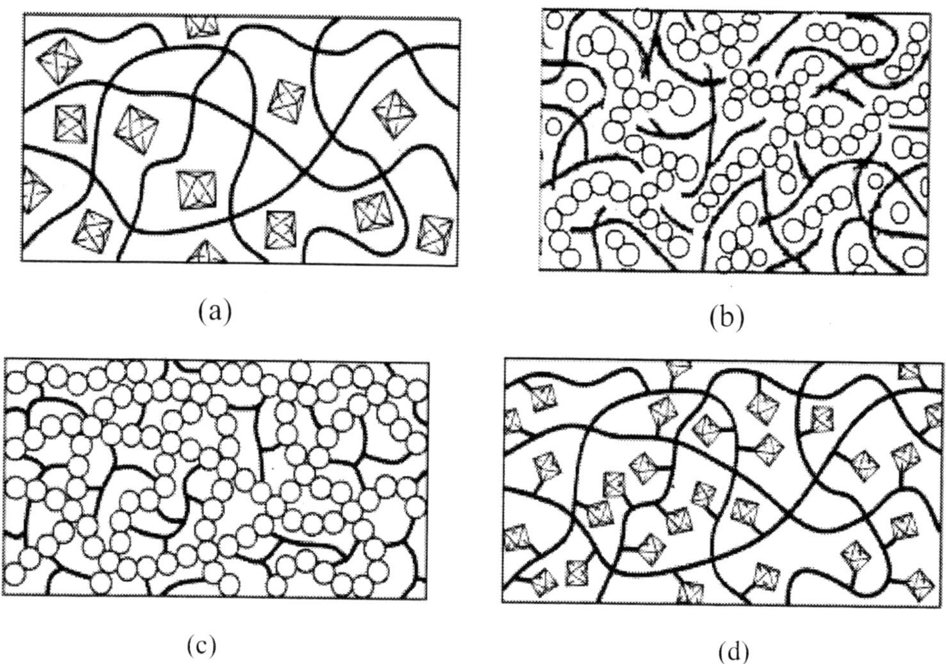

Figure 1. Different kinds of organic–inorganic hybrid materials. (a) embedding of the inorganic moiety into the organic polymer, without covalent bonds; (b)embedding of the organic moiety into the inorganic oxide network, without covalent bonds; (c) incorporation of organic groups by bonding to the inorganic network and (d) incorporation of inorganic groups by bonding to the organic polymer backbone. Adapted from ref. [18]

Besides the above classifications and nomenclatures, some other nomenclatures are used for hybrid materials with particular structures. For examples, in early study, the hybrid materials from sol-gel process were considered to be the modification of conventional polymers or ceramics, and were termed "ceramers" by Wilkes et al. [20] and "ormosils" or "ormocers" by Schmidt et al. [21]. The term "ormocer" has been used in some later research reports [22].

Hybrid membranes can be prepared from hybrid materials and therefore can be classified in similar ways. The classification method based on the nature of the host and guest phases is more commonly used and so in the research publications, the nomenclatures "organic-inorganic hybrid membranes" or "inorganic-organic hybrid membranes" are frequently seen, though sometimes also with some extent of ambiguity. In this book chapter, both organic-inorganic and inorganic-organic hybrid membranes are considered. The saying of "organic/inorganic hybrid membranes", rather than "organic-inorganic and inorganic-organic hybrid membranes" in the title is just for convenience.

Some other classification and nomenclatures as to hybrid membranes with particular structures are also used. For instances, the hybrid membrane composed of organic polymer and inorganic particle phase is called "mixed matrix membrane (MMM)" [4, 23-26]; "supported" [27-28] or "unsupported" [29-30], "homogeneous" or "heterogeneous" [31-32] hybrid membranes can be identified according to their physical or chemical structures.

Among the different hybrid membranes, one kind is special which is prepared starting from inorganic or organic substrates or membranes. Especially, surface of inorganic substrates can be grafted with organic components [33-35]. This kind of membranes is more like the traditional "composite membranes" since inorganic and organic layers can be distinguished. However, chemical interaction, including hydrogen, ionic interactions, covalent or iono-covalent bonds can be formed between the different layers. So they can also be accepted as a kind of hybrid membranes.

3. Historic Development of Hybrid Membranes

The early origin of hybrid materials and membranes may trace back to more than a century ago. For examples, sol-gel processable monomers containing two or more trichlorosilyl or trialkoxysilyl groups have been known for over 60 years [36-37]. Prior to the late 1980s, virtually all of these compounds were used as coupling agents, surface modifiers, or coatings and as components of adhesive formulations. Among these, the surface modification of inorganic substrates would form the inorganic-organic hybrid membranes [38].

Despite these early trials, the real beginning of hybrid materials and membranes was from 1980s. Only from then, the hybrid materials and membranes were prepared at a larger scale and the words "organic-inorganic" and "inorganic-organic" were frequently used and gradually accepted. Some pioneering works that should be mentioned include the first report of mixed matrix membranes (MMMs) for gas separation by Paul and Kemp [39] and initiation of sol-gel hybrid materials by Schmidt [21, 40-42] and Wilkes [43-45].

In the work of Paul and Kemp [39], they added 5Å zeolite into rubbery polymer polydimethyl siloxane (PDMS) and found very large increase in the diffusion time lag but only minor effects on the steady-state permeation.

In Schmidt's work, organically modified alkoxysilanes $R'_nSi(OR)_{4-n}$ (n=1-3, R= alkyl, R'=organic groups) [21, 40-42] was directly used as the precursor of the sol-gel process. In Wilkes', chain ends of polymers or oligomers (such as poly(dimethylsiloxane) (PDMS) and poly(tetramethylene oxide) (PTMO)) were functionalized with silanol (-SiOH) or alkoxysilane (-SiOR) groups [43-45]. In both cases, the -SiOH or -SiOR groups underwent the hydrolysis and condensation reactions to form inorganic Si-O-Si bonding, while organic groups R' or organic polymer chains remained attached to Si without changes, since the Si-C bond will only be broken by oxidation at temperature higher than 600 °C [46-47]. So the organic groups can be incorporated into a silica glass and this gives birth to a kind of hybrid materials.

Following these early works, researchers have made enormous efforts in the particular field of hybrid materials and membranes. Large numbers of hybrid materials/membranes have been fabricated and various application fields found. Their preparation, characterization, and applications have become a fast expanding area of research. Searching with the help of Web of Science (http://portal.isiknowledge.com), the publications with the topic of "organic-inorganic" or "inorganic-organic" have exceeded 5,700 in the past 15 years.

The high activity of research in hybrid materials/membranes mainly origins from their excellent potential and opportunity in keeping and enhancing the best properties of both organic and inorganic components [19]. For example, characterizations of organic polymers such as functionalization, ease of processing at low temperatures, toughness and those of inorganic materials such as hardness, chemical and thermal stability, transparency, electrical and magnetic activity may be united in hybrid materials/membranes [22]. Not only this, new/synergetic properties not accessible otherwise with classical composites can also be generated [19].

"Synergy" might sound like a modern word, but it was used by the ancient Greeks and the concept it conveys is as old as human civilization. Indeed, a primitive precursor of hybrid composites with synergic properties can even be found among ancient materials, such as adobe which is a mixture of clay and straw. The addition of fibrous straw helps to prevent cracks, specially during the curing process, thus leading to a material with properties superior to the sum of the properties of its components; a synergic material. In the hybrid materials/membranes, organic and inorganic components are mixed at micro scale, often at nano- or molecular scale. So the interphase in composite mixtures plays vital role and the concept of synergy can take on a chemical dimension. Research results have shown that through suitable hybridism method, the hybrid materials/membranes can show improved optical [48-50], and electrical [51] properties, luminescence [52-53], ionic conductivity [54-58], and selectivity [30, 59-60], as well as chemical [61-62] or biochemical [63-65] activity.

Taking account of the attractive properties and rapid development of hybrid materials/membranes, the general introduction and tracking would be very necessary. Publications of the hybrid membranes were enormous. Reviews as to some particular hybrid membranes have been published and can be referred to [4, 31, 66-70]. The following sections of this book chapter are not intended to be comprehensive reviews of the many respects of the hybrid membranes. Instead, they aim to make readers understand the primary ideas for the preparation of the hybrid membranes and appreciate some of their unique properties and potential applications. Since the authors' research experience mainly focus on the sol-gel hybrid membranes, sol-gel process and the membranes thus fabricated will be paid special attention to. Especially, reports of ion exchange hybrid membranes of the authors' fabrication will be cited as illustration. Actually, the sol-gel process was, and remains to be the most important and commonly used method for the hybrid membrane preparation.

4. Preparation of Hybrid Membranes

Preparation of hybrid membranes can be carried out by different means. In the following, we'd introduce the important and commonly used preparation methods. It should be noted that the classification of these methods as listed above is just general and mainly for the convenience of discussion. In many cases, different methods can be utilized in one preparation reaction system.

4.1. Sol-Gel Process

4.1.1. General Background of Sol-Gel Process

For the preparation of hybrid materials, the difference between organic and inorganic component in thermal stability should be taken into account. While inorganic systems are thermally quite stable and are often formed at high temperature, most organic polymers have an upper temperature limit of around 250°C. Therefore, the synthesis of hybrid systems requires a strategy wherein the formation of the components is well-suited to each, e.g. the use of a low temperature formation procedure [18]. The sol-gel process is an ideal procedure for such requirement, since its processing temperature is lower than classical high temperature solid-state reactions for the preparation of the inorganic component. Other advantages of sol-gel process include high purity of reactants and possibility to mix the precursors of different phases on molecular level [71]. So the sol-gel process has been most commonly utilized in the preparation of hybrid materials [17, 72-74].

Sols are dispersions of colloidal particles in a liquid, while a gel is a porous 3-dimensionally interconnected solid network that expands in a stable fashion throughout a liquid medium and is only limited by the size of the container [75]. The sol-gel processing has been developed since the mid-1800s and was initially used for the preparation of inorganic ceramic and glass. In the sol-gel process, alkoxide precursors $M(OR)_n$ (M= Si, Ti, Zr, Al, B, etc.) are usually used as starting materials. They can undergo the hydrolysis and polycondensation processes in the presence of water, as represented in Scheme 1 with silicone alkoxide as the example [74]. The hydrolysis and polycondensation reactions initiate at numerous sites within the $Si(OR)_4$ $+H_2O$ solutions. When sufficient interconnected Si-O-Si bonds are formed in a region, they respond cooperatively as colloidal (submicrometer) particles or a sol. With time, the colloidal particles and condensed silica species link together to become a three-dimensional network and gelation occurs [71].

Subsequent steps after gelation may include aging, drying, dehydration (or chemical stabilization) and densification to obtain dense gel-silica [71]. The step of aging involves maintaining the gel for hours to days completely immersed in liquid, while drying of a gel removes the liquid from the interconnected pore network. Dehydration (or chemical stabilization) step remove surface silanol (Si-OH) bonds from the pore network, and the densification process at high temperature eliminates the pores, so density of the gel ultimately becomes equivalent to fused quartz.

It should be noted that not all the steps are needed for the preparation of inorganic oxides or hybrid materials. For example, to get porous silicone oxides, the densification at high temperature (up to 1000°C) can be avoided. Also in some cases, part of the M-OH groups bonds from the pore network can be maintained, so the dehydration or chemical stabilization process is not necessarily needed. Therefore, the processing temperature can be much lower than the conventional fusing method. Actually, for the hybrid membrane materials preparation, generally only hydrolysis and polycondensation at temperature < 100 °C, and aging and drying at temperature < 200 °C are needed.

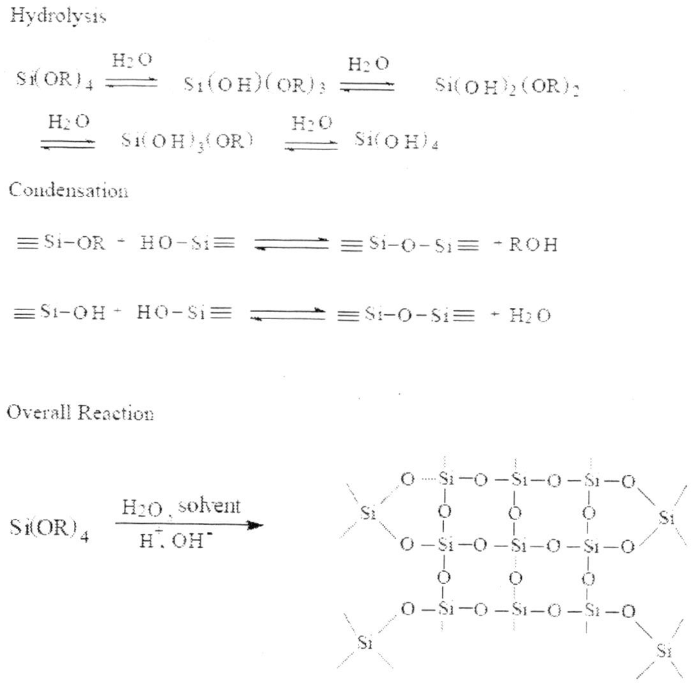

Hydrolysis

$$Si(OR)_4 \overset{H_2O}{\rightleftharpoons} Si(OH)(OR)_3 \overset{H_2O}{\rightleftharpoons} Si(OH)_2(OR)_2 \overset{H_2O}{\rightleftharpoons} Si(OH)_3(OR) \overset{H_2O}{\rightleftharpoons} Si(OH)_4$$

Condensation

$$\equiv Si{-}OR + HO{-}Si\equiv \;\rightleftharpoons\; \equiv Si{-}O{-}Si\equiv + ROH$$

$$\equiv Si{-}OH + HO{-}Si\equiv \;\rightleftharpoons\; \equiv Si{-}O{-}Si\equiv + H_2O$$

Overall Reaction

$$Si(OR)_4 \xrightarrow[H^+,\,OH^-]{H_2O,\ \text{solvent}}$$

Scheme 1. The hydrolysis and polycondensation of silicone alkoxysilane $Si(OR)_4$. Adapted from ref [74].

It is researched that the structure of a gel is established at the time of gelation and the subsequent steps such as aging and drying depend upon the gel structure [71]. So it is quite helpful to understand and control the kinetics of the hydrolysis and condensation reactions that the precursors undergo during the early stage of the sol-gel process. Among the different alkoxide precursors, silicone alkoxides are the most commonly used and studied since they are much less reactive than the other ones such as Ti, Zr, Al, and B. The sequence of reactivity is expressed as follows [76-77]:

$$Zr(OR)_4,\ Al(OR)_3 > Ti(OR)_4 > Sn(OR)_4 >> Si(OR)_4$$

Actually, Zr, Al, Ti and Sn alkoxides are so sensitive to moisture, even in the absence of a catalyst, that precipitation of the oxide will generally occur as soon as water is present [74]. Hydrolysis and condensation of silicone alkoxide precursors are usually much slower and in most cases, acidic or basic catalysts are added into the reaction system.

As mentioned in section **3**, Schmidt and Wilkes initiated the sol-gel hybrid materials in 1980s [21, 40-45]. They used organically modified alkoxysilanes $R'_nSi(OR)_{4-n}$ (n=1-3, R= alkyl, R'=organic group) or alkoxysilane functionalized polymers as the sol-gel precursors, so that short organic chain or long organic polymer chain can be covalently linked with the inorganic silica network through the Si-C bonding. The sol-gel reaction of triethoxysilane-functionalized polytetramethylene oxide (PTMO) and tetraethoxysilane (TEOS) can be shown as an example in Scheme 2 [74].This idea of preparing hybrid membrane materials is still being widely utilized at present. Besides, some other sol-gel processing methods have also been developed, as will be discussed in the following.

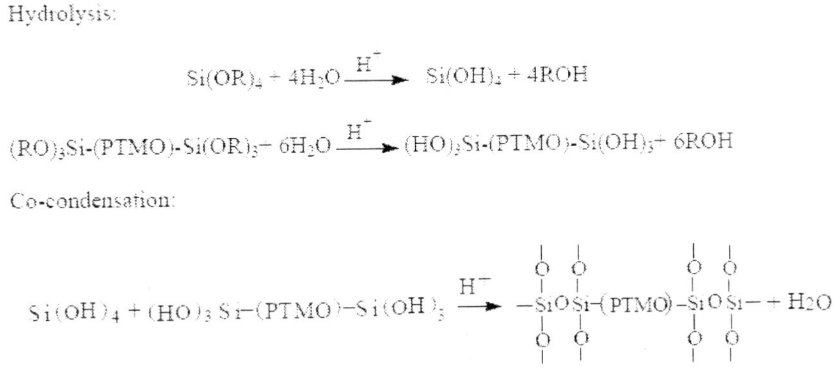

Hydrolysis:

$$Si(OR)_4 + 4H_2O \xrightarrow{H^+} Si(OH)_4 + 4ROH$$

$$(RO)_3Si\text{-}(PTMO)\text{-}Si(OR)_3 + 6H_2O \xrightarrow{H^+} (HO)_3Si\text{-}(PTMO)\text{-}Si(OH)_3 + 6ROH$$

Co-condensation:

$$Si(OH)_4 + (HO)_3Si\text{-}(PTMO)\text{-}Si(OH)_3 \xrightarrow{H^+} -SiOSi\text{-}(PTMO)\text{-}SiOSi- + H_2O$$

Scheme 2. The sol-gel reaction of triethoxysilane-functionalized PTMO and tetra-alkoxysilane. Adapted from Ref [44, 74].

4.1.2. Sol-Gel of Alkoxysilanes $R'_nSi(OR)_{4-n}$ or $(RO)_3Si\text{-}R'Si(OR)_3$

Similar to Schmidt's method, organically modified alkoxysilanes $R'_nSi(OR)_{4-n}$ (n=1-3, R= alkyl, R'=organic group) with various organic groups R' have been developed to undergo the sol-gel process together with or without other precursors [22]. According to the applicational needs, $R'_nSi(OR)_{4-n}$ with functional R' group can be chosen, so that H^+ [78-82] or OH^- conductivity [83-84], ion exchange capacity [85-87], special selectivity [88-89] or optical properties [22, 49], and etc can be induced into the resulted hybrid membranes. Length, rigidity and hydrophilicity of the R' groups can influence the properties of the hybrid membranes to a great extent. Some commonly used $R'_nSi(OR)_{4-n}$ are listed in table 1 for illustration [29-30, 78-82, 85-93]. The chemical structure of the silica network from sol-gel reaction of diphenyldimethoxysilane or phenyltrimethoxysilane with tetramethoxysilane (TMOS) is shown in figure 2 as an example [90].

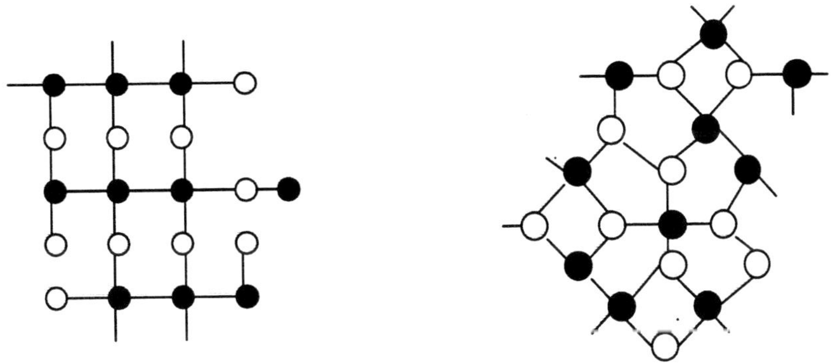

Figure 2. Schematic of the siloxane structure of phenyltrimethoxysilane (PTMOS) - tetramethoxysilane (TMOS) (a) and diphenyldimethoxysilane (DPMOS) - TMOS (b) membrane materials (black ball represents TMOS silicon atoms, while ball represents an PTMOS or DPMOS silicon atom). Adapted from ref. [90].

Table 1. Some of the organically modified alkoxysilane $R'_nSi(OR)_{4-n}$ for the preparation of hybrid membranes

Organically modified alkoxysilanes or chlorosilanes	Other precursors	Other components	Intended application	ref
methyltriethoxysilane (MTES)	Tetraethoxysilane (TEOS)	-	Gas separation	[29-30]
diphenyldimethoxysilane or phenyltrimethoxysilane	Tetramethoxysilane (TMOS)	-	Gas separation	[90]
3-(sulfonic acid) propyl triethoxysilane, 2-(4-sulfonic acid phenyl) ethyltrimethoxysilane or 2-(4-sulfonamide phenyl) ethyltrimethoxysilane, 3-(glycidoxypropyl) methyldimethoxysilane, γ-glycidyloxypropyltrimethoxysilane (GLYMO), 3-(methacryloxypropyl) trimethoxysilane) (MEMO)	-	-	proton-conducting membrane fuel cell (PMFC)	[78]
sulfonated allyltrimethoxysilane(ALLMO), GLYMO, MEMO	TEOS	-	PMFC	[79]
sulfonated MEMO or sulfonated ALLMO, GLYMO, MEMO or ALLMO	TEOS	-	PMFC	[80]
sulfonated ALLMO, p-vinylphenylmethyl diethoxysilane, N-(3-triethoxysilylpropyl)-4,5-dihydroimidazole	-	-	PMFC	[81-82]
MEMO	-	AlOOH nanoparticles	coating	[91]
N-octadecyldimethyl[3-(trimethoxysilyl) propyl]ammonium chloride	TEOS	-	sensor	[88]
MEMO, GLYMO	TEOS	$AgNO_3$	coating	[92]
MTES, octyltriethoxysilane, octadecyltrimethoxysilane or phenyltriethoxysilane	-	-	reverse osmosis	[93]
2-(trimethoxysilyl)ethyl-2-pyridine or N-trimethoxysilylpropyl-N,N,Ntrimethylammonium chloride	TEOS	-		[85]
3-(mercaptopropyl) trimethoxysilane	-	-	ultrafiltration and nanofiltration	[86-87]
aminopropyle-triethoxysilane	TEOS	-	electrochemical analysis	[89]

Among the different kinds of $R'_nSi(OR)_{4-n}$, the ones containing a polymerizable R' group (such as an acrylate [78-80,91-92], epoxy [78-80,92] or styrene group [81-82, 94]) are of special interest since they can be polymerized to form organic polymer network after the sol-gel process. So such precursors are called as dual network precursors [18].

The sol-gel process of alkoxysilanes is always catalyzed by acid (hydrochloric acid [29-30, 86-87], nitric acid [93] or acetic acid [85]). Ethanol [85-87] or methanol [90] is commonly used as the solvent. After reaction for a certain time, the sol-gel solution can be cast [90, 93], spin-coated [88, 91] or dip-coated [85-878, 92], then dried and heated to form the hybrid membranes. In some cases, certain tricks are needed to get the desirable properties of the membranes. For example, the H_2O/Si ratio can be lower if sol-gel solution with lower viscosity and polymerization degree is wanted [78-79, 81-82, 92]. When dual network precursor is used, thermal or UV curing can be followed after the casting or coating of the sol-gel solution, so that organic polymer network can be formed and the chemical structure of hybrid membranes strengthened [78-82, 91-92, 94].

In the author's lab, inorganic alumina substrates were dip-coated with the sol-gel solution from 3-(mercaptopropyl) trimethoxysilane (MPTS). After drying and heating processes, hybrid membranes were formed with the alumina plates as supports. Then the membranes were oxidized with hydrogen peroxide to be transformed to negatively-charged ones [86-87]. The whole procedure for the preparation of the membranes could be schemed as in Scheme 3. Ion exchange capacities (IECs) of the membranes increased with an increase of the coating times within a range of $(1.0\text{-}2.3)\times 10^{-2}$ mmol cm^{-2} for 1-4 coating times. The thermal stability was confirmed by TGA results, which showed that the membranes could endure a temperature as high as 250°C in N_2 atmosphere.

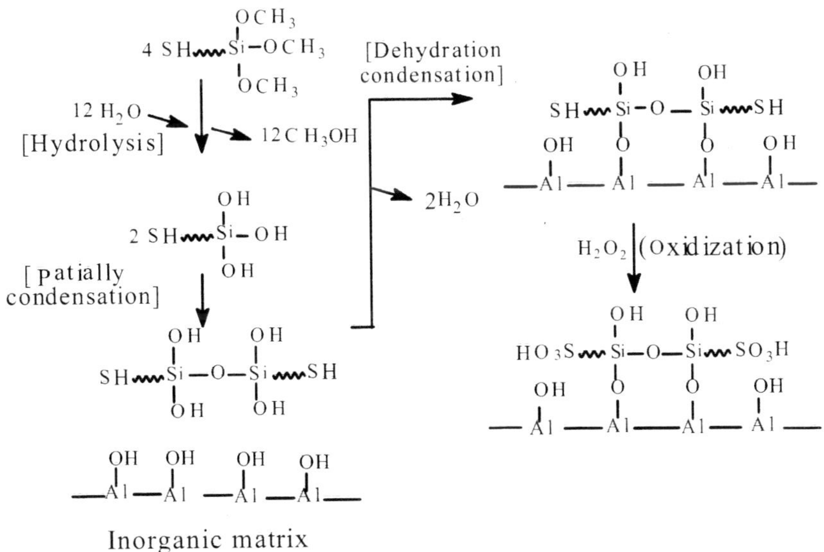

Scheme 3. Mechanism of sol-gel and oxidization processes of MPTS from ref [86-87].

In our lab, dual network precursors, γ-methacryloxypropyltrimethoxysilane (γ-MPS) and γ-glycidoxypropyltrimethoxysilane (GPTMS) were also used for sol-gel process together with tetraethoxysilane (TEOS) and N-triethoxysilylpropyl-N,N,N-trimethylammonium (TESPA (+)) [95-96]. In some cases, organic monomer, glycidylmethacrylate (GMA) was also added. After the sol-gel process of the mixture of the precursors at room temperature for 2 days, photoinitiator (Darocur 1173) was added. The mixture was cast onto Teflon plate, solvent evaporated, and then cured by UV and heat treatment in sequence. During the above preparation procedures, the acrylate or epoxy groups of γ-MPS or GPTMS could form organic network after the UV and thermal curing, while the TESPA (+) could render the hybrid membranes anion-exchangeable. The whole preparation processes can be shown in Scheme 4.

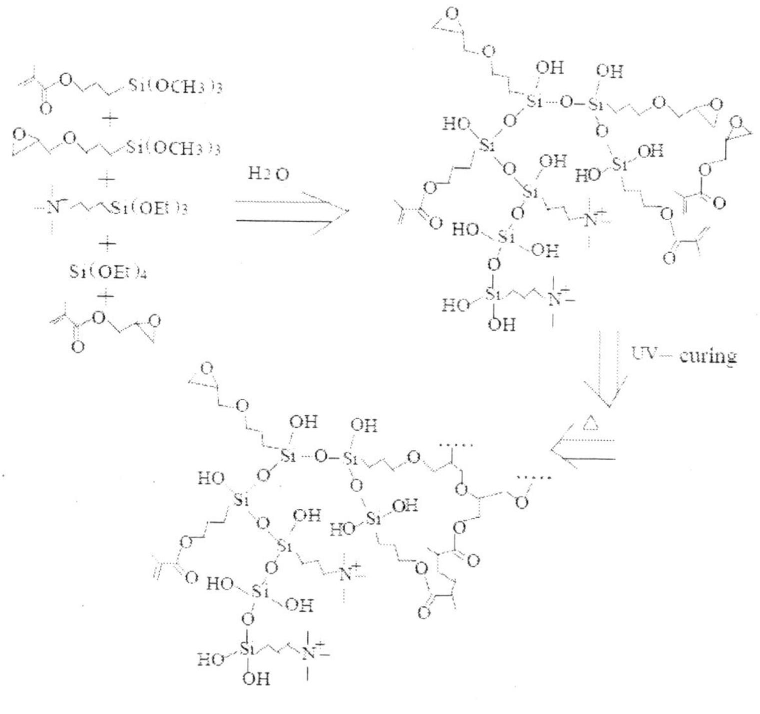

Scheme 4. The preparation of the anion-exchange hybrid membranes from γ-methacryloxypropyltrimethoxysilane (γ-MPS), γ-glycidoxypropyltrimethoxysilane (GPTMS), tetraethoxysilane (TEOS), N-triethoxysilylpropyl-N,N,N-trimethylammonium (TESPA (+)) and glycidylmethacrylate (GMA) [95-96].

Hybrid membranes can also be formed via the co-condensation of relatively low molecular weight polyfunctional alkoxysilane $(RO)_3Si\text{-}R'Si(OR)_3$ with or without other metal alkoxides. The organic group R' can be varied in length, rigidity, geometry of substitution, and functionality. Because the organic group remains an integral component of the hybrids, this variability provides an opportunity to modulate bulk properties such as porosity, thermal stability, refractive index, optical clarity, chemical resistance, hydrophobicity, and dielectric constant [97-98]. For instances, the R' can be such functional organic groups as arylene, alkylene, alkenylene, and alkynylene groups [97-98, 74]. For special application, R' with functional groups can also be synthesized. In Barboiu's work, $(RO)_3Si\text{-}R'Si(OR)_3$ in which R' contains fixed sulphide and amino binding sites [99] or crown ether [100-103] (figure 3) was prepared from several steps of reaction. Organic-inorganic hybrid membranes from the sol-gel process of such precursor can be used for the facilitated transport of ions (silver), amino acids or organic acids.

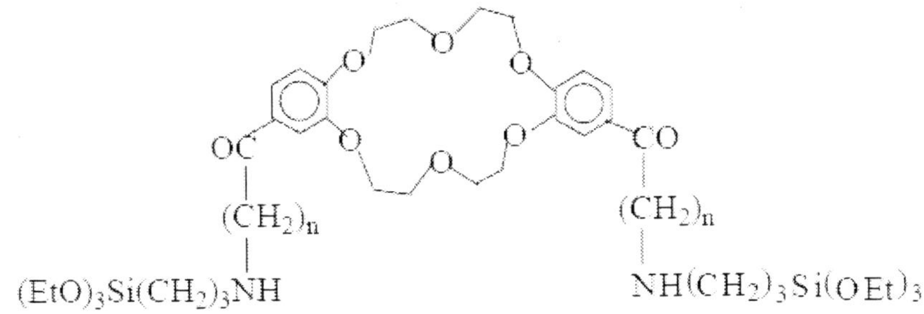

Figure 3. The polyfunctional alkoxysilane $(RO)_3Si\text{-}R'Si(OR)_3$ containing dibenzo 18-crown-6 synthesized. Adapted from ref. [100].

4.1.3. Sol-Gel of Alkoxysilane or Silanol Functionalized Polymers/Oligomers

Similar to Wilkes' method, varieties of polymers/oligomers were functionalized to undergo sol-gel process in the absence or presence of other precursors. Specifically, when the side or the end of polymers/oligomers chains contains such reactive groups as hydroxy (-OH) [54,104-106], ester group (-C(=O)O-) [60, 107], or amine (-NH- or $-NH_2$) [108] groups, they can be functionalized with alkoxysilane groups after reaction with organically modified alkoxysilanes. More recently, $SiCl_4$ was used to react with sulfonated polyetheretherketone (SPEEK) after treatment of SPEEK with BuLi at low temperature. Then ethanol was added, so that silanol (-SiOH) groups was covalently connected with SPEEK [109].

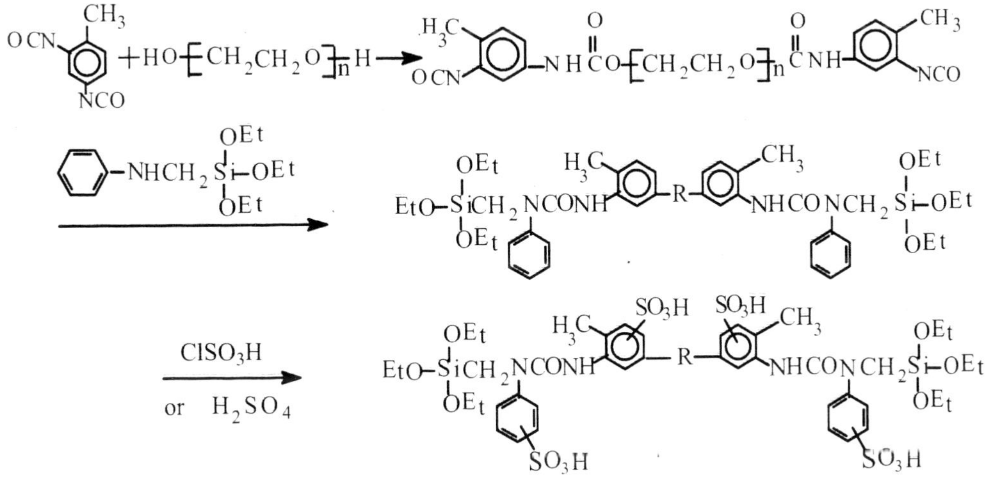

Scheme 5. Preparation of negatively charged alkoxysilane functionalized polyethylene oxide (PEO) from ref [110-111].

In our lab, different polymers, such as polyethylene oxide (PEO) [83-84, 110-114], poly(methyl acrylate) (PMA) [107], or poly(2,6-dimethyl-1,4-phenylene oxide) (PPO) [115-

117] was functionalized with alkoxysilane and used as sol-gel precursors. The -OH groups at the end of PEO chain, ester groups (-C(=O)-OCH_3) of PMA chains or -CH_2Br groups of brominated BPPO can be used for the reactions with organically modified alkoxysilane. The synthesis of alkoxysilane functionalized PEO included the following three steps, as illustrated in Scheme 5 [110-111]: the endcapping of PEO with tolylene 2,4-diisocyanate (TDI), linkage of phenylaminomethyl triethoxysilane (ND-42) to the endcappped PEO and the sulfonation reaction with chlorosulfonic acid or fuming sulfuric acid. This synthesized precursor then underwent hydrolysis and polycondensation with HCl as the catalyst. After casting or dip-coating of the sol-gel solution on Teflon plate or alumina plates and drying up to 100-110°C, non-supported or supported cation exchange hybrid membranes were obtained.

Since the functionalized polymers/oligomers contain relatively long polymer chains, sol-gel process may be different from the conventional sol-gel process of metal alkoxide or organically modified alkoxysilane. Wilkes et al. proposed a highly schematic model of hybrid materials from sol-gel process of alkoxysilane or silanol terminated polymer/oligormers [118]. This model structure is represented in figure 4. The crowded areas represent the condensed species while the coiled interconnecting lines represent the oligomeric species. Other research reports also confirmed the existence of micro-phase separation in the hybrid materials from such precursors: a Si-O domain and the polymer/oligormer domain [119-120]. Surivet et al. tried to apply the organic condensation polymerization theories to the sol-gel process of alkoxysilane terminated PEO [119]. Some efforts to investigate the sol-gel process of alkoxysilane functionalized PEO and organically modified alkoxysilane (*N*-[3-(trimethoxysilyl) propyl] ethylene diamine (A-1120)) have also been conducted by the authors [121]. Amine groups in A-1120 rendered the pH value of the reaction solution quite high, so that the polycondensation reaction and the formation of cyclic and branched agglomerate were quite rapid for A-1120. Polycondensation reaction was much slower for PEO-$[Si(OCH_3)_3]_2$ due to high steric hindrance. Pre-reaction of either PEO-$[Si(OCH_3)_3]_2$ or A-1120 would deteriorate the homogeneity of the hybrid and so the compatibility of the materials was the highest when PEO-$[Si(OCH_3)_3]_2$ and A-1120 co-reacted from the same time. This observation result deviates from the conventional acknowledge that pre-reaction of precursors which is less reactive should increase the compatibility of the final materials.

Dual network precursor, i. e. the organically modified alkoxysilane with acrylate, epoxy or styrene groups can also be used to copolymerize with organic monomers. Side chain of the resulted copolymers contains alkoxysilane groups and can be used for sol-gel process with or without other precursors. A series of hybrid membranes, such as polyacrylate -TiO_2-SiO_2 [48], polyacrylate-SiO_2 [122], poly (styrene-co-methacrylate)-silica [123], etc., have been prepared through sol-gel reaction of such precursors.

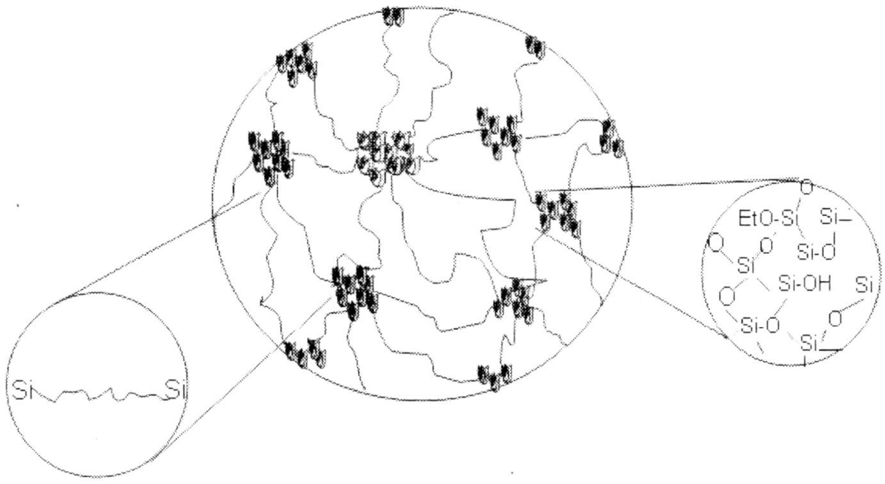

Figure 4. Model structure of ceramers proposed by Wilkes et al. Adapted from ref. [118].

In our lab, dual network precursor, γ-methacryloxypropyl trimethoxy silane (γ-MPS) was copolymerized with glycidylmethacrylate (GMA) [122]. The epoxy groups which were contained in the copolymers were then converted to quaternary amine groups through ring-opening reaction with amines, so that positively charged polymer precursors could be obtained. Supported hybrid membranes were prepared through the sol-gel process of these precursors. The membranes preparation procedures, including the copolymerization of GMA and γ-MPS, ring-opening reaction and sol-gel process can be shown as in Scheme 6. Some fundamental properties of the hybrid membranes are shown in table 2. It can be seen from Table 2 as the γ-MPS component increased in the copolymer, *Td* value and water contact angle of the hybrid membranes generally increased, while the anion exchange capacity (IEC) and water uptake (W_R) decreased. The charge transition point of the hybrid membranes deduced from their streaming potential (SP) behavior decreased from pH>12 to pH=7-8 as the γ-MPS component increased.

Table 2. Thermal properties, anion-exchange capatities (IEC) and hydrophilicity of hybrid membranes from γ-methacryloxypropyl trimethoxy silane (γ-MPS) and glycidylmethacrylate (GMA) from Ref [122]

Membrane compositions (GMA/ γ-MPS)	*Td* (°C)	IEC(mmol/g)	Water contact angle	Water uptake (W_R) (%)
83/17	228	1.36	83.4°	141.4 (±5.2)
71/29	223	1.10	84.2°	134.87(± 1.5)
56/44	241	1.00	87.0°	82.96 (±2.4)
35/65	251	0.83	88.2°	79.35(±2.2)

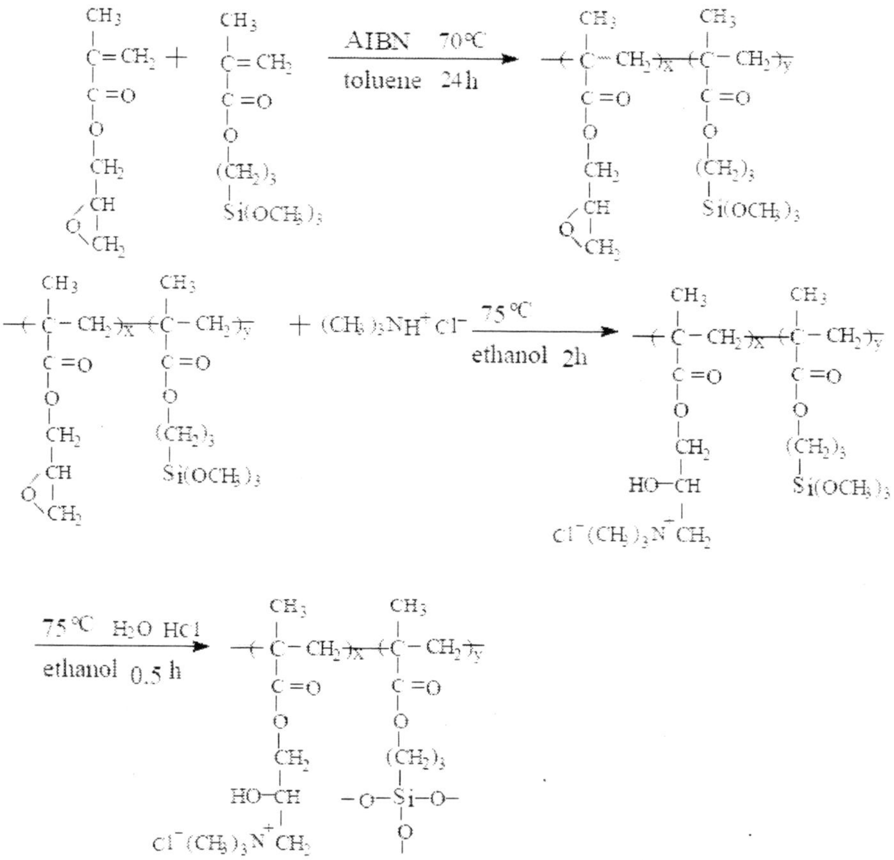

Scheme 6. The preparation of anion-exchange hybrid membranes from γ-methacryloxypropyl trimethoxy silane (γ-MPS) and glycidylmethacrylate (GMA) from ref [122].

4.1.4. Sol-Gel Process of Metal Alkoxides in the Matrix of Polymers

Sol-gel process of metal alkoxides can be carried out in the matrix of polymers. This is an important method to fabricate the class I hybrid materials. When polymers with -OH [120, 124-128], amine (-NH- or $-NH_2$ [128-129]) groups are used, hydrogen bonding can be formed between the -M-OH groups and the polymer chains, so that the homogeneity can be improved. In some cases, organically modified alkoxysilanes are also used [125-128]. The organic groups thus connected to the inorganic phases can help improve the uniformity of its dispersal in the organic polymer phase. For example, when small amount of 3-aminopropyltrimethoxysilane was used together with TEOS for the in-situ sol-gel reaction in the solution of poly(ether imide), dispersion of the formed inorganic phase was in the nanoscale [130].

For the actual processing, sol-gel precursors can be mixed with the solution of polymers. Water and catalyst are added and the reaction proceeds for a certain time, followed by casting, drying and heating [124-132]. Many factors may influence the dispersal of inorganic particles in the polymer matrix, such as the species of the precursors and solvent, the composition, the processing procedures. In the work of Silveira et.al [131], TEOS underwent sol-gel process in presence of poly(methyl methacrylate) (PMMA) and the mechanism of phase separation was investigated. Spinodal decomposition (SD) was confirmed for systems with intermediate PMMA/TEOS compositions occurring with a simultaneous viscosity increase. For systems with low PMMA content, a behaviour typical of nucleation and growth was detected. In another case by Zoppi et.al. [132], different metal alkoxide, i. e. tetraethoxysilane (TEOS) or titanium tetraisopropoxide (TiOP) was compared as the sol-gel precursor in presence of poly(ethylene oxide-b-amide-6) (PEBAX). Rates of hydrolysis and condensation reactions of TiOP were much higher than TEOS, giving rise to more obvious phase separation and more rigid structure of the hybrid membranes. The hybrid membranes from TEOS were very homogeneous even for inorganic contents as high as 80 wt%. In a third case, the influence of different solvents for preparation of poly(amide-imide) /TiO_2 hybrid membranes was mentioned [129]. The use of dimethylacetamide (DMAC) generally leads to a phase-separated membrane, because DMAC can form hydrogen bonding with the polymer's amide group, hindering the molecular interaction between the polymer and the TiO_2 domains and leading to a phase separated system. Tetrahydrofuran (THF) can favor the formation of optically transparent membranes, since only weak polar interactions between the THF molecules and the polymer repeat units exit, favoring interaction between the polymer and the TiO_2 domains.

Besides the reaction of precursor in the solution of polymer, precursor can also be penetrated into the polymer membranes. For practical processing, cross-linked, ionomeric, or crystalline polymeric host is at first swelled with a compatible solution containing metal alkoxides. Then the mixture of the sol-gel precursors, solvent, and sometimes also catalyst is penetrated into the polymer membrane matrix. Thus inorganic species, generally in the form of particles with a characteristic size of a few hundred angstroms, can be generated *in situ* within the polymers. In the work of Siuzdak [133] or Apichatachutapan [134], poly[ethyleneco-methacrylic acid] or Nafion which contains ionic clusters on the order of a few nanometers was used as the polymer membrane matrix. These ionic clusters were considered preferred sites in the ionomeric form for inorganic oxide nanoparticle formation via *in situ* sol-gel reactions. The *in-situ* formation of TiO_2 in Nafion membrane can be shown in figure 5 [134].

4.2. Surface Grafting of Inorganic Substrates with Organically Modified Alkoxysilane or Chlorosilane

As has been introduced above, efforts in modification of inorganic substrates through grafting of organically modified alkoxysilane or chlorosilane have begun more than a century ago. Researches have continued over the several decades. Some modification examples of inorganic substrates by alkoxysilanes or chlorosilanes were listed and summarized in the work of Javaid [5].

Different from sol-gel process, the surface grafting method does not go through the "sol" to "gel" route. Rather, the silane compounds will react with the surface hydroxyl (-OH) groups of the inorganic supports. No colloidal particles, no porous 3-dimensionally interconnected solid network will be formed.

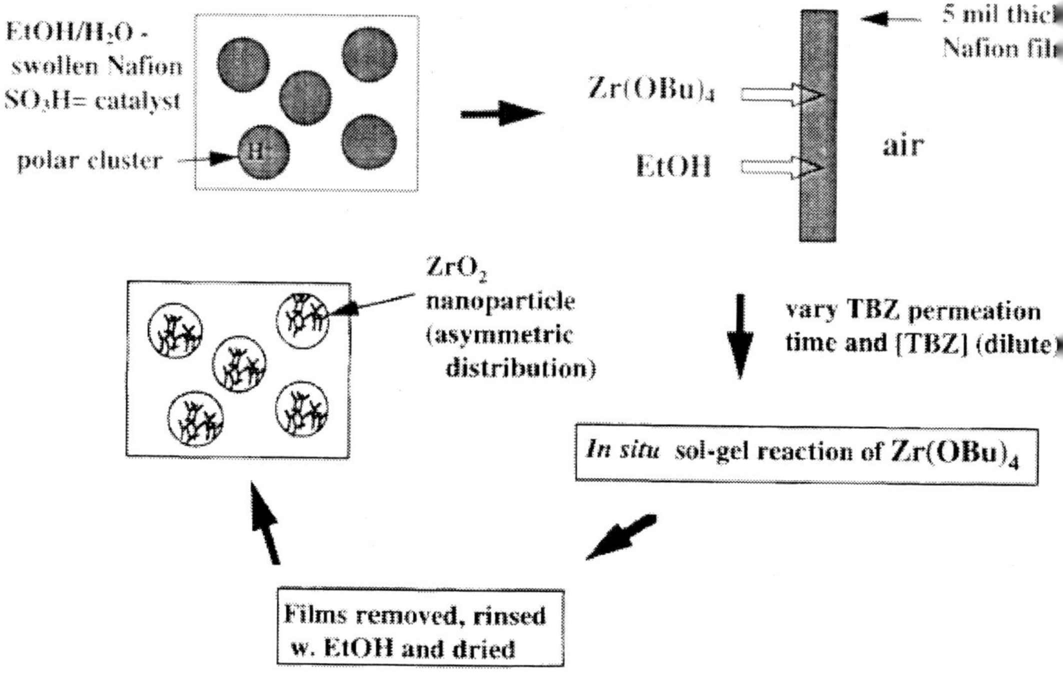

Figure 5 Schematic depiction of Nafion/[zirconium oxide] nanocomposite formulation via in-situ sol-gel reactions that initiate within nanometers in-size EtOH+H_2O containing sulfonic acid clusters. Adapted from ref. [134].

In the case of trichlorosilanes, due to the highly reactive trichloro-head group (R-$SiCl_3$), the reaction with -OH groups of the substrates takes place at room temperature without any catalyst. It is assumed that the silane molecules are first hydrolyzed by the trace quantities of water present either on the surface of the support or in the solvent followed by the formation of a covalent bond with the surface. All three hydroxyl groups (Si-OH) can react either with other surface silanols or with the silanols present on adjacent silane molecules forming a cross-linked polymer layer [135-136].

Due to the complexity of the reactions, it is agreed that even under ideal conditions the deposition of a monolayer (composed of close-packed, trans-extended, vertical chains) is difficult and often irreproducible [34, 137]. To depress the formation of a polymer layer, the amount of water present should be carefully controlled or a monochlorosilane, rather than trichlorosilane can be used as the grafting agent. For example, when monochlorosilane (R-Si$(CH_3)_2$Cl) is used, its reaction with the surface silanols is much more involved than that of the trichlorosilanes. Due to the steric hindrance of bulky methyl groups, there is either no or a very sluggish reaction of chloro and surface hydroxyl groups at ambient conditions [136]. This reaction can be promoted by using a base catalyst. The reaction of dimethyloctadecylchlorosilane (OCS) and octadecyltrichlorosilane (ODS) with surface hydroxyl (-OH) groups of inorganic substrates is shown in Scheme 7 [136].

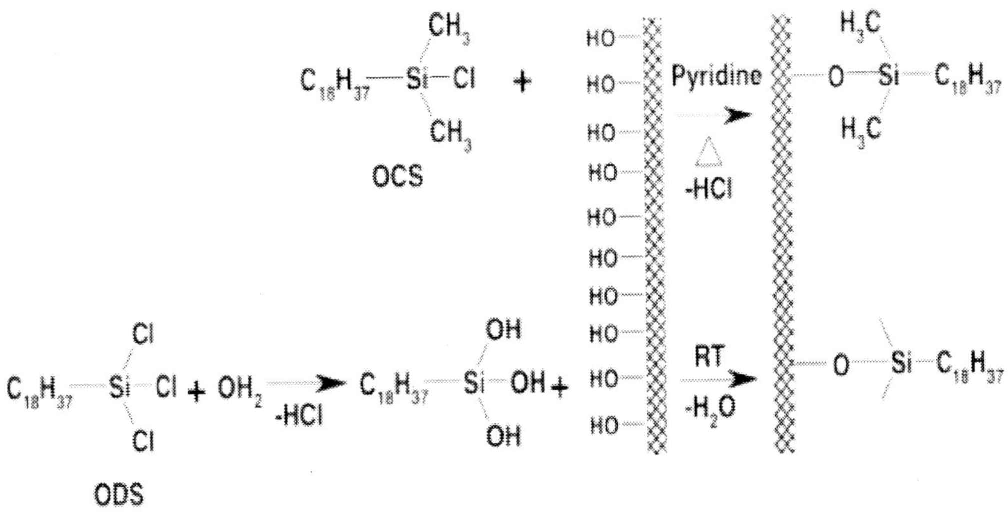

Scheme 7. Reaction mechanisms of the dimethyloctadecylchlorosilane (OCS) and octadecyltrichlorosilane (ODS) molecules with the surface hydroxyl groups. From ref. [136] with the permission of the authors

In actual preparation process, some handling methods can be utilized for avoiding undesirable side actions. For examples, to avoid the reaction of silane molecules with the atmospheric water, silanization can be carried out under flowing in a glovebag [138-139]. Pre-treatment of the substrates with alcohol, hydrogen peroxide (H_2O_2), or water is necessary for cleaning or hydroxylate of the surface [136, 138-139]..

In some cases, after the grafting of the organically modified silane to the substrates, polymerization can be followed. This can be classified into the so-called "grafting-from" method in polymerization science. For example, in the work of Faibish [140], free-radical graft polymerization of vinylpyrrolidone onto the membrane surface was conducted after the treatment of the surface with vinyltrimethoxysilane, as shown in Scheme 8. Since the vinyltrimethoxysilane was covalently linked to the surface, linkage of the polymer chains, at least partially, to the surface could be guaranteed. In the work of Ohya [141], a so-called "nanotechnological copolymerization method" was applied. That is, an organically modified alkoxysilane such as 3-aminopropyltriethoxysilane was firstly grafted onto the porous ceramic support. The amino groups of the alkoxysilane could act as a functional group for further co-polymerization. Then two organic monomers, monomer 1 (such as tetracarboxylic dianhydrides) and monomer 2 (such as diamine) were applied onto the support in sequence repeatedly. By changing the number of repeating unit in the polymer, the length of the grafted polymer chain could be effectively controlled. Scheme of such nanotechnological technique was shown in Scheme 9 [141].

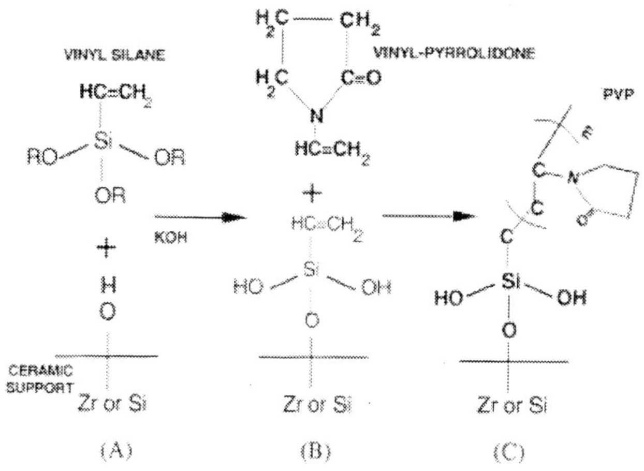

Scheme 8. Steps in vinylpyyrolidone (VP) free-radical graft polymerization onto zirconia or silica surfaces. From ref. [140] with the permission of the authors.

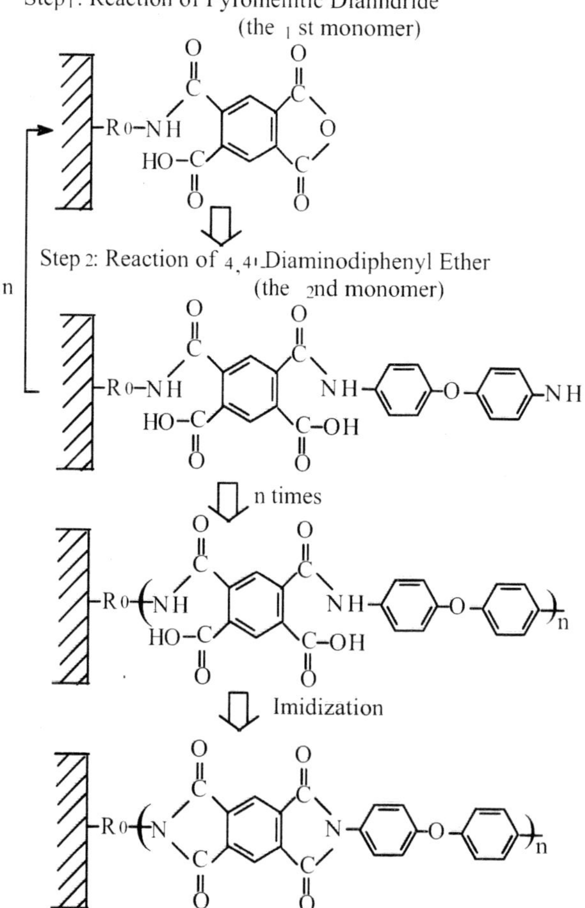

Scheme 9. The formation of polyimide on the surface of inorganic substrates. Adapted from ref. [141].

4.3. Impregnation or Mixing Methods

For the preparation of the class I hybrid materials/memebranes, organic component can be impregnated into the voids of oxide gels, or inorganic particles are just dispersed through mixing into organic polymers [142]. Mixing in the solution is commonly utilized [143-145]. In some cases, melt blending method is also used [146]. The organic component may be conventional low molecular weight organics or organic polymers [143-145]. It may also be other active substances such as enzymes, proteins or various organic dyes [147]. In some cases, they may also be organic monomers and can undergo polymerization thereafter.

Such impregnation or mixing method is often used to fabricate the "mixed matrix membranes" (MMMs) [25,143]. MMM constists of organic polymer and inorganic particle phase, such as zeolite, molecule sieves, porous silica, carbon nanotubes or non-porous nano-size particles [4, 25, 142, 146, 148-150]. MMMs have the potential to achieve higher selectivity, permeability, or both relative to the existing polymeric membranes, resulting from the addition of the inorganic particles with their inherent superior separation characteristics. At the same time, the fragility inherent in the inorganic membranes may be avoided by using a flexible polymer as the continuous matrix. So they have been commonly explored for separation applications.

It should be noted that the transport properties of MMMs are strongly dependent on the nanoscale morphology of the membranes. The morphology of the interface is a critical determinant of the overall transport property. So interface voids between the particles and the polymer matrix, or blocking of the inorganic particles pores by polymer chains should be avoided.

To achieve more homogeneous dispersal of the inorganic components in the organic matrix, inorganic particles can be modified by organic groups through the use of organically modified alkoxides $R'_nSi(OR)_{4-n}$ [151] or polyfunctional alkoxysilane $(RO)_3Si$-$R'Si(OR)_3$ [152]. Or organic polymers containing such groups as -OH, $-NH_2$ can be chosen to form hydrogen linkage with the -MOH groups of the inorganic component. In the work of Pechar and et al [151], zeolite was incorporated into polyimide to form MMM. Interfacial contact between the two phases relied on introducing amine functional groups on the zeolite surface and covalently linking them with carboxylic acid groups present along the polyimide backbone. In some other cases, the sol from metal alkoxide or organically modified metal alkoxide was used for impregnation into the polymer solutions [130, 153]. The following drying and heating processes after film-casting can accelerate the further growth of the inorganic particles. This method can be regarded as a utilization of both the in-situ sol-gel reaction method and the mixing/integration method.

4.4. Other Preparation Methods

Beside the above listed preparation methods, other ones for preparing hybrid membranes have also been used, such as partial pyrolysis [154-155], chemical vapor deposition (CVD) [156-158] and plasma polymerization of monomers on the surface of inorganic substrates [159-164].

"Pyrolysis" of polymer materials is their degradation process at high temperature. When the temperature is as high as 700-1000°C, generally complete pyrolysis will occur, all the unstable covalent bonding being removed and unstable elements evaporated. When the temperature is controlled to be lower, or the heating time shorter, the polymer materials can only be partially pyrolyzed. Some organic components will remain in the inorganic network, and so the product can be considered as a kind of hybrid materials. Partial pyrolisis of silicones to form microporous silica has been practiced as a method for forming thermally stable membranes in 1980s [165]. It was shown that microporous membranes containing both inorganic silicon dioxide and organic silicone phases were introduced. In Stewens et al's work [155], partial pyrolysis of poly(dimethyl siloxane) (PDMS) was followed by an oxidative stabilization step to produce porous hybrid membrane (Scheme 10). The partial pyrolysis could enhance thermal stability and create a micropous structure, while the oxidation process could promote mechanical strength and flexibility.

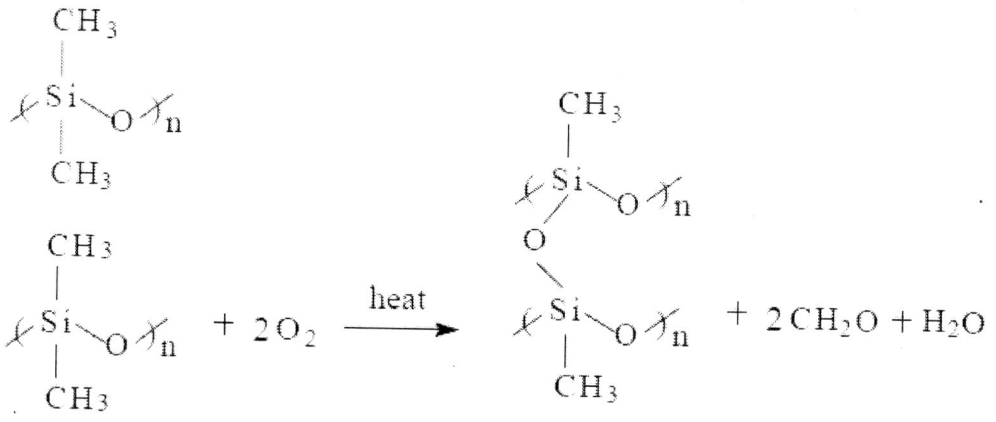

CH_3
Si O n
CH_3

CH_3
Si O n + $2O_2$ —heat→ CH_3 Si O n / O / Si O n CH_3 + $2CH_2O + H_2O$
CH_3

Scheme 10. Mechanism for degradation and oxidation of poly(dimethyl siloxane). Adapted from ref[155].

Chemical vapor deposition process (CVD) is a gas process generally used for preparation of inorganic materials/membranes. When organically modified alkoxysilane is used as the starting materials and the CVD processing conditions controlled to be mild, some C-Si bonding can be remained thereafter and so organic groups will be covalently connected to the inorganic substrates to form a kind of hybrid membranes. In the work of Ida et al [157-158], surface corona discharge-induced plasma chemical process-chemical vapor deposition (SPCP-CVD) of (γ-aminopropyltriethoxysilane (γ-APTES) was applied on ceramic membranes. Surface of the obtained hybrid membranes contained $-NH_2$ groups, which could be used for enzyme immobilization (Scheme 11).

Plasma-induced graft polymerization technique is well known as a surface modification method. In some cases, organic monomers can be introduced into the surface of inorganic substrates, then induced by plasma for polymerization to form organic polymer layer. The resulted membrane is more like the traditional inorganic-organic composite membranes, since generally only weak linkage exists between the inorganic substrate and the organic layer. In the work of H. Yasuda [159], it was stated that "transport characteristics of many plasma polymers are not typical of polymers but are in between those of polymers and of inorganic materials". The interested readers may refer to some other intense researches [160-164] for this kind of materials/membranes.

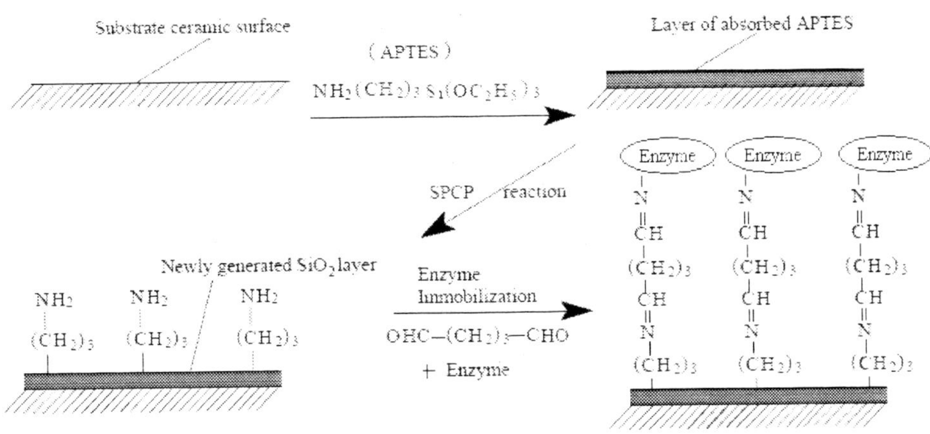

Scheme 11. Schematic diagram of the surface modification treatment by SPCP method and following enzyme immobilization. Adapted from ref. [157-158].

5. Applications of Hybrid Membranes

As has been stated in the above sections, the hybrid membranes are rapidly-developing fields and their applications are wide. Not only this, new application fields are being continuously found and so it is hard to give an overview. In this section, we'd just give some typical application examples, mainly to show the great potential and excellent merits of hybrid membranes. However, it should be born in mind that not all the hybrid membranes can show the best qualities of both the organic components and inorganic components. Optimization of membrane properties in some respects is accompanied with sacrifice to some extent of some other properties in many cases. However, there are indeed many hybrid membranes which show properties superior upon the original organic or inorganic materials. For design of needed hybrid membranes, the choosing of suitable starting materials and suitable preparation method is vital.

5.1. Gas Separation

Among the hybrid membranes for gas separation, the hybrid membranes from organically modified inorganic substrates and mixed matrix membranes (MMMs) attract the highest attention. When organic component is grafted onto the surface of inorganic substrate, the grafting conditions of the organic components play important role in the properties of the membranes. If the organic components block or fill the substrates micropores, the pemeability of the hybrid membranes will decrease to a large extent, sometimes by several orders of magnitude [34, 166]. The ideal condition should be mono-layer, unpacked crosslinkage, the so called brush-shape. For example, in the work of Castro et al. [167-168], polyvinylpyrrolidone chains were grafted onto silica supports through a two-step graft polymerization process. It was supposed that polymer brush-layer was formed on the surface of the support. Such brush-layer increased selectivity without decreasing permeability and reduced fouling during the cross-flow filtration of an oil/water emulsion.

Selectivity over gases, for example, will be changed as compared to the original inorganic substrates [169]. Sometimes, reverse selectivities over gases as compared with the original unmodified substrates can be observed [139, 170]. For example, when heptadecafluoro-1,1,2,2-tetrahydrodecyltrichlorosilane was grafted to porous Vycor glass, permeance of CO_2 at low pressure was enhanced, due to the existence of the glassy polymer chains [33]. In Paterson et al.'s work [166], 5 nm alumina membranes were surface derivatized with PDMS silicone oil. The modified membrane in some cases(e.g.CO_2/N_2) exhibited selectivities similar to a dense PDMS film, while in other cases (e.g. CH_4/N_2) showed significantly higher selectivity. So the membrane after modification no longer remained porous and gas permeation was occurring predominantly due to the solution-diffusion mechanism.

For MMMs, comprehensive review has been made, including the development, the preparation and gas separation properties [4, 143, 171]. Idea for designing of suitable MMMs for certain gas separation has also been pointed out: In the cases where it is more sensible to allow the smaller component to pass through, inorganic fillers with molecular sieving characteristics and polymers based on the size selection should be combined to produce MMMs; on the other hand, when the selective transport of more condensable molecules through the membrane is more economical, the MMMs may include microporous media that favor a selective surface flow mechanism and polymers that separate the mixtures by solubility selectivity. More recently, non-porous nano-size fillers have also been used, whose function is to systematically manipulate the molecular packing of the polymer chains, hence enhancing the separation properties of glassy polymeric membranes [148-150].

Besides the modified inorganic substrates and MMMs, another kind of membrane that may be of interest for gas separation is hollow hybrid membrane based on the sol-gel reaction of organically modified alkoxysilanes. In Ballweg et al's work [172], dual network precursor was used. After the build up of the inorganic network in sol-gel reactions, spinnable resin could be formed, which was then extruded through a ring shaped nozzle to form a hollow thread. The final curing was started by UV-light initiating the polymerization of organic groups. The fibers showed good mechanical properties (tensile strength up to 110 MPa, Young's modulus up to 2 GPa with elongation at rupture between 5 and 30 %) and easy processability. Oxygen permeability could be up to 130 000 cm^3/m^2 d bar. The geometrical dimension of the fibers (outer fiber diameter 90 mm to 5 mm with wall thickness from 15 mm to 2 mm) could be controlled by various processing parameters.

5.2. Pervaporation

Pervaporation, as an environment-benign and energy-saving technology, has attracted considerable research interest to separate azeotropic mixtures or close boiling points mixtures. Many reports can be found in seeking robust membrane materials with higher permeability and permselectivity. Despite of concentrated efforts to tailor polymer structure to improve separation properties, current polymeric membrane materials have seemingly reached a limit in the tradeoff between permeation flux and selectivity [11]. To circumvent this hurdle, much research interest has focused on exploration of organic-inorganic hybrid membrane materials. Incorporation of absorbents (such as zeolites, carbon molecule sieves) into organic polymers has been extensively investigated, in hope of combining the excellent size-sieving capacity of absorbents with good membrane-forming properties of polymers. For such mixed matrix membranes (MMMs), nonselective and impermeable voids at the interface of absorbent particles and polymer matrix should be carefully restrained or avoided.

Meanwhile, hybrid membranes through sol-gel process of (organically modified) metal alkoxide in the matrix of organic polymers have also been developed. The in-situ growth of Si-O-Si network in the polymers can result into silica particles more homogeneously dispersed in the hybrid membranes. Poly(vinyl alcohol) (PVA), chitosan and etc which contain a large amount of -OH or -NH_2 groups are commonly used as the polymer matrix. In Liu's work, a series of PVA or chitosan/SiO_2 hybrid membranes were prepared for pervaporation. For example, hybrid membranes from the sol-gel reaction of γ-aminopropyltriethoxysilane (APTEOS) in the solution of PVA were investigated for pervaporation separation of ethanol/water mixtures [127]. The amorphous region of the hybrid membranes increased with increasing APTEOS content, and both the free volume and the hydrophilicity of the hybrid membranes increased when APTEOS content was less than 5 wt%. The swelling degree of the hybrid membranes was restrained in an aqueous solution owing to the formation of hydrogen and covalent bonds in the membrane matrix. Permeation flux increased remarkably with APTEOS content increasing, and water permselectivity increased at the same time. The trade-off between the permeation flux and water permselectivity of the hybrid membranes was broken. The hybrid membrane containing 5 wt% APTEOS has highest separation factor of 536.7 at 50 °C and permeation flux of 0.0355 kg m^{-2} h^{-1} in pervaporation separation of 5 wt% water in the feed [127]. Chitosan-silica hybrid membranes (CSHMs) through similar preparation method were investigated in pervaporation of methanol/dimethyl carbonate (MeOH/DMC) mixtures [128]. Thermal stability of the hybrid membranes was enhanced, degree of swelling was greatly depressed, the selectivity of solubility and diffusion were greatly improved over the original chitosan membrane. The hybrid membrane presented superior separation behaviors with a flux of 1265 g/(h m^2) and separation factor of 30.1 in pervaporation separation of 70 wt% MeOH in feed at 50°C.

5.3. Ultrafiltration (UF), Nanofiltration (NF) and Reverse Osmosis(RO)

The reports of hybrid membranes for pressure-driven membrane separation processes such as UF, NF or RO application are relatively less common, mainly because the conventional organic or inorganic membranes have achieved great success in these fields. Also, increasing of price and higher complexity of preparation procedures render the exploration of hybrid membranes not so attractive. However, the great variety and high potential of hybrid membranes in modifying the conventional membranes should not be ignored [173]. Especially, for applications under more harsh circumstances [174], such as higher temperature, stronger acid, basic or oxiding conditions, development of hybrid membranes with higher thermal and chemical stabilities is highly desirable.

In the authors' lab, supported negatively or positively charged hybrid membranes were prepared using 3-(mercaptopropyl) trimethoxysilane (MPTS) or alkoxysilane functionalized PEO as the sol-gel precursors [86-87, 111-113]. The membranes can be thermally stable up to 200°C under N_2 atmosphere and their pore diameter can be varied from around 1μm to several nanometers. Therefore, these membranes can be potentially used for ultrafiltration and nanofiltration. The negative or positive charge of the active hybrid membrane layer was useful to reject charged species in the feed or reduce the fouling [175-177]. In Nunes's work, organic-inorganic hybrid membranes from sol-gel of alkoxysilane functionalized PEO was evaluated for nanofiltration. Cut offs down to 860g/mol could be got, with a water flux of 2.5l/h m^2 bar [108].

Lee et al coated the polyamide TFC RO membranes with sol-gel solution of organically modified alkoxysilanes, then cured at 70°C [93,178]. When long-chain alkyl (such as octadecyl) substituted alkoxysilane or rigid phenyl substituted alkoxysilane was used, results showed that the relative flux of the hybrid membranes decreased, but their selectivity toward NaCl increased to 99.6%. In particular, the membrane from phenylalkoxysilane could show 99.6% salt rejection at a water permeability of 0.673m^3/(m^2 day). MMM containing NaA zeolite nanoparticles dispersed within 50-200 nm thick polyamide films was investigated for RO[179]. At the high nanoparticle loadings, hybrid membranes could show pure water permeability nearly double that of the original polyamide membranes with equivalent solute rejections.

5.4. Modification of Electrodes for Analytical Purpose

The chemical modification of conventional electrodes has attracted much attention during the past decades because it provides a powerful means to bring new qualities to the electrode surface. Among the wide range of electrode modifiers, the sol-gel hybrid materials have been the focus of attention because of advantageous features such as mechanical stability and durability, two- or three dimensional rigid structure, possibility for molecular recognition or discrimination, or intrinsic catalytic properties [180-184].

The electrodes modified with hybrid materials have found important applications in analytical field. For examples, target analyte can be accumulated by the modified electrodes. The accumulation may be through adsorption or complexation. In the work of Walcarius

[89], carbon paste electrodes modified with aminated silicas were prepared using aminopropyle-triethoxysilane as one of the precursors. The electrodes can be used for the accumulation of copper (II) due to the existence of the amine ligands. The accumulation of target analyte may also be through ion exchange. In Heineman' work [185-186], the analytical utilities of electrodes modified with hybrid membranes containing long-chain polymers were explored. It was demonstrated that cationic analytes (as methyl viologen or $[Re^{I}(DMPE)_3]^+$ with DMPE=1,2-bis-(dimethylphosphino)ethane) partitioned much more quickly in a sol-gel-Nafion composite electrode compared to the corresponding pure Nafion-modified electrode. This improvement was ascribed to the rigid three dimensional open structure of the composite membrane, which was induced by its inorganic component (silica network), leading to enhanced mass transfer to and from the ion exchange sites located inside the membrane. Enhancing diffusion processes in Nafion by dispersing the polymer into a sol-gel-derived silica matrix was otherwise evidenced by electrochemical impedance spectroscopy and other techniques [187-188]. This behavior was exploited for electrochemical sensing of $Re^{I}(DMPE)_3]^+$ at both a planar graphite electrode and carbon-fiber microelectrode covered with a sol-gel-Nafion membrane, lowering by 10000 times the detection limit compared to that of unmodified electrodes because of the substantial preconcentration of the analyte in the membrane [186,189]. The use of other polymers embedded in a sol-gel glass resulted in similar improvements for the analysis of both cations and anions ($Fe(CN)_6^{3-/4-}$, $Ru(NH_3)_6^{3+}$, and $Ru(bpy)_3^{2+}$) [190].

5.5. Proton Conducting Membranes for Fuel Cells

Polymer electrolyte fuel cells are promising candidates as power generators for zero-emission vehicles [191]. Polymer electrolytes are also expected to play an important role in the development of improved energy sources, which is required in parallel to the huge development of cellular phones, book-type computers, and many personal digital assistants [192]. However, the organic polymer electrolyte membranes usually lack thermal stability, resulting in narrow operational temperature windows. Even for the expensive perfluorosulfonic polymer membranes (Nafion@, Flemion@, Aciplex@ and Dow@), their thermal and chemical stabilities need to be enhanced on some application occasions. Some other limitations of these perfluorosulfonic polymer membranes are the inability to retain water at high temperatures and the high methanol permeation that limits their application in direct methanol fuel cell (DMFC).

Exploration of ion conducting hybrid membranes to overcome the shortcomings of organic ones is mainly through two different ways. The first one concentrates on improving the performances of the perfluorosulfonic ionomeric membranes by incorporating inorganic or hybrid component. For example, Nafion/SiO_2 hybrid membranes, obtained through the *in situ* sol-gel reaction of tetra-alkoxysilane in Nafion matrix, can show similar proton conductivity, higher water uptake and lower methanol crossover rate as compared to pure Nafion membranes when the SiO_2 content is within a certain limit. Using the hybrid membranes for DMFC, the current density can be as high as $650mA/cm^2$ at 0.5V (125°C) [193].

The other way is to explore new and improved charged hybrid membranes to replace the perfluorosulfonic ionomeric ones. The new membranes may be from the incorporation of inorganic particles or inorganic solid acid into the organic polymer matrix. For example, the SiO_2/polyethylene oxides (PEO), polypropylene oxide (PPO) or olytetramethylene oxide (PTMO) hybrid membranes doped with monododecylphosphate (MDP) or 12-phosphotungstic acid (PWA) were found to be thermally stable up to 250°C and possess protonic conductivities of approximately 10^{-4}S/cm at temperature windows from room temperature to 160°C and relative humidity [194]. Homogeneous ion-conducting hybrid membranes have also been explored for energy technological applications. In the work of Poinsignon et al, poly(benzyl sulfonic acid)siloxane [195] and poly(benzyl sulfonic acid-diethylbenzenehexyl) siloxane [196] were obtained from the sol-gel process of organically modified alkoxysilanes and following chemical treatments. Chemical structures of these two polysiloxanes are as shown in figure 6. The latter product has thermal stability up to 250°C as well as a high ionic conductivity (about 1.6×10^{-2} S cm^{-1}) at room temperature and so is a promising solid electrolyte for practical fuel cell applications. In the work of Depre et al [78], new inorganic-organic sulfonic acid/sulfonamide functionalized protonic conductors for application as coatings, layers or separators in film batteries, as well as membranes in fuel cells under humid conditions were developed. Their thermal stability up to 200°C is limited by the oxidizing potential of the sulfonyl group. Cells reflecting the typical behavior of protonic conductors under humid and dry conditions were assembled. Under similar humid conditions these new membranes show remarkable high conductivities within the same range as Nafion®.

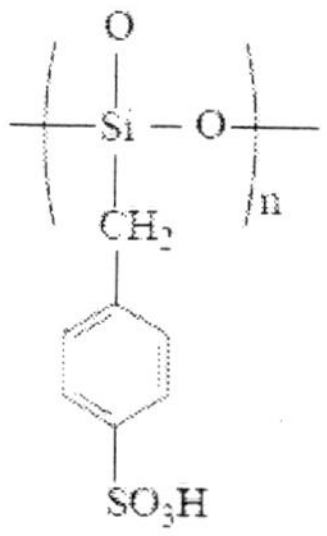

(a)

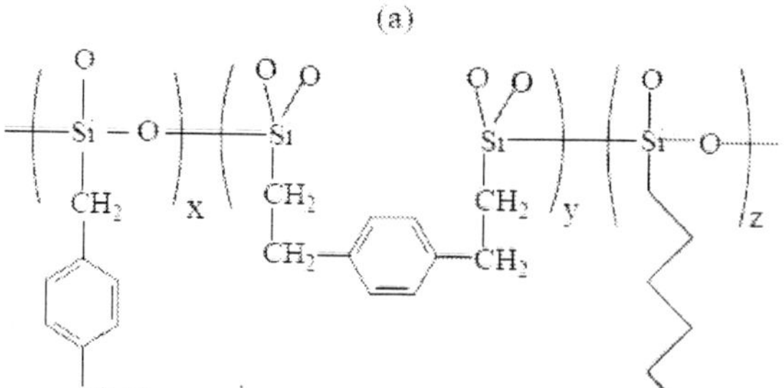

(b)

Figure 6. The chemical structure of poly(benzyl sulfonic acid)siloxane (a) and poly(benzyl sulfonic acid-diethylbenzenehexyl) siloxane (b) prepared by Poinsignon et al. Adapted from ref. [195-196].

5.6. Other Applications

Beside the above separation applications, there are also some other separations utilizing hybrid membranes, such as membrane affinity chromatography, electrochromic or optical displays, chemical sensors, diffusion dialysis, electro-driven separation techniques, membrane catalysis, membrane reactor and etc. [27,32, 70, 197-203]. As examples, in Ruckenstein's work [70], affinity membranes were prepared from the modifying of glass fiber filters using polyfunctional alkoxysilane: bis[3-(tri- methoxysilyl)-propyl] amine, 1,2-bis(triethoxysilyl)-ethane, or 1,3-diethoxy-1,1,3,3- tetramethyldisiloxane. Such kind of hybrid membranes could combine the biocompatibility of the organic component with the physical properties of the glass, such as porosity, rigidity, thermal stability, durability with organic solvents and oxidizing agents. The supported hybrid membranes prepared by Shi et al from the sol-gel process of TEOS in the presence of charged organic polymers were used to develop spectroscopic sensors for pH and transition metal ions, as well as spectroelectrochemical sensors for transition metal complexes [197-199]. In Barboiu's work [100], new heteropolysiloxane (HPS) materials containing the 2,3,11,12-bis[4-(10-aminodecylcarbony1)]-benzo-18-crown-6 and 2,3,11,12-bis[4-(2- aminoethylcarbonyl)] -benzo-18-crown-6 compounds chemically linked in a reticulate silica matrix were synthesized by the sol-gel process. From these materials, dialysis membranes used for the facilitated competitive transport of silver/copper ions were obtained. The selectivity of the membranes was improved by the incorporation of the complexing molecules which allowed a better solubility of silver ions in the membrane.

6. Conclusions and Remarks

In this chapter, the concept, general development history, preparation and application of hybrid membranes are briefly summarized. Considering the large amount of publications that have emerged in this field during the recent years, the illustration provided in this chapter is brief and thus quite limited. Even so, one could be surprised by the high level of diversity in both the applications and the production of this kind of membranes. The high interests in the hybrid membranes arise mainly from the synergy exhibited by the organic-inorganic hybrids resulting from the combination in one material of the intrinsic properties of their two components (rigid inorganic matrix with tailored structure and organic groups or macromolecules with specific functionalities).

Despite the great success that has been achieved in preparing different hybrid membranes for different application fields, important problems still exist and should be resolved for their better developments. Firstly, the formation process of the hybrid membranes need more research. Especially for the sol-gel process, the reaction kinetics of different precursors, including metal alkoxides, organically modified alkoxysilanes and alkoxysilane functionalized polymers/oligomers need to be explored further. As has been mentioned above, the sol-gel reaction is a quite complex process involving multi-steps. Especially, during the early steps before gelation occurred, different kinds of hydrolysis and polymerization reactions may take place at numerous sites within the precursors under different reaction conditions. Researches have shown that the actual sol-gel reaction kinetics will influence the properties of the obtained hybrids to a large extent. So an investigation of the sol-gel kinetics under different reaction conditions is quite necessary before design and controll of the properties of the charged hybrid membrane can be achieved. Up to now, the reaction kinetics of the most commonly used precursors, such as tetra-alkoxysilane are still not quite distinct, let along that of other precursors with more complex chemical structures. The slow development is mainly due to the complexity of the sol-gel reaction kinetics. To the authors' opinion, more attention should be paid and more analytical instrument/tools should be developed and used. Besides, researchers engaging in different fields, such as physics, chemistry, materials science and engineering should work together for quicker development in this aspect.

Another problem that should be resolved is the insufficient in the understanding of the hybrid membranes structure-morphology-property relation at microcosmic scale. As has been researched, the properties of the hybrid membranes are influenced greatly by the hybrism fashion between organic/inorganic components. So it would be quite desirable to understand the structure-morphology-property relation of the membranes and design different hybrid membranes for different application fields accordingly. Various instrument and tools can help in this respect. Specifically, use of mechanical, thermal analysis, X-ray scattering, and microscopy methods in conjunction with other standard tools such as NMR can be invaluable in leading to a truer understanding of how these membranes function in light of their chemistry and morphological features.

7. Acknowledgments

Financial supports from the Natural Science Foundation of China (No. 20576130, 20636050, 20604027), National Basic Research Program of China (No. 2009CB623403) and Foundations of Educational Committee of Anhui Province (ZD2008002, KJ2008A072, KJ2009A003) are greatly appreciated. We would also like to thank the authors of the papers cited in this review for their original contributions.

References

[1] Glarer, J. The early history of reverse osmosis membrane development. *Desalination.* 117 (1998) 297-309

[2] Humphrey, J. L. Separation processes - playing a critical role. *Chemical Engineering Progress.* 91 (10) (1995) 31-41.

[3] Matsuura, T. Recent progresses in membrane separation science and technologies: a review. pp. 983-1004. Chapter 30 of Comprehensive Analytical Chemistry XX/XVII. Editor: *J. Pawliszyn.* 2002 Elsevier Science B.V.

[4] Chung, T. S.; Jiang, L. Y.; Li, Y.; Kulprathipanja, S. Mixed matrix membranes (MMMs) comprising organic polymers with dispersed inorganic fillers for gas separation. *Prog. Polym. Sci.* 32 (2007) 483-507.

[5] Javaid, A. Membranes for solubility-based gas separation applications. *Chemical Engineering Journal.* 112(2005) 219-226.

[6] Smitha, B.; Sridhar, S.; Khan, A. A. Solid polymer electrolyte membranes for fuel cell applications-a review. *J. Membr. Sci.* 259 (2005) 10-26.

[7] Seiler, M. Hyperbranched polymers: Phase behavior and new applications in the field of chemical engineering. *Fluid Phase Equilibria.* 241 (2006) 155-174.

[8] Lin, Y. S. Microporous and dense inorganic membranes: current status and prospective. *Sep. Purif. Technol.* 25 (2001) 39-55.

[9] McLeary, E. E.; Jansen, J. C.; Kapteijn, F. Zeolite based films, membranes and membrane reactors: Progress and prospects. *Microporous and Mesoporous Materials.* 90 (2006) 198-220.

[10] Lu, G. Q.; Costa, J. C. D.; Duke, M.; Giessler, S.; Socolow, R.; Williams, R. H.; Kreutz, T. Inorganic membranes for hydrogen production and purification: A critical review and perspective. *J. Colloid Interface Sci.* 314 (2007) 589-603.

[11] Robeson, L. M. Correlation of separation factor versus permeability for polymeric membranes. *J. Membr. Sci.* 62 (1991) 165-185.

[12] Singh, A. S.; Koros, W. J. Air separation properties of flat sheet homogeneous pyrolytic carbon membranes. *J. Membr. Sci.* 174 (2000) 177-188

[13] Kim, Y. K.; Lee, J. M.; Park, H. B.; Lee, Y. M. The gas separation properties of carbon molecular sieve membranes derived from polyimides having carboxylic acid groups. *J. Membr. Sci.* 235 (2004) 139-146.

[14] Saracco, G.; Neomagus, H. W. J. P.; Versteeg, G. F.; Van Swaaij, W. P. M. High-temperature membrane reactor: potential and problems. *Chem. Eng. Sci.* 54 (1999) 1997-2017.

[15] Caro, J.; Noack, M.; Kolsch, P.; Schaefer, R. Zeolite membrane-state of their development and perspective. *Micropor. Mesopor. Mater.* 38 (2000) 3-24.

[16] Judeinstein, P.; Sanchez, C. Hybrid organic-inorganic materials: A land of multi-disciplinarity. *J. Mater. Chem.* 6 (1996) 511-525.

[17] Sanchez, C.; Ribot, F. Design of hybrid organic-inorganic materials synthesized via sol-gel chemistry. *N. J. Chem.* 18 (1994) 1007-1047.

[18] Kickelbick, G. Concepts for the incorporation of inorganic building blocks into organic polymers on a nanoscale. *Prog. Polym. Sci.* 28 (2003) 83-114.

[19] Pedro, G. R. Hybrid Organic-Inorganic Materials-In Search of Synergic Activity. *Adv. Mater.* 13 (2001) 163-174.

[20] Wilkes, G. L.; Orler, B.; Huang, H. *Polym. Prepr.* 1985, 26, 300.

[21] Schmidt, H. New type of non-crystalline solids between inorganic and organic materials. *J. Non-Cryst. Solids.* 73 (1985) 681-691.

[22] Haas, K. H. Hybrid inorganic-organic polymers based on organically modified Si-alkoxides. *Advanced engineering materials.* 2(9) (2002) 571-582.

[23] Hosseini, S. S.; Li, Y.; Chung, T. S.; Liu, Y. Enhanced gas separation performance of nanocomposite membranes using MgO nanoparticles. *J. Membr. Sci.* 302 (1-2) (2007) 207-217

[24] Snyder, M. A.; Tsapatsis M. Hierarchical nanomanufacturing: From shaped zeolite nanoparticles to high-performance separation membranes. *Angewandte Chemie-international edition.* 46 (40) (2007) 7560-7573.

[25] S. Husain, W. J. Koros. Mixed matrix hollow fiber membranes made with modified HSSZ-13 zeolite in polyetherimide polymer matrix for gas separation. *J. Membr. Sci.* 288(2007) 195-207.

[26] Liu, B. B.; Cato, Y. M.; Wang, T. H.; Yuan, Q. Preparation of novel ZSM-5 zeolite-filled chitosan membranes for pervaporation separation of dimethyl carbonate/methanol mixtures. *J. Appl. Polym. Sci.* 106 (3) (2007) 2117-2125.

[27] Xi, F. N.; Wu, J. M.; Lin, X. F. Novel nylon-supported organic-inorganic hybrid membrane with hierarchical pores as a potential immobilized metal affinity adsorbent. *J. Chromatogr.* A 1125 (1) (2006) 38-51.

[28] Tezuka, T.; Tadanaga, K.; Matsuda, A.; Hayashi, A.; Tatsumisago, M. Utilization of glass papers as a support for proton conducting inorganic-organic hybrid membranes from 3-glycidoxypropyltrimethoxysilane, tetraalkoxysilane and orthophosphoric acid. *Solid state ionics.* 176 (39-40) (2005) 3001-3004.

[29] Dirè, S.; Pagani, E.; Babonneau, F.; Ceccato, R.; Carturana, G. Unsupported SiO_2-based organic-inorganic membranes Part 1.-Synthesis and structural characterization. *J. Mater. Chem.* 7(1) (1997) 67-73.

[30] Dirè, S.; Pagani, E.; Ceccato, R.; Carturan, G. Unsupported SiO_2-based organic-inorganic membranes. Part 2: Surface features and gas permeation. *J. Mater. Chem.* 7 (1997) 919-922.

[31] Wu, C. M. Xu, T. W.; Liu, J. S. "Charged Hybrid Membranes by the Sol-gel Approach: Present States and Future Perspectives ". Chapter 1, In "Focus on Solid State Chemistry", Edited by Arte M. Newman, ISBN: 1-60021-265-4, Nova Science Publishers Inc., USA, NY 11788, 2006, pp 1-44.

[32] Nagarale, R.K.; Shahi, V. K.; Rangarajan, R. Preparation of polyvinyl alcohol-silica hybrid heterogeneous anion-exchange membranes by sol-gel method and their characterization. *J. Membr. Sci.* 248 (2005) 37-44.

[33] Singh, R. P.; Way, J. D.; McCarley, K. C. Development of a model surface flow membrane by modification of porous vycor glass with a fluorosilane. *Ind. Eng. Chem. Res.* 43 (2004) 3033-3040.

[34] McCarley, K. C.; Way, J. D. Development of a model surface flow membrane by modification of porous-alumina with octadecyltrichlorosilane. *Sep. Purif. Technol.* 25 (2001) 195-210.

[35] Kuraoka, K.; Chujo, Y.; Yazawa, T. Hydrocarbon separation via porous glass membranes surface-modified using organosilane compounds. *J. Membr. Sci.* 182 (2001) 139-149.

[36] Barry, A. J.; Depree, L.; Hook, D. E. *Br. Patent.* 635,645, 1944.

[37] Shea, K. J.; Loy, D. A. Bridged Polysilsesquioxanes. Molecular-Engineered Hybrid Organic-Inorganic Materials. *Chem. Mater.* 13 (2001) 3306-3319.

[38] Boldebuck, E. N. *U.S. Patent.* 2,551,924, 1951.

[39] Paul, D. R.; Kemp, D. R. The diffusion time lag in polymer membranes containing adsorptive fillers. *J. Polym. Sci: Polym. Phys.* 41 (1973) 79-93.

[40] Schmidt, H; Philipp, G. New materials for contact lenses prepared from Si- and Ti-alkoxides by the sol-gel process. *J. Non-Cryst. Solids.* 63 (1984) 283-292.

[41] Schmidt, H; Scholze, H; Kaiser, H. Principles of hydrolysis and condensation reaction of alkoxysilanes. *J. Non-Cryst. Solids.* 63 (1984) 1-11.

[42] Schmidt, H.; Scholze, H.; Tunker, G. Hot melt adhesives for glass containers by the sol-gel process. *J. Non-Cryst. Solids.* 80 (1986) 557-563.

[43] Huang, H. H.; Orler, B.; Wilkes, G. L. Structure-property behavior of new hybrid materials incorporating oligomeric species into sol-gel glasses. *Macromolecules.* 20 (1987) 1322-1330.

[44] Wilkes, G. L.; Brennan, A. B.; Huang, H. H.; Rodrigues, D.; Wang, B. The synthesis structure and property behavior of inorganic-organic hybrid network materials prepared by the sol-gel process. *Mater. Res. Soc. Symp. Proc.* 171 (1990) 15-29.

[45] Wilkes, G. L.; Huang, H.; Glaser, R. H. Silicon-Based Polymer Science: A Comprehensive Resource; Advances in Chemistry Ser. 224; American Chemical Society: Washington, DC, 1990; p 207.

[46] Kamiya, K.; Yoko, T.; Sano, T.; Tanaka, K. Distribution of carbon particles in carbon/SiO_2 glass composites made from $CH_3Si(OC_2H_5)_3$ by the sol-gel method. *J. Non-Cryst. Solids.* 119 (1990) 14-20.

[47] Bommel, M. J.; Bernards, T. N. M.; Boonstra, A. H. The influence of the addition of alkyl-substituted ethoxysilane on the hydrolysis-condensation process of TEOS. *J. Non-Cryst. Solids.* 128 (1991) 231-242.

[48] Lee, L. H.; Chen, W. C. High-Refractive-Index Thin Films Prepared from Trialkoxysilane-Capped Poly(methyl methacrylate) *Titania Materials Chem. Mater.* 13 (2001) 1137-1142.

[49] Harreld, J. H.; Dunn, B.; Zink, J. I. Effects of organic and inorganic network development on the optical properties of ORMOSILs. *J. Mater. Chem.* 7 (1997) 1511-1517.

[50] Kim, H. K.; Kang, S. J.; Choi, S. K.; Min, Y. H.; Yoon, C. S. Highly efficient organic/inorganic hybrid nonlinear optic materials via sol-gel process: Synthesis, optical properties, and photobleaching for channel waveguides. *Chem. Mater.* 11 (1999) 779-788.

[51] Mager, M.; Jonas, F.; Eiling, A.; Guntermann, U. German Patent 19 650 147 A1, 1998.

[52] Seo, S. J.; Zhao, D.; Suh, K.; Shin, J. H.; Bae B. S. Synthesis and luminescence properties of mesophase silica thin films doped with in-situ formed europium complex. *J. Lumin. 128* (2008) 565-572.

[53] Cordoncillo, E.; Viana, B.; Escribano, P.; Sanchez, C. Room temperature synthesis of hybrid organic-inorganic nanocomposites containing Eu^{2+}. *J. Mater. Chem.* 8 (1998) 507-509.

[54] Nakajima, H.; Honma, I. Proton-conducting hybrid solid electrolytes for intermediate temperature fuel cells. *Solid State Ionics.* 148 (2002) 607-610.

[55] Rikukawa, M.; Sanui, K. Proton-conducting polymer electrolyte membranes based on hydrocarbon polymers. *Prog. Polym. Sci.* 25 (2000) 1463-1502.

[56] Kalappa, P.; Lee, J. H. Proton conducting membranes based on sulfonated poly(ether ether ketone)/TiO_2 nanocomposites for a direct methanol fuel cell. *Polym. Int.* 56 (2007) 371-375.

[57] Jannasch, P. Recent developments in high-temperature proton conducting polymer electrolyte membranes. *Curr. Opin. Colloid Interface Sci.* 8 (2003) 96-102.

[58] Li, L.; Yuxin Wang, Y. X. Proton conducting composite membranes from sulfonated polyethersulfone Cardo and phosphotungstic acid for fuel cell application. *J. Power Sources.* 162 (2006) 541-546.

[59] Joly, C.; Goizet, S.; Schrotter, J. C.; Sanchez, J.; Escoubes, M. Sol-gel polyimide-silica composite membrane: gas transport properties. *J. Membr. Sci.* 130 (1997) 63-74.

[60] Cornelius, C. J.; Marand, E. Hybrid silica-polyimide composite membranes: gas transport properties. *J. Membr. Sci.* 202 (2002) 97-118.

[61] Battioni, P.; Cardin, E.; Louloudi, M.; Schollhorn, B.; Spyroulias, G. A.; Mansuy, D.; Traylor, T. G. Metalloporphyrinosilicas: A new class of hybrid organic-inorganic materials acting as selective biomimetic oxidation catalysts. *Chem. Commun.* 17 (1996) 2037-2038.

[62] Ciuffi, K. J.; Sacco, H. C.; Valim, J. B.; Manso, C. M. C. P.; Serra, O. A.; Nascimento, O. R.; Vidoto, E. A.; Iamamoto, Y. Polymeric organic-inorganic hybrid material containing iron(III) porphyrin using sol-gel process. *J. Non-Cryst. Solids.* 247 (1999) 146-152.

[63] Jesus, D. S.; Couto, C. M. C. M.; Araújo, A. N.; Montenegro, M. C. B. S. M. Amperometric biosensor based on monoamine oxidase (MAO) immobilized in sol-gel film for benzydamine determination in pharmaceuticals. *J. Pharm. Biomed. Anal.* 33 (2003) 983-990.

[64] Maines, A.; Ashworth, D.; Vadgama, P. Diffusion restricting outer membranes for greatly extended linearity measurements with glucose oxidase enzyme electrodes. *Analytica Chimica Acta.* 333 (1996) 223-231.

[65] Wang, B. Q.; Zhang, J. Z.; Cheng, G. J.; Dong, S. J. Amperometric enzyme electrode for the determination of hydrogen peroxide based on sol-gel/hydrogel composite film. *Analytica Chimica Acta.* 407 (2000) 111-118.

[66] C. Guizard, A. Bac, M. Barboiu, N. Hovnanian. Hybrid organic-inorganic membranes with specific transport properties Applications in separation and sensors technologies. *Sep. Purif. Technol.* 25 (2001) 167-180.

[67] Bouclé, J.; Ravirajan, P.; Nelson, J. Hybrid polymer-metal oxide thin films for photovoltaic applications. *J. Mater. Chem.* 17 (2007) 3141-3153.

[68] Licoccia, S.; Traversa, E. Increasing the operation temperature of polymer electrolyte membranes for fuel cells: From nanocomposites to hybrids. *J. Power Sources.* 159 (2006) 12-20.

[69] Barboiu, M.; Cazacu, A. ; Michau, M. ; Caraballo, R. ; Herault, C. A. ; Banu, A. P. Functional organic-inorganic hybrid membranes. *Chem. Eng. Process.* 2007, doi:10.1016/j.cep.2007.07.018

[70] Ruckenstein, E.; Guo, W. Cellulose and glass fiber affinity membranes for the chromatographic separation of biomolecules. *Biotechnol. Prog.* 20 (2004) 13-25.

[71] Hench, L. L.; West, J. K. The sol-gel process. *Chem. Rev.* 90 (1990) 33-72.

[72] Jose, N. M.; Prado, L. A. S. D. Hybrid organic-inorganic materials: Preparation and some applications. *Quimica Nova.* 28 (2) (2005) 281-288.

[73] Ogoshi, T.; Chujo, Y. Organic-inorganic polymer hybrids prepared by the sol-gel method. *Composite Interfaces.* 11 (8-9) (2005) 539-566.

[74] Wen, J.; Wilkes, G. L. Organic/inorganic hybrid network materials by the sol-gel approach. *Chem. Mater.* 8 (1996) 1667-1681.

[75] Pierre, A. C. Introduction to sol-gel processing. Kluwer academic publishers. 1998. p.1-2

[76] Bradley, D. C.; Mehrotra, R. C.; Gaur, D. P. Metal alkoxides. Academic: London. 1978.

[77] Ribot, F.; Toledano, P.; Sanchez, C. Hydrolysis condensation process of beta-diketonates-modified cerium(IV) isopropoxide. *Chem. Mater.* 3 (1991) 759-764.

[78] Depre, L.; Ingram, M.; Poinsignon, C.; Popall, M. Proton conducting sulfon/sulfonamide functionalized materials based on inorganic-organic matrices. *Electrochim. Acta.* 45 (2000) 1377-1383.

[79] Depre, L.; Kappel, J.; Popall, M. Inorganic-organic proton conductors based on alkylsulfone functionalities and their patterning by photoinduced methods. *Electrochim. Acta.* 43 (10-11) (1998) 1301-1306.

[80] Popall, M.; Du, X. M. Inorganic-organic copolymers as solid state ionic conductors with grafted anions. *Electrochim. Acta.* 40(13-14) (1995) 2305-2308.

[81] Jacob, S.; Cochet, S.; Poinsignon, C.; Popall, M. Proton conducting inorganic-organic matrices based on sulfonyl- and styrene derivatives functionalized polycondensates via sol-gel processing. *Electrochim. Acta.* 48 (2003) 2181-2186.

[82] Jacob, S.; Poinsignon, C.; Popall, M. Inorganic-organic hybrid protonic polymeric materials for fuel cells based on polycondensed and organically cross-linked sulfonyl- and styrene-functionalized alkoxysilanes. *Electrochim. Acta.* 50 (2005) 4022-4028.

[83] Wu, Y. H.; Wu, C. M.; Yu, F.; Xu, T. W.; Fu, Y. X. Free-standing anion-exchange PEO-SiO_2 hybrid membranes. *J. Membr. Sci.* 307 (2008) 28-36.

[84] Xu, T. W.; Wu, Y. H.; Wu, C. M. Preparation methodologies of the organic-inorganic hybrid anion-exchange membranes. Chinese Patent, Apply Number: 200610097187.7

[85] Kogure, M.; Ohya, H.; Paterson, R.; Hosaka, M.; Kim, J. J.; McFadzean, S. Properties of new inorganic membranes prepared by metal alkoxide methods Part II: New inorganic-organic anion-exchange membranes prepared by the modified metal alkoxide methods with silane coupling agents. *J. Membr. Sci.* 126 (1997) 161-169.

[86] Wu, C. M.; Xu, T. W.; Yang, W. H. A new inorganic-organic negatively charged membrane: membrane preparation and characterizations. *J. Membr. Sci..* 224 (2003) 117-125.

[87] Wu, C. M.; Xu, T. W.; Yang, W. H. A method to prepare a hybrid negatively charged membrane, Chinese Patent Number: ZL 03 1 31571.2

[88] Liu, A.; Zhou, H. S.; Honma, I. Electrochemical investigation of the permselectivity of a novel positively-charged sol-gel silicate prepared from tetraethyloxysilane and N-octadecyldimethyl-[3- (trimethoxysilyl)propyl] ammonium chloride. *Electrochemistry Communications.* 7 (2005) 1-4.

[89] Walcarius, A.; Lüthi, N.; Blin, J. L.; Su, B. L.; Lamberts, L. Electrochemical evaluation of polysiloxane-immobilized amine ligands for the accumulation of copper(II) species. *Electrochim. Acta.* 44 (1999) 4601-4610.

[90] Smaïhi, M.; Jermoumi, T.; Marignan, J.; Noble, R. D. Organic-inorganic gas separation membranes: preparation and characterization. *J. Membr. Sci.* 116 (1996) 211-220.

[91] Sepeur, S.; Kunze, N.; Werner, B.; Schmidt, H. UV curable hard coatings on plastics. *Thin Solid Films.* 351 (1999) 216-219.

[92] De, G.; Kundu, D. Silver-nanocluster-doped inorganic-organic hybrid coatings on polycarbonate substrates. *J. Non-Cryst. Solids.* 288 (2001) 221-225.

[93] Kim, N.; Shin, D. H.; Lee, Y. T. Effect of silane coupling agents on the performance of RO membranes. *J. Membr. Sci.* 300 (2007) 224-231.

[94] Kato, M.; Sakamoto, M.; Yogo, T. Synthesis of proton-conductive sol-gel membranes from trimethoxysilylmethylstyrene and phenylvinylphosphonic acid. *J. Membr. Sci.* 303 (2007) 43-53.

[95] Wu, C. M.; Wu, Y. H.; Xu, T. W.; Fu, Y. X. Novel anion-exchange organic-inorganic hybrid membranes prepared through sol-gel reaction and UV/thermal curing. *J. Appl. Polym. Sci.* 107 (3) (2008) 1865-1871.

[96] Xu, T. W.; Wu, C. M.; Wu, Y. H. Preparation methodologies of the anion-exchange hybrid membranes through sol-gel and UV/thermal curing. Chinese Patent, Apply Number: 200710023886.1

[97] Shea, K. J.; Loy, D. A. Aryl-bridged polysilsesquioxanes--new microporous materials. *Chem. Mater.* 1 (1989) 572-574.

[98] Loy, D. A.; Shea, K. J. Bridged polysilsesquioxanes-Highly porous hybrid organic-inorganic materials. *Chem. Rev.* 95 (1995) 1431-1442.

[99] Villamo, O.; Barboiu, C.; Barboiu, M.; Yau-Chun-Wan, W.; Hovnanian, N. Hybrid organic-inorganic membranes containing a fixed thio ether complexing agent for the facilitated transport of silver versus copper ions. *J. Membr. Sci.* 204 (2002) 97-110.

[100] Barboiu, M.; Luca, C.; Gmzard, C.; Hovnanian, N.; Cot, L.; Popescu, G. Hybrid organic-inorganic fixed site dibenzo 18-crown-6 complexant membranes. *J. Membr. Sci.* 129 (1997) 197-207.

[101] Barboiu, M.; Guizard, C.; Luca, C.; Albu, B.; Hovnanian, N.; Palmeri, J. A new alternative to amino acids transport: facilitated transport of L-phenylalanine by hybrid siloxane membranes containing a fixed site macrocyclic complexant. *J. Membr. Sci.* 161 (1999) 193-206.

[102] Barboiu, M.; Guizard, C.; Hovnanian, N.; Palmeri, J.; Reibel, C.; Cot, L.; Luca, C. Facilitated transport of organics of biological interest I. A new alternative for the separation of amino acids by fixed-site crown-ether polysiloxane membranes. *J. Membr. Sci.* 172 (2000) 91-103.

[103] Barboiu, M.; Guizard, C.; Luca, C.; Hovnanian, N.; Palmeri, J.; Cot, L. Facilitated transport of organics of biological interest II. Selective transport of organic acids by macrocyclic fixed site complexant membranes. *J. Membr. Sci.* 174 (2000) 277-286.

[104] Honma I.; Takeda, Y.; Bae, J. M. Protonic conducting properties of sol-gel derived organic / inorganic nanocomposite membranes doped with acidic functional molecules. *Solid State Ionics.* 120 (1999) 255-264.

[105] Chen, J.; Chareonsak, R.; Puengpipat V.; Marturunkakul S. Organic/inorganic composite materials for coating applications. *J. Appl. Polym. Sci.* 74 (1999) 1341-1346.

[106] Lu, Z. H.; Liu, G. J.; Duncan, S. Poly(2-hydroxyethyl acrylate-co-methyl acrylate)/SiO_2/ TiO_2 hybrid membranes. *J. Membr. Sci.* 221(2003) 113-122.

[107] Wu, C. M.; Xu, T. W.; Yang, W. H. Synthesis and characterizations of novel positively charged PMA-SiO_2 nanocomposites. *Eur. Polym. J.* 41 (2005) 1901-1908.

[108] Sforca, M. L.; Yoshida, I. V. P.; Nunes, S. P. Organic-inorganic membranes prepared from polyether diamine and epoxy silane. *J. Membr. Sci.* 159 (1999) 197-207.

[109] Vonaa, M. L. D.; Marania, D.; D'Epifanioa, A.; Traversaa, E.; Trombettab, M.; Licoccia, S. A covalent organic/inorganic hybrid proton exchange polymeric membrane: synthesis and characterization. *Polymer.* 46 (2005) 1754-1758.

[110] Wu, C. M.; Xu, T. W.; Yang, W. H. Synthesis and characterizations of new negatively charged organic-inorganic hybrid materials: effect of molecular weight of sol-gel precursor. *Journal of Solid State Chemistry.* 177 (2004) 1660-1666.

[111] Wu, C. M.; Xu, T. W.; Gong, M.; Yang, W. H. Synthesis and characterizations of new negatively charged organic-inorganic hybrid materials. Part II. Membrane preparation and characterizations. *J. Membr. Sci.* 247 (2005) 111-118.

[112] Wu, C. M.; Xu, T. W.; Yang W. H. Fundamental studies of a new hybrid (inorganic-organic) positively charged membrane: membrane preparation and characterizations. *J. Membr. Sci.* 216 (2003) 269-278.

[113] Xu, T. W.; Wu, C. M.; Yang, W. H. A method to prepare a hybrid positively charged membrane, Chinese Patent Number: ZL 03 1 32271.9.

[114] Liu, J. S.; Xu, T. W.; Gong, M.; Yu, F.; Fu, Y. X. Fundamental studies of novel inorganic-organic charged zwitterionic hybrids: 4. New hybrid zwitterionic membranes prepared from polyethylene glycol (PEG) and silane coupling agent. *J. Membr. Sci.* 283 (2006) 190-200.

[115] Zhang, S. L.; Wu, C. M.; Xu, T. W.; Gong, M.; Xu, X. L. Synthesis and characterizations of anion exchange organic-inorganic hybrid materials based on Poly(2,6-dimethyl-1,4-phenylene oxide) (PPO). *Journal of Solid State Chemistry.* 178 (2005) 2292-2300.

[116] Zhang, S. L.; Xu, T. W.; Wu, C. M. Preparation and characterizations of new anion exchange hybrid membranes based on Poly (2,6-dimethyl-1,4-phenylene oxide) (PPO). *J. Membr. Sci.*269 (2006) 142-151.

[117] Xu, T.W., Zhang S.L. A method to prepare a hybrid positively charged membrane. *Chinese Patent ZL.* 2004 1 0065737.8 .

[118] Huang, H. H.; Wilkes, G. L. Structure-property behavior of new hybrid materials incorporating oligomeric poly(tetramethylene oxide) with inorganic silicates by a sol-gel process .3. effect of oligomeric molecular-weight. *Polym. Bull.* 1987, 18, 455-462.

[119] Surivet, F.; Lam, T. M.; Pascaultj, J. P.; Phami, Q. T. Organic-Inorganic Hybrid Materials. 1. Hydrolysis and Condensation Mechanisms Involved in Alkoxysilane-Terminated Macromonomers. *Macromolecules.* 25 (1992) 4309-4320.

[120] Yano, S.; Iwata, K.; Kurita, K. Physical properties and structure of organic-inorganic hybrid materials produced by sol-gel process. *Mater. Sci. & Eng.* C 6 (1998) 75-90.

[121] Wu, C. M.; Xu, T. W.; Gong, M.; Yang, W. H. Multi-step sol-gel process and its effect on the morphology of polyethylene oxide (PEO)/SiO_2 anion-exchange hybrid materials. *Eur. Polym. J.* 43 (2007) 1573-1579.

[122] Wu, Y. H.; Wu, C. M.; Gong, M.; Xu, T. W. New anion exchanger organic-inorganic hybrid materials and membranes from a copolymer of glycidylmethacrylate and γ-methacryloxypropyl trimethoxy silane. *J. Appl. Polym. Sci.* 102 (2006) 3580-3589.

[123] Aparicio, M.; Castro, Y.; Duran, A. Synthesis and characterisation of proton conducting styrene-co-methacrylate-silica sol-gel membranes containing tungstophosphoric acid. *Solid State Ionics.* 176 (2005) 333-340.

[124] Kim, J. H.; Lee, Y. M. Gas permeation properties of poly(amide-6-b-ethylene oxide)-silica hybrid membranes. *J. Membr. Sci.* 193 (2001) 209-225.

[125] Guo, R. L.; Hu, C. L; Pan, F. S.; Wu, H.; Jiang, Z. Y. PVA-GPTMS/TEOS hybrid pervaporation membrane for dehydration of ethylene glycol aqueous solution. *J. Membr. Sci.* 281 (2006) 454-462.

[126] Peng, F. P.; Lu, L. Y.; Sun, H. L.; Wang, Y. Q.; Wu, H.; Jiang, Z. Y. Correlations between free volume characteristics and pervaporation permeability of novel PVA-GPTMS hybrid membranes. *J. Membr. Sci.* 275 (2006) 97-104.

[127] Zhang, Q. G.; Liu, Q. L.; Jiang, Z. Y.; Chen, Y. Anti-trade-off in dehydration of ethanol by novel PVA/APTEOS hybrid membranes. *J. Membr. Sci.* 287 (2007) 237-245.

[128] Chen, J. H.; Liu, Q. L.; Fang, J.; Zhu, A. M.; Zhang, Q. G. Composite hybrid membrane of chitosan-silica in pervaporation separation of MeOH/DMC mixtures. *J. Colloid Interface Sci.* 316 (2007) 580-588

[129] Hu, Q.; Marand, E.; Dhingra, S.; Fritsch, D. Poly(amide-imide)/TiO_2 nano-composite gas separation membranes: Fabrication and characterization. *J. Membr. Sci.* 135 (1997) 65-79.

[130] Zhong, S. H.; Li, C. F.; Xiao, X. F. Preparation and characterization of polyimide-silica hybrid membranes on kieselguhr-mullite supports. *J. Membr. Sci.* 199 (2002) 53-58.

[131] Silveira, K. F.; Yoshida, I. V. P.; Nunes, S. P. Phase separation in PMMA/silica sol-gel systems. *Polymer.* 36(7) (1995) 1425-1434.

[132] Zoppi, R. A.; Neves, S.; Nunes, S. P. Hybrid films of poly(ethylene oxide-*b*-amide-6) containing sol-gel silicon or titanium oxide as inorganic fillers: effect of morphology and mechanical properties on gas permeability. *Polymer.* 41 (2000) 5461-5470.

[133] Siuzdak, D. A.; Start, P. R.; Mauritz, K A. Surlyn®/Silicate hybrid materials. I. Polymer in situ sol-gel process and structure characterization. *J. Appl. Polym. Sci.* 77 (2000) 2832-2844.

[134] Apichatachutapan, W.; Moore, R. B.; Mauritz, K. A. Asymmetric Nafion/(zirconium oxide) hybrid membranes via in situ sol-gel chemistry. *J. Appl. Polym. Sci.* 62 (1996) 417-426.

[135] Fadeev, A. Y.; McCarthy, T. J. Self-Assembly is not the only reaction possible between alkyltrichlorosilane and surfaces: monomolecular and oligomeric covalently attached layers of dichloro- and trichlorosilanes on silicon. *Langmuir.* 16 (2000) 7268-7274.

[136] Singh, R. P.; Way, J. D.; Dec, S. F. Silane modified inorganic membranes: Effects of silane surface structure. *J. Membr. Sci.* 259 (2005) 34-46.

[137] Brzoska, J. B.; Azouz, I. B.; Rondelez, F. Silanization of solid substrates - a step toward reproducibility. *Langmuir*. 10 (1994) 4367-4373.

[138] Javaid, A.; Ford, D. M. Solubility-based gas separation with oligomer-modified inorganic membranes Part II. Mixed gas permeation of 5 nm alumina membranes modified with octadecyltrichlorosilane. *J. Membr. Sci.* 215 (2003) 157-168

[139] Javaid, A.; Hughey, M. P.; Varutbangkul, V.; Ford, D. M. Solubility-based gas separation with oligomer-modified inorganic membranes. *J. Membr. Sci.* 187 (2001) 141-150.

[140] R. S. Faibish, Y. Cohen. Fouling-resistant ceramic-supported polymer membranes for ultrafiltration of oil-in-water microemulsions. *J. Membr. Sci.* 185 (2001) 129-143.

[141] Okazaki, H. Ohya, S. I. Semenova, S. Kikuchi, M. Aihara, Y. Negishi, Nanotechnological method to control the molecular weight cut-off and/or pore diameter of organic-inorganic composite membrane. *J. Membr. Sci.* 141(1998) 65-74.

[142] Jang, M. Y.; Yamazaki, Y. Preparation and characterization of composite membranes composed of zirconium tricarboxybutylphosphonate and polybenzimidazole for intermediate temperature operation. *J. Power Sources.* 139 (2005) 2-8.

[143] T. T. Moore, W. J. Koros. Non-ideal effects in organic-inorganic materials for gas separation membranes. *J. Mol. Struct.* 739 (2005) 87-98.

[144] Bottino, A.; Capannelli, G.; D'Asti, V.; Piaggio, P. Preparation and properties of novel organic-inorganic porous membranes. *Sep. Purif. Technol.* 22-23 (2001) 269-275.

[145] Qrinn, R.; Laciak, D. V.; Pez, G. P. Polyelectrolyte-salt blend membranes for acid gas separations. *J. Membr. Sci.* 131 (1997) 61-69.

[146] Wu, C. S.; Liao H. T. Study on the preparation and characterization of biodegradable polylactide/multi-walled carbon nanotubes nanocomposites. *Polymer*. 48 (2007) 4449-4458.

[147] Avnir, D.; Braun, S.; Lev, O.; Ottolenghit, M. Enzymes and Other Proteins Entrapped in Sol-Gel Materials. *Chem. Mater.* 6 (1994) 1605-1614.

[148] Merkel, T. C.; Freeman, B. D.; Spontak, R. J.; He, Z.; Pinnau, I. Ultrapermeable, reverse-selective nanocomposite membranes. *Science*. 296 (2002) 519-522.

[149] He, Z.; Pinnau, I.; Morisato, A. Nanostructured poly(4-methyl-2-pentyne)/silica hybrid membranes for gas separation. *Desalination*. 146 (2002) 11-15.

[150] Merkel, T. C.; He, Z.; Pinnau, I.; Freeman, B. D.; Meakin, P.; Hill, A. J. Sorption and transport in poly (2,2-bis(trifluoromethyl)- 4,5-difluoro-1,3-dioxole-co-tetrafluoro-ethylene) containing nanoscale fumed silica. *Macromolecules*. 36 (2003) 8406-8414.

[151] T. W. Pechar, S. Kim, B. Vaughan, E. Marand, M. Tsapatsis, H. K. Jeong, C. J. Cornelius. Fabrication and characterization of polyimide-zeolite L mixed matrix membranes for gas separations. *J. Membr. Sci.* 277 (2006) 195-202.

[152] Khayet, M.; Villaluenga, J.P.G.; Valentin, J.L.; LÓpez-Manchado, M.A.; Mengual, J.I.; Seoane, B. Filled poly(2,6-dimethyl-1,4-phenylene oxide) dense membranes by silica and silane modified silica nanoparticles: characterization and application in pervaporation. *Polymer*. 46 (2005) 9881-9891

[153] Li, C. N.; Sun, G. Q.; Rena, S. Z.; Liu, J. Wang, Q.; Wu, Z. M.; Sun, H.; Jin, W. Casting Nafion-sulfonated organosilica nano-composite membranes used in direct methanol fuel cells. *J. Membr. Sci.* 272 (2006) 50-57.

[154] Shelekhin A. B.; Grosgogeat, E. J.; Hwang, S. T. Gas separation properties of a new polymer inorganic composite membrane. *J. Membr. Sci.* 66 (1992) 129-141.
[155] Stevens N. S. M. Formation of hybrid organic/inorganic composite membranes via partial pyrolysis of poly(dimethyl siloxane). *Chem. Eng. Sci.* 53 (1998) 1699-1711.
[156] Wróbel A. M.; Wickramanayaka S.; Nakanishi, Y.; Hatanaka, Y.; Pawlowski, S.; Olejniczak, W. Atomic hydrogen-induced chemical vapor deposition of a Si:C:H thin film materials form alkylsilane precursors. *Diamond and related materials.* 6 (1997) 1081-1091.
[157] Ida, J.; Matsuyama T.; Yamamoto, H. Surface modification of a ceramic membrane by the SPCP-CVD method suitable for enzyme immobilization. *J. Electrostatics.* 49 (2000) 71-82.
[158] Ida, J.; Matsuyama, T.; Yamamoto, H. Immobilization of glucoamylase on ceramic membrane surfaces modified with a new method of treatment utilizing SPCP-CVD. *Biochem. Eng. J.* 5 (2000) 179-184.
[159] H. Yasuda. Plasma polymerization for protective coatings and composite membranes. *J. Membr. Sci.* 1984 (18) 273-284.
[160] Nishimiya, K.; Haraguchi, T.; Ide, S.; Hatanaka, C.; Kajiyama, T.; Goto, M. Modification of solid surface by plasma-graft polymerization. Kitakyushu Kogyo Koto Senmon Gakko Kenkyu Hokoku 24 (1991) 139.
[161] Haraguchi, T.; Nishimiya, K.; Ide, S.; Hatanaka, C.; Isomura, K.; Nakashio, F.; Goto, M.; Kajiyama, T. Application of hollow fiber modified by plasma-graft polymerization to rare-earth metal extractor. Kitakyushu Kogyo Koto Senmon Gakko Kenkyu Hokoku 25 (1992) 97.
[162] Kai, T.; Yamaguchi, T.; Nakao, S. Preparation of organic/inorganic composite membranes by plasma-graft filling polymerization technique for organic-liquid separation. *Ind. Eng. Chem. Res.* 39 (2000) 3284-3290.
[163] Nehlsen, S.; Hunte, T.; Mtfller, J. Gas permeation properties of plasma polymerized thin film siloxane-type membranes for temperatures up to 350°C. *J. Membr. Sci.* 106 (1995) 1-7.
[164] Shi, D. L.; He, P.; Wang, S. X.; van Ooij, W. J.; Wang, L. M.; Zhao, J. G.; Yu, Z. Interfacial particle bonding via an ultrathin polymer film on Al2O3 nanoparticles by plasma polymerization. *J. Mater. Res.* 17 (5) (2002) 981-990.
[165] Lee, K. H.; Khang, S. J. High temperature membrane. *US patent.* 4,640,901 (1987)
[166] Leger, C.; Lira, H. D. L.; Paterson, R. Preparation and properties of surface modified ceramic membranes. Part II. Gas and liquid permeabilities of 5 nm alumina membranes modified by a monolayer of bound polydimethylsiloxane (PDMS) silicon oil. *J. Membr. Sci.* 120 (1996) 135.
[167] Castro, R. P.; Cohen, Y.; Monbouquette, H. G. Silica-supported polyvinylpyrrolidone filtration membranes. *J. Membr. Sci.* 115 (1996) 179-190.
[168] Castro, R. P.; Cohen, Y.; Monbouquette, H. G. The permeability behavior of polyvinylpyrrolidone- modified porous silica membranes. *J. Membr. Sci.* 84 (1993) 151-160.
[169] Singh, R. P.; Jha, P.; Kalpakci, K.; Way, J. D. Dual-Surface-Modified Reverse-Selective Membranes. *Ind. Eng. Chem. Res.* 46 (2007) 7246-7252.

[170] Randon, J.; Paterson, R. Preliminary studies on the potential for gas separation by mesoporous ceramic oxide membranes surface modified by alkyl phosphonic acids. *J. Membr. Sci.* 134 (1997) 219-223.

[171] Pal, R. Permeation models for mixed matrix membranes. *J. Colloid Interface Sci.* 317 (1) (2008) 191-198.

[172] Ballweg, T.; Wolter, H.; Storch, W.; Königsreuther, P. Poster Jahrestagung der Deutschen Keramischen Gesellschaft DKG, Freiberg/Sachsen, 5-7.10.1999, Deutsche Keramische Gesellschaft 1999.

[173] Guizard, C.; Julbe, A.; Larbot, A.; Cot, L.; Nanostructures in sol-gel derived materials: application to the elaboration of nanofiltration membranes. *J. Alloys Compd.* 188 (1992) 8-13.

[174] Koter, S.; Piotrowski, P.; Kerres, J. A. Comparative investigation of ion exchange membranes. *J. Membr. Sci.* 153 (1999) 83-90.

[175] Ma, H. M.; Hakim, L. F.; Bowman, C.N.; Davis R. H. Factors affecting membrane fouling reduction by surface modification and backpulsing. *J. Membr. Sci.* 189 (2) (2001) 255-270.

[176] Dai, Y.; Jian, X.; Zhang, S. H.; Guiver, M. D. Thermostable ultrafiltration and nanofiltration membranes from sulfonated poly(phthalazinone ether sulfone ketone). *J. Membr. Sci.* 188 (2001) 195-203.

[177] Xu, T. W.; Yang, W. H. A novel positively charged composite membranes for nanofiltration prepared from PPO by in situ amines crosslinking. *J. Membr. Sci.* 215 (2003) 25-32.

[178] Shin, D. H.; Kim, N. W.; Lee, Y. T. Surface modification of reverse osmosis membrane skin layer by silane compound. *Membr. J.* 16 (2006) 106.

[179] Jeong, B. H.; Hoek, E. M. V.; Yan, Y. S.; Subramani, A.; Huang, X. F.; Hurwitz, G.; Ghosh, A. K.; Jawor, A. Interfacial polymerization of thin film nanocomposites: A new concept for reverse osmosis membranes. *J. Membr. Sci.* 294 (1-2) (2007) 1-7.

[180] Walcarius, A. Electrochemical Applications of Silica-Based Organic-Inorganic Hybrid Materials. *Chem. Mater.* 13 (2001) 3351-3372.

[181] Lev, O.; Wu, Z.; Bharathi, S.; Glezer, V.; Modestov, A.; Gun, J.; Rabinovich, L.; Sampath, S. Sol-gel materials in electrochemistry. *Chem. Mater.* 9 (1997) 2354-2375.

[182] Walcarius, A. Analytical applications of silica-modified electrodes - A comprehensive review. *Electroanalysis.* 10 (1998) 1217-1235.

[183] Walcarius, A. Electroanalysis with pure, chemically modified, and sol-gel-derived silica-based materials. *Electroanalysis.* 13 (2001) 701-708.

[184] Wang, J. Sol-gel materials for electrochemical biosensors. *Anal. Chim. Acta.* 399 (1999) 21-27.

[185] Barroso-Fernandez1, B.; Lee-Alvarez, M. T.; Seliskar, C. J.; Heineman, W. R. Electrochemical behavior of methyl viologen at graphite electrodes modified with Nafion@ sol-gel composite. *Anal. Chim. Acta.* 370 (1998) 221-230.

[186] Hu, Z.; Seliskar, C. J.; Heineman, W. R. Voltammetry of [Re(DMPE)(3)](+) at ionomer-entrapped composite-modified electrodes. *Anal. Chem.* 70 (1998) 5230-5236.

[187] Mauritz, K. A. Organic-inorganic hybrid materials: perfluorinated ionomers as sol-gel polymerization templates for inorganic alkoxides. *Mater. Sci. & Eng.* C 6 (1998) 121-133.

[188] Zoppi, R. A.; Nunes, S. P. Electrochemical impedance studies of hybrids of perfluorosulfonic acid ionomer and silicon oxide by sol-gel reaction from solution. *J. Electroanal. Chem.* 445 (1998) 39-45.

[189] Hu, Z.; Heineman, W. R. Oxidation state speciation of [Re-I(DMPE)3](+)/[Re-II(DMPE)$_3$](2+) by voltammetry with a chemically modified microelectrode. *Anal. Chem.* 72 (2000) 2395-2400.

[190] Petit-Dominguez, M. D.; Shen, H.; Heineman, W. R.; Seliskar, C. Electrochemical behavior of graphite electrodes modified by spin-coating with sol-gel entrapped ionomers. *J. Anal. Chem.* 69 (1997) 703-710.

[191] Scrosati, B.; Neat, R. J. In Applications of electroactive polymers; Scrosati, B., Ed.; Chapman and Hall: London, 1993.

[192] Gray, F. M. Solid Polymer Electrolytes, Fundamentals and Technological Applications; VCH Publishers: New York, 1991.

[193] Jung, D.H.; Cho, S.Y.; Peck, D.H.; Shin, D.R.; Kim, J.S. Performance evaluation of a Nafion/silicon oxide hybrid membrane for direct methanol fuel cell. *J. Power Sources.* 106 (2002) 173-177.

[194] Honma, I.; Nomura, S.; Nakajim. H. Protonic conducting organic/inorganic nanocomposites for polymer electrolyte membrane. *J. Membr. Sci.* 185 (2001) 83-94.

[195] Gautier-Luneau, I.; Denoyelle, A.; Sanchez, J. Y.; Poinsignon, C. Organic-inorganic protonic polymer electrolytes as membrane for low-temperature fuel cell. *Electrochim. Acta.* 37 (9) (1992) 1615-1618.

[196] Sanchez, J. Y.; Denoyelle, A.; Poinsignon, C. Poly(benzylsulfonic acid)siloxane as proton-conducting electrolyte. *Polym. Adv.Technol.* 4 (1993) 99-105.

[197] Shi, Y. N.; Seliskar, C. J.; Heineman, W. R. Spectroelectrochemical Sensing Based on Multimode Selectivity Simultaneously Achievable in a Single Device. 2. Demonstration of Selectivity in the Presence of Direct Interferences. *Anal. Chem.* 69(23) (1997) 4819-4827.

[198] Shi, Y. N.; Seliskar, C. J. Optically Transparent Polyelectrolyte-Silica Composite Materials: Preparation, Characterization, and Application in Optical Chemical Sensing. *Chem. Mater.* 9 (3) (1997) 821-829.

[199] Shi, Y. N. Sol-gel derived polyelectrolyte-silica composite materials: preparation, characterization and applications in spectroscopic and spectroelectrochemical sensing. Doctororal dissertation. College of Arts and Sciences. 1998

[200] Honma, I.; Hirakawa, S.; Yamada, K.; Bae, J.M. Synthesis of organic / inorganic nanocomposites protonic conducting membrane through sol-gel processes. *Solid State Ionics.* 118 (1999) 29-36.

[201] Mauritz, K.A. Nanophase separated ionomers as sol-gel polymerization templates for inorganic alkoxides: Applications as proton conducting membranes, Conference on electroactive membranes: From fundamentals to applications, Monte Verita, Switzerland, 1999.

[202] Mehendale, N. C.; Bezemer, C.; Walree, C. A.; Gebbink, R. J. M. K.; Koten, G. Novel silica immobilized NCN-pincer palladium(II) and platinum(II) complexes: Application as Lewis acid catalysts. *J. Mol. Catal.* A 257 (2006) 167-175.

[203] Zhong, S. H.; Sun, H. W.; Wang, X. T.; Shao, H. Q.; Guo, J. B. Preparation of hybrid membrane reactors and their application for partial oxidation of propane with CO_2 to propylene. *J. Membr. Sci.* 278 (2006) 212-218.

In: Advances in Membrane Science and Technology
Editor: Tongwen Xu
ISBN: 978-1-60692-883-7

Chapter V

Pervaporation Membranes for Organic Separation

Xiangyi Qiao, Lan Ying Jiang, Tai-Shung Chung[*] and Ruixue Liu

National University of Singapore, Department of Chemical & Biomolecular Engineering, 10 Kent Ridge Crescent, Singapore 119260, Singapore

Abstract

Pervaporation is a promising and unique membrane separation process for liquid separation, especially for azeotropic mixtures or close boiling point mixtures that are difficult to be separated by conventional distillation or adsorption. In this chapter, an overview on pervaporation membrane separation process was briefly summarized. Fundamentals of pervaporation process were included but emphasis was given to solution-diffusion model. The discrepancy between using permeance & selectivity and flux & separation factor for analyzing membrane separation performance versus operating conditions was discussed. In order to elucidate the basic criteria for materials selection for pervaporation membranes, efforts were given to correlate with important factors that governing permeant transportation through pervaporation membranes such as interactions between permeant & permeant, interactions between permeant & membrane, and kinetic aspects. The structure engineering section illustrated major considerations involved in developing pervaporation membranes for a specific application. Examples were highlighted to recent progress of pervaporation membranes in materials, membrane morphology and membrane separation properties in three major application areas; namely, organic dehydration, separation of organic from dilute aqueous solution, and organic-organic separation. The organic solvent dehydration by pervaporation has already been well developed industrially. Organophilic membranes have shown significant potential to recover biofuels from fermentation broths; however, its large scale application relies on improved energy efficiency, reduction of membrane cost and improvement in membrane performance. Organic-organic separation by pervaporation

[*] Correspondence concerning this chapter should be addressed to Tai-Shung Chung at chencts@nus.edu.sg

has the broadest potential application in chemical industry; but it remains the most challenging research area which needs breakthroughs in new membrane material development and in-depth understanding on each specific separation process.

Keyword: Pervaporation, Solution-diffusion model, Membrane structure, Hydrophilic membranes, Organophilic membranes, Organoselective membranes

5.1. Introduction

Pervaporation distinctively differentiates itself from other membrane separation processes with a phase change across the membrane. In pervaporation, the liquid mixture to be separated is placed in contact with one side of a membrane and the permeated product is removed as a low-pressure vapor from the other side; the components partially evaporate inside and selectively transport through a homogeneous, nonporous membrane, which leads to separation. Based on the different approaches to achieve the partial pressure difference across the membrane, pervaporation can be classified into vacuum pervaporation, sweep gas pervaporation and thermopervapaoration [1, 2].

Pervaporation separation has demonstrated incomparable advantages in separation of azeotropic, close boiling and heat sensitive mixtures. As a comparison, distillation utilizes the composition difference between liquid and vapor phases to achieve separation, and no adequate separation can be performed if the components in the mixture have close-boiling points or have the same composition in liquid and vapor phase (i.e. azeotropes) [3]. Alternatively, pervaporation provides an effective way to separate close-boiling point liquids and azeotropic mixtures; this is because it is not limited by relative volatility of the components. Baker et al. [1] illustrated a typical pervaporation application in ethanol dehydration where water was preferentially transported through a polyvinyl alcohol (PVA) pervaporation membrane especially at the azeotropic concentration range and therefore pure ethanol was produced.

Pervaporation is ranked the third highest among the 31 techniques under evaluation as a potential energy saving process in fluid separation techniques [4]. It is well known that phase change from liquid to vapor in pervaporation is energy-intensive. However, pervaporation survives this challenge by two properties. Firstly, pervaporation purposely allows the minor components pass through the membranes; this is because only the component passing through the membrane consumes the latent heat involved in the phase change [5]. Another feature is that pervaporation utilizes the most selective membranes to minimize the energy waste on the components which are supposed to be maintained in the retentate. As a consequence, the pervaporation process is more economical in removal of small amount of impurities from a bulk solution. The best application of pervaporation is to combine it with a distillation system to replace the energy-intensive azeotropic distillation or the costly molecular sieve adsorption. In other words, distillation is employed to purify alcohol at high water content, while pervaporation is utilized to remove the rest of water at around or above azeotropic point. As Pearce showed in his report [6], the overall production cost for 99wt% purity ethanol was reduced by more than 30% with an energy saving of 60% for a hybrid distillation and pervaporation system compared to conventional distillation technology. As shown by numerous examples, the pervaporation process can also be considered as one of so-called 'clean technologies', especially well suited for the treatment and recycling of volatile organic compounds and pollution prevention. This is because no additional species are introduced to the feed stream and it can be operated at ambient temperature.

Among the many aspects influencing the pervaporation efficiency, the membrane is the key element. This review analyzes the recent progress of pervaporation membranes. Aspects such as fundamental transport mechanisms, important factors governing permeant transportation, materials selection and membrane formation, and membrane separation properties will be covered.

5.2. Fundamentals of Pervaporation Separation Process

5.2.1. Solution-Diffusion Model

Two basic models have been developed to describe the mass transport in pervaporation; namely, the solution-diffusion model and the pore flow model. The solution-diffusion model has been widely adopted by most of pervaporation membrane researchers due to the good agreement between theory and experiments [2, 7]. The mass transfer in pervaporation membrane is described as a three-step process: (i) the permeant is dissolved in the feed side of the membrane; (ii) the permeant diffuses though the membrane; and (iii) the permeant evaporates as vapor at the downstream side of the membrane. Figure 5.1 illustrates the solution-diffusion model. For a pervaporation process, the transportation rates of molecules from the bulk feed liquid mixture to the membrane surface and the removal of vapors at the downstream side also play important roles to the overall mass transport. The accumulation of less permeable substances in the feed side lead to concentration polarization, while the inefficient removal of permeant at the downstream influence the productivity.

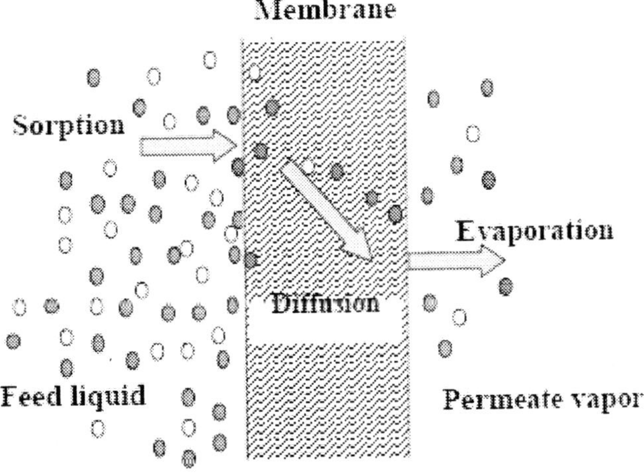

Figure 5.1. Schematic diagram of solution-diffusion model [2].

The solution-diffusion model is applicable to non-porous polymeric membranes in which the transport of permeating molecules relies on the thermally agitated motion of polymer chain segments. Based on the solution-diffusion model, the permeability coefficient P is given by the product of diffusion coefficient D and sorption coefficient S:

$$P = D\,S \tag{5.1}$$

For a binary pervaporation system consisting of components A and B, the diffusion selectivity (D_A/D_B) depends greatly on (1) the penetrant size and shape, (2) the mobility of polymer chains, (3) the interstitial space between polymer chains, and (4) the interactions between penetrant and penetrant plus interactions between penetrant and membrane material [8]. The sorption selectivity (S_A/S_B) prefers more condensable molecules or molecules which have special interaction with membrane materials [9].

The driving force for mass transport through a pervaporation membrane is the chemical potential gradient, i.e. partial pressure gradient (fugacity). The transport equation based on solution-diffusion mechanism for pervaporation can be derived as follows [7, 8]. The flux J_i of one component is proportional to its driving force, i.e. the chemical potential gradient across the membrane ($d\mu_i/dl$)

$$J_i = -L_i \frac{d\mu_i}{dl} \tag{5.2}$$

where L_i is a proportionality coefficient. Under isothermal conditions (constant temperature, T), the chemical potential (μ_i) in pressure and concentration driven processes is

$$\mu_i = \mu_i^0 + RT\ln(c_i) + v_i(p - p_i^0) \tag{5.3}$$

where c_i is molar concentration of a component, v_i is the molar volume, p is pressure, and superscript 0 refers to standard state of a component. The chemical potential gradient $d\mu_i$ is then given as

$$d\mu_i = RTd\ln(c_i) + v_i dp \tag{5.4}$$

The pressure within the membrane is assumed to be constant in the solution-diffusion model; therefore, combining Equations (5.2) and (5.4) gives

$$J_i = -\frac{RTL_i}{c_i}\frac{dc_i}{dl} \tag{5.5}$$

If RTL_i/c_i is replaced by diffusion coefficient D_i, the integration of Equation (5.5) across the membrane provides

$$J_i = \frac{D_i(c^f_{i(m)} - c^p_{i(m)})}{l} \tag{5.6}$$

where l is the membrane thickness, $c_{i(m)}$ represents the concentration of component i inside the membrane, and the superscripts f and p represent the feed and permeate side, respectively.

By assuming the chemical potential equilibrium at the liquid mixture/membrane feed interface and a hypothetical vapor state in equilibrium with the liquid mixture, one can obtain

$$c^f_{i(m)} = \frac{\gamma^f_{i.G}}{\gamma^f_{i(m)} \cdot p^s_i} p^f_i = S_i \cdot p^f_i \tag{5.7}$$

where γ is the activity coefficient, superscript s indicates the saturated state, subscripts G represents the hypothetical vapor phase, and S_i is defined as the sorption coefficient in a gas phase. Similarly, the equilibrium at the permeate gas and permeate membrane surface gives

$$c^p_{i(m)} = \frac{\gamma^p_{i.G}}{\gamma^p_{i(m)} \cdot p^s_i} p^p_i = S_i \cdot p^p_i \tag{5.8}$$

Finally, Equation (5.6) can be rewritten as follows:

$$J_i = \frac{D_i S_i (p^f_i - p^p_i)}{l} = \frac{P_i(p^f_i - p^p_i)}{l} \tag{5.9}$$

where P_i, the membrane permeability as a product of diffusion coefficient (D_i) and sorption coefficient (S_i), can be obtained from the above equation.

Conventionally, D_i and S_i are considered to be constant. In pervaporation, the membrane may often be seriously swollen due to the much complicated and strong physicochemical interaction between the highly condensable permeant molecules and the membrane material. This would lead to the changes of sorption characteristics and diffusion properties of the permeants in the membrane [10]. Different from air separation membranes where the selectivity obtained from mixed gases is not much different from the pure gas tests, the selectivity of pervaporation for a specific mixture is not only far lower than the permeability ratio of the pure components but also varies as a function of the feed composition.

5.2.2. Performance Parameters: Flux and Separation Factor

The performance of a pervaporation membrane is typically characterized by flux (J) and separation factor (α), as defined by the following equations:

$$J = \frac{Q}{At} \tag{5.10}$$

$$\alpha_{2/1} = \frac{y_2 / y_1}{x_2 / x_1} \tag{5.11}$$

where, Q is the total mass transferred over time t, A is the membrane area, x_2 and y_2 are the mole fractions of one component in the feed and permeate, respectively, and x_1 and y_1 are the mole fractions of the other component in the feed and permeate, respectively. Flux is obtained from the amount of permeant collected from a laboratory setup at a certain time interval divided by membrane area, as defined in Equation (5.10); while separation factor is defined as the concentration ratio of two components in a binary system, as defined in Equation (5.11). For very dilute feed solutions, an enrichment factor is often used to represent membrane selectivity which is the ratio of concentrations of the preferentially permeating component in the permeate and in the feed, respectively [11].

Because of the existence of a trade-off relationship between flux and separation factor, that is, the flux and separation factor usually perform in the opposite way, Huang and his coworkers [12] introduced pervaporation separation index (PSI) to evaluate the overall performance of a membrane. PSI was originally defined as a product of permeation flux and separation factor:

$$PSI = J_t \cdot \alpha \tag{5.12}$$

where J_t is the total permeation flux. However, in this definition, the PSI can be large if the membrane has a high flux even when α is equal to 1. Therefore, the definition of PSI was later modified as a product of J_t and $(\alpha - 1)$ [13].

Flux and separation factor are direct performance data which are easy to use and compare. However, from the standpoint of investigating intrinsic membrane properties, these two parameters are not the pertinent guidelines for membrane materials comparison and membrane development. This is due to the fact that these two parameters combine both membrane properties and operating conditions, thus one cannot discern the true and individual effects of operating conditions and membrane's intrinsic properties on system performance. Therefore, this intermingled effect creates difficulties for membrane scientists to compare membrane materials for pervaporation.

5.2.3. Performance Parameters: Permeance and Selectivity

Wijmans and Baker [14] were the pioneers proposing the use of permeance and selectivity instead of flux and separation factor to investigate pervaporation membrane performance and properties. Recently, Wijmans [15] re-emphasized the importance of using permeance and selectivity, while Guo et al. [16] and Qiao et al. [17] elaborated their differences and importance for performance interpretation by detailed examples on the comparison of flux vs. permeance and separation factor vs. selectivity.

Equation (5.9) gives the relationship between flux, permeability, and driving force of vapor pressure. The partial vapor pressure (fugacity) of each component on the feed side in this equation can be calculated based on its concentration in the feed liquid mixture as follows:

$$p_i^f = x_i \gamma_i p_i^s \tag{5.13}$$

Saturated vapor pressure p_i^s and activity coefficient γ_i can be obtained from the vapor-liquid equilibrium data with the aid of the HYSYS DISTIL software (version 5.0, provided by Hyprotech Ltd, Canada) or the Antoine equation and Wilson equation, respectively. By substituting Equation (5.13) to Equation (5.9), one can obtain permeance ($\overline{P_i}$) based on the following equation

$$\overline{P_i} = P_i / l = J_i / (x_i \gamma_i p_i^s - p_i^p) \tag{5.14}$$

Permeance is more convenient for an asymmetric/composite membrane where the dense selective layer thickness is not readily available, while permeability (*P*) is usually related to a dense membrane. Both are direct indicators of the intrinsic properties of a membrane, and can be determined directly from experiments with the help of the above equations. The membrane selectivity is defined as the ratio of permeability or permeance of two permeating components.

Through investigating the dehydration of aqueous butanol mixtures across PVA membranes, Guo et al. [16] found that using permeance and selectivity could clarify and quantify the contribution by the nature of the membrane to the separation performance. For example, for the dehydration of aqueous butanol mixtures, water flux increased rapidly with increasing feed temperature; however, their corresponding water permeance showed an opposite trend. Traditionally, the increased water flux at a higher temperature was explained by the increased thermal motion of polymer chains and the expansion of the free volume. However, the declining trend of permeance on temperature revealed that the driving force also played an important role. In addition, the negative temperature effect on sorption should also be accounted.

Furthermore, Qiao et al. [17] investigated the dehydration of isopropanol and butanol. In contrast to the results obtained from water flux, a comparison of water permeance from different alcohol systems revealed that the mass transport of water inside the membrane was actually not strongly affected by the different alcohols. Similarly, based on permeance, the important physicochemical properties which influenced alcohol transport were mainly attributed to molecular linearity and their affinity to water and PVA as reflected by solubility and polarity parameters. In addition, the investigation on the effect of feed water content on separation factor change may mislead the analysis of water influence on membrane performance and exaggerate the plasticization phenomenon. The insight information would be very hard to be discovered if one only examines membrane performance by flux and separation factor.

To give an example, the following equation describes the relation between separation factor (α) and selectivity (β) for water to isopropanol in the dehydration of an aqueous isopropanol system when the permeate side pressure p^p is very small and negligible [18]:

$$\alpha_{2/1} = \frac{J_2/J_1}{x_2/x_1} = \frac{x_1}{x_2}\frac{\overline{P}_2(\gamma_2 x_2 p_2^s - y_2 p^p)}{\overline{P}_1(\gamma_1 x_1 p_1^s - y_1 p^p)} = \frac{x_1}{x_2}\frac{\overline{P}_2\gamma_2 x_2 p_2^s}{\overline{P}_1\gamma_1 x_1 p_1^s} = \frac{\overline{P}_2\gamma_2 p_2^s}{\overline{P}_1\gamma_1 p_1^s} = \beta_{2/1}\frac{\gamma_2 p_2^s}{\gamma_1 p_1^s} \tag{5.15}$$

Where β is the mole-based membrane selectivity, J are the molar flux, x and y are the mole fractions at the feed and permeate side, respectively, $\overline{P}$ is the permeance based on mole, and subscripts 1 and 2 represent isopropanol and water, respectively. β, the mole-based membrane selectivity, is defined as

$$\beta_{2/1} = \overline{P}_2 / \overline{P}_1 \tag{5.16}$$

Obviously, the mole-based selectivity and separation factor are related by a term of $\gamma_2 P_2^s / \gamma_1 P_1^s$. Table 5.1 lists the calculation of the ratio $\gamma_2 P_2^s / \gamma_1 P_1^s$ for a solution containing 85wt% (63mol %) isopropanol in water. The data in this table illustrates that the ratio of $\gamma_2 P_2^s / \gamma_1 P_1^s$ decreases with increasing temperature; therefore the difference between membrane selectivity and separation factor increases with increasing temperature according to Equation (5.15). Furthermore, since γ_2 / γ_1 is a concentration dependent term, the membrane selectivity as defined in Equation (5.16) shows less dependence on feed water content compared to the separation factor.

Table 5.1. The calculations of $\gamma_2 P_2^s / \gamma_1 P_1^s$ [18]

T (°C)	γ_1	γ_2	P_1^s (kPa)	P_2^s (kPa)	$\gamma_1 P_1^s$ (kPa)	$\gamma_2 P_2^s$ (kPa)	$\gamma_2 P_2^s / \gamma_1 P_1^s$
100	1.1448	1.8802	141.98	70.565	162.538	132.676	0.816
80	1.1618	1.9436	67.708	34.109	78.663	66.295	0.843
60	1.1813	2.0106	28.748	14.849	33.96	29.855	0.879

5.3. Factors Influencing Pervaporation Membrane Performance

It is self-evident that membrane separation performance depends on the membrane materials, the structure of the membrane, and the interactions between permeant-permeant and permeant-membrane. The factors involving for understanding the pervaporation separation mechanism, i.e. interaction between permeant and membrane, interaction between permeant and permeant, and kinetic aspects of permeant transportation, are critical for material screening and structure optimization of pervaporation membranes.

5.3.1. Interaction between Permeant and Membrane

As aforementioned, membrane performance is mathematically determined by selective sorption and/or selective diffusion. However, these two theoretical parameters are actually not constant during the separation process and may vary significantly with physicochemical properties of membranes and penetrants. The most important factor which determines the selective sorption may be the interaction between penetrant and membrane. This interaction can be qualitatively described and analyzed from three aspects; namely, solubility parameter approach, polarity parameter approach, and Flory-Huggins interaction parameter approach.

Solubility parameter represents the nature and magnitude of the interaction force between different molecules. It is frequently used to qualitatively describe the affinity between solvent and polymer, the interactions between polymer and permeant, and permeant and permeant [10, 19, 20]. If a solvent has a solubility parameter close to that of a polymer, it is regarded as a good solvent for the polymer. The solubility parameter is calculated by Equation (5.17)

$$\delta_{sp}^2 = \delta_d^2 + \delta_p^2 + \delta_h^2 \tag{5.17}$$

where δ_d, δ_p, and δ_h represent dispersion force, polar force and hydrogen bonding component of the solubility parameter, respectively. The solubility parameters of solvents are available in literature [21], while the solubility parameters of polymers can be estimated by group contribution methods [20].

The Flory-Huggins interaction parameter describes the sorption of a penetrant in a polymer material or the thermodynamic interaction between a solvent and a polymer [22]. Although the assumption that the interaction parameter χ is a constant over the entire activity range remains questionable, it can be determined in a convenient way by a single sorption experiment in a pure liquid according to the following equation [23]:

$$\chi = -\frac{\ln\phi + (1-\phi)}{(1-\phi)^2} \tag{5.18}$$

where ϕ is the penetrant volume fraction in the membrane. A lower interaction parameter indicates a higher affinity between polymer and penetrant.

Polarities of the membrane and penetrants also have strong relation to permeant transport. Thereby polarity parameters (E_T) are used to describe polymer/liquid interactions as well. Yoshikawa et al. [24] investigated effects of polymer polarity on pervaporation separation performance for ethanol/water mixtures. When the polarity parameter of the membrane deviated from that of water, the selectivity of water decreased. The polarity parameter is particularly useful when a polymer structure is not available. Jonquieres et al. [25] showed a linear relationship between sorption and liquid solubility parameters. Later they found a good correlation of flux and polarity parameter of alcohols for the pervaporation of alcohol/ETBE mixtures [26]. For a given solvent, the definition of polarity parameters $E_T(30)$ is given as the transition energy for the longest wavelength visible absorption band of the pyridinium-N-phenoxide betaine called Reichardt's dye. The polarity parameters for more than 270 organic solvents are collected by Reichardt [27].

However, the real affinity and complicated interaction between the feed component and the membrane may be beyond the prediction by means of solubility parameter, polarity parameter and Flory-Huggins interaction parameter because of the evolution of membrane structure with time during the tests. For example, highly hydrophilic materials such as polyvinyl alcohol (PVA), polyacrylic acid (PAA), agarose, alginate and chitosan may swell excessively and exhibit poor selectivity in aqueous organic mixtures due to the strong interaction between the feed water and the materials. Therefore, one needs to maintain a proper balance of hydrophilicity and hydrophobicity for the development of pervaporation dehydration membranes. Another example is that the physical structure of polyimide membranes may be influenced by operating conditions with the sorption of water and organic solvents, and thus the pervaporation performance is affected. Guo and Chung [28] demonstrated that the Matrimid® polyimide membranes had performance change after a thermal cycle process, due to the interactions between feed molecules and membrane, the plasticization of membrane and the non-equilibrium nature of the dense-selective skin in an asymmetric membrane. It is therefore important to identify a proper conditioning procedure to maintain an optimum and stable performance.

5.3.2. Interactions between Permeant and Permeant

Not only the interaction between permeant and membranes affects pervaporation performance, the interaction between permeant and permeant also influences their transport properties. This is so-called the "coupling effect" which has often been observed in the mass transfer through pervaporation membranes. Thermodynamically, the interactions between permeant molecules always exist. For example, the permeate molecules could form clusters in the hydrophilic environment in a membrane whereby induced thermodynamic coupling [29, 30]. On the other hand, the mutual drag between the diffusion of permeant molecules induces kinetic coupling. According to Drioli et al. [31], the dissolved permeant component may cause change in polymer free volume and membrane morphology, and thus result in facilitated or inhibited transport of other component through the membrane. Schaetzel et al. [32] investigated dehydration of aqueous ethanol solution through a PVA-PAA-comaleic acid membrane and suggested that coupling effect for the flux of ethanol (the slow permeant) was more significant than that for the flux of water (the fast permeant). Qiao et al. [17] observed a strong coupling between isopropanol and water through a PVA membrane and illustrated that the degree of coupling not only depended on the membrane properties, such as the degree of cross-linking, affinity to water and structure responses on temperature rise, but also relied on the physicochemical properties of penetrants, such as molecular linearity (or the aspect ratio) of penetrant molecules, and their solubility parameters and polarity parameters.

5.3.3. Kinetic Aspects

According to the solution-diffusion mechanism, high selectivity can be obtained if one component has strong affinity to the membrane and can diffuse faster in the membrane than other components. From the kinetic aspects, the movements of molecules from the upstream to the downstream side depend on the interstitial space created by the thermally agitated motion of polymer chains to allow molecule jump from one site to another site [2, 33]. As a result, the polymer chain stiffness and packing, which are determined by the monomer composition, chain configuration and functional groups, also affect the performance of a pervaporation membrane. The chain stiffness tends to hinder molecule diffusion and lowers permeability; while polymer chains consists of a larger interstitial space and a bigger free volume tends to have a higher permeation rate. Polyimide and polyamide are typical examples which show high selectivity but low permeability because of their rigid chain backbones and low free volumes. For this kind of materials, a high diffusion selectivity of water over other organic solvents can be achieved by tailoring the packing density of polymer chains. Peng et al. [34, 35, 36] demonstrated that the fractional free volume of PVA membrane can be increased and the free volume cavity size can be tuned by the incorporation of graphite and carbon nanotube into the PVA matrix.

The thermodynamic and kinetic influences can be potentially elucidated with the aid of molecular simulation and modeling. The molecular simulation makes it possible to predict the membrane performance of a designed material for a specific application and therefore reduce the tedious experimental testing during membrane development. The molecular dynamic simulation can be carried out by Materials Studio (Discover and Amorphous Cell modulus) or other software. For example, Peng et al. [36] employed Materials Studio to predict the fractional free volume change of the PVA membranes after various modifications. They also illustrated that the simulation results provided good explanations for the observation from experiments. Adoor et al. [37] observed greater affinity between silicalite-1 particles and the sodium alginate (NaAlg) polymer matrix than that between the particles with the PVA polymer matrix. This observation was later confirmed by the greater interaction energy between NaAlg and zeolite calculated using a COMPASS (condensed-phase optimized molecular potentials for atomistic simulation studies) force field. A more rigorous and detailed thermodynamic model using the UNIFAQ-FV group contribution calculation was also proposed to study polymer blend formulations and their membrane performance [38]. Nevertheless, there still lacks a universal model that can quantitatively predict the actual membrane performance for pervaporation because of the complicated thermodynamic interactions and sophisticated mass transports in pervaporation membranes.

5.4. Structure Engineering of Pervaporation Membranes

The membrane in a pervaporation process has to differentiate liquid molecules in angstrom scale (Å). For example, the difference in kinetic diameters between a water molecule (3.0Å) and an isopropanol molecule (4.7Å) is only 1.7Å. Although the difference is much bigger than the size difference between O_2 and N_2 molecules (kinetic diameter: 3.46 vs. 3.64Å) in gas separation, it is still necessary to use a dense membrane to achieve the separation. In general, in order to develop a suitable membrane for pervaporation, one should screen and choose appropriate materials to construct the crucial dense-selective layer which not only have superior separation performance (i.e., high flux and separation factor), but also possess good mechanical strength, chemical resistance, thermal stability, and minimal solvent-induced swelling [39, 40]. Koros et al. [41] illustrated the fundamental factors controlling material selection for polymeric membranes used for solution-diffusion based permeation separations. An entire hierarchy including four structure levels was summarized. That is, the chemical composition of the polymer which construct the selective layer, the steric relationship of polymer chains in the selective skin, the morphology of the selective skin and the structure integrity between skin and porous support. Because operation at high temperatures was unavoidable in pervaporation, the durability of the membrane material and the sealing material of modules also presented great challenges [41, 42].

Usually, dense membranes are prepared to obtain the intrinsic sorption and transport property of a membrane material. If the separation performance generated from dense membranes meets the requirements, asymmetric or composite membranes made of the same polymeric material will be developed. The development of asymmetric or composite membranes consisting of an ultra-thin dense selective layer represents the real interest in industrial applications. However, it is not an easy task because the resultant membranes may perform differently from its dense films as well as from each other if the fabrication parameters are changed.

A composite membrane consists of a selective layer and a substrate made from different materials. Its flux (or permeance), separation factor (or selectivity) and membrane mechanical strength may be a combination of different layers [9]. Composite membranes are typically prepared by first casting the microporous support, followed by deposition, coating or in situ polymerization of the selective dense layer on top of the support. Ideally, composite membranes are capable to combine the properties of different materials and offer a higher permeation rate than dense films due to the reduced thickness of the selective layer. However, the resistance and the separation properties of the sublayer may interfere with the overall membrane performance or even dominate the separation process [43]. Similar to gas separation membranes, a resistance model has been derived for pervaporation membranes to describe the relationship between the overall permeation resistance and the resistances from skin and sublayer. A model analogous to an electric flow in a circuit is shown in Figure 5.2. The overall permeation resistance can thus be determined by the following equation [2]

$$R_t = R_1 + R_2 R_3 /(R_2 + R_3) \tag{5.19}$$

where R_1 represents the resistance of the skin layer, R_2 and R_3 are the resistances of the pores and the polymer matrix in the porous substrate, respectively. Huang and Feng illustrated that the selectivity achievable in an asymmetric pervaporation membrane was determined by the resistances in both skin layer and substrate [44]. Liu et al. [40] also showed that the performance enhancement of dual-layer P84 co-polyimide/polyethersulfone (PES) hollow fibers upon heat treatment was restricted by the drastical increase in the resistance of the PES inner layer.

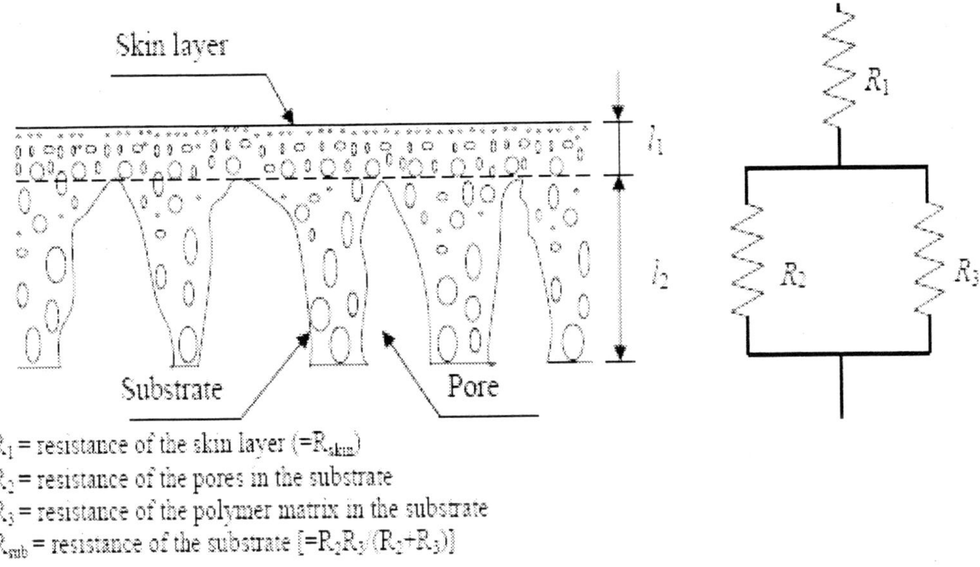

Figure 5.2. The resistance model of mass transport through an asymmetric membrane [2].

The integrity of composite membranes is also of great concern when the top layer and the substrate exhibit different degrees of swelling. It may thus produce a big stress at the interface which consequently causes instability and delaminates the membrane structure [5]. Traditionally, this problem was solved by crosslinking the top layer, inserting a gutter layer between the skin layer and the substrate, and in-situ polymerization. Dual-layer hollow fibers made by co-extrusion of two polymeric solutions provide the opportunity to form a good adhesion among inner and outer layers by dry-jet wet spinning [45, 46, 47]. Li et al. [46] demonstrated conceptually that delamination-free dual-layer hollow fibers could be produced by varying outer- and inner-layer spinning dope composition, bore fluid composition as well as post-treatment conditions. In the air gap region, molecular diffusion may take place at the interface and create an interlocking structure at the interface [48]. The inter-layer molecular diffusion can be further enhanced with an indent spinneret design [49]. For pervaporation, Liu et al. [40] were the pioneer developing BTDA-TDI/MDI co-polyimide (P84)/PES dual-layer hollow fibers with good integrity for the dehydration of isopropanol/water mixtures, while Wang et al. [50] were the pioneer developing polybenzimidazole (PBI)/P84 dual-layer hollow fibers for the dehydration of tetrafluoropropanol (TFP). The interface can be further improved if the polymeric materials in the inner and outer layers have strong interactions. Novel PBI/polyetherimide (PEI) dual-layer hollow fibers have been developed for the dehydration of isopropanol [51]. Because of hydrogen bonding between PBI and PEI, the dual layer hollow fibers exhibited excellent integration at the interface. Meanwhile, the integrity of outer and inner layers was found beneficial to the overall performance.

The traditional single-layer asymmetric membrane consisting of an integrally skinned selective layer and a porous support layer made from the same material is still a promising structure due to the easy and simple fabrication procedure. These membranes are prepared by the phase inversion technique whereby a homogenous polymer solution is transformed into a three dimensional macromolecular network in flat sheet or hollow fiber geometry [52]. However, the reproducibility is very sensitive to the process conditions. Using the same polymer, significant differences in morphology and performance between asymmetric membranes under different preparation conditions have been observed [41]. Some key factors influence the formation of phase inversion flat sheet membranes include: polymer dope composition, the volatility of solvent and non-solvent additives, chemistry of the quench medium, cast thickness, evaporation time, drying process and post-treatment conditions [53, 54, 55].

The formation mechanism for hollow fibers is more complicated than that for flat sheet membranes and additional controlling factors are involved, such as air gap distance, fiber take-up rate, dope and bore fluid flow rates, bore fluid chemistry, and spinneret dimension [56, 57, 58]. Table 5.2 briefly summarizes the effect of some controlling factors in hollow fiber fabrication process. Since the invention of phase-inversion technology by Leob and Sourirajan [59], it was generally admitted that the formation of a defect-free membrane with an ultra-thin skin layer is difficult due to the complexity in the phase-inversion process. As a result, various post treatments such as thermal treatment, crosslinking and coating are exploited to eliminate the defects and to improve the separation performance at a compensation of increasing processing time, cost and reduction of flux. Recent advances in membrane fabrication proved that through the proper control and adjustment of the preparation conditions in the phase inversion process, defect-free asymmetric membranes may be obtainable without the aid of post treatments [48, 58, 60, 61, 62]. Pinnau and Koros [60] fabricated defect-free asymmetric flat membranes by casting a dope contained solvent and non-solvent followed by forced convective evaporation of solvent though a dry/wet phase inversion process. Chung et al. [61] developed defect free 6FDA-durene hollow fibers from a modified Lewis acid: base complex dope. Clausi and Koros [62] produced defect-free Matrimid hollow fibers with a thin skin layer by spinning dopes comprising volatile and non-volatile solvents, and a polymer. Chung et al. [63] also proposed that the spinning dope should exhibit significant chain entanglement in order to produce hollow fibers with an ultra-thin skin layer and minimum defects. Peng and Chung [58] demonstrated that by varying spinneret dimension with a proper take-up rate and air gap distance, defect-free Torlon® hollow fibers with ultra-thin dense layers can be formed from a polymer/solvent binary system. Although defect-free membranes were mainly investigated for gas separation, the principles developed in these studies were applicable for the future development of pervaporation membranes. It is believed that one can obtain a defect-free pervaporation membrane through properly controlled parameters in the fabrication process. Recently developed PBI dual-layer hollow fibers are excellent examples [50, 51].

Table 5.2. Effects of some controlling factors in hollow fiber fabrication process [57, 58, 64]

Factors	Major effects
Solvent and non-solvent	Solvent exchange rate, phase inversion process
Air gap distance	Macrovoid / defects / orientation
Gravitation force	Orientation and defects
Take-up rate / draw ratio	Polymer chain orientation / elongation
Dope concentration / viscosity	Fiber morphology and defects
Bore fluid chemistry	Inner membrane morphology
Outer coagulant chemistry	Outer membrane morphology
Temperature	Affect most of the above
Spinneret design	Shear rate / polymer chain orientation / membrane morphology
Post-treatment (i.e. thermal treatment, crosslinking, drying method and coating, etc)	Membrane morphology / defects control

5.5. Recent Research Progress of Pervaporation Membranes

Many reviews have summarized the performance and characteristics of membrane materials in various pervaporation applications [2, 5, 39, 65, 66, 67, 68, 69]; therefore this section will only focus on the recent research progress and future trends.

5.5.1. Hydrophilic Membranes

The broadest industrial application of pervaporation is for the dehydration of organic solvents especially alcohols. Polymers which contain hydrophilic groups such as hydroxyl (-OH), carboxyl (-COOH), carbonyl (-CO) and amino (-NH_2) groups are intensively studied for pervaporation dehydration process.

Highly Hydrophilic Materials

As mentioned before, highly hydrophilic materials such as poly(vinyl alcohol) (PVA) and poly(acrylic acid) (PAA) have strong affinity to water but exhibit excessive swelling in aqueous solutions and this leads to drastic loss of selectivity. Cross-linked poly (vinyl alcohol) (PVA) is the most popular material for pervaporation dehydration. PVA membranes have been successfully commercialized by Gesellschart für Trenntechnik (GFT, now Sulzer Chemtech) after extensive researches to improve its permselectivity and stability. Crosslinking and grafting are commonly used to make PVA membranes more stable and selective; however, decreased permeability is often observed after crosslinking. The crosslinking agents that have been studied include [70, 71]: fumaric acid, glutaraldehyde

(GA), HCl, citric acid, maleic acid, formic acid, amic acid, sulfur-succinic acid, and formaldehyde.

Blending is another economical and effective approach to suppress swelling and to enhance performance. Namboodiri and Vane studied blending of poly (allylamine hydrochloride) (PAAHCl) and PVA for ethanol and isopropanol dehydration, and found that both water flux and selectivity were increased with the addition of PAAHCl and the performance was tunable by varying blend composition and curing conditions [72, 73].

Incorporation of nanoparticles especially zeolite molecular sieves into the polymer matrix is also very promising to improve the stability and enhance the separation performance. Despite some trade-offs between permeability and selectivity obtained by the early attempts of embedding zeolite 3A, 4A, 5A and 13X into PVA membranes [74], Guan et al. [71] successfully developed multilayer PVA and zeolite 3A mixed matrix membranes with the selective layer crosslinked by fumaric acid. Both flux and separation factor for ethanol dehydration were enhanced significantly after the incorporation of zeolite particles. The key factors of fabricating a successful mixed matrix membrane for gas separation are also applicable to the development of pervaporation membranes, i.e., the choices of appropriate polymer and filler, as well as the controlled interstitial defects between the polymer phase and the zeolite phase [75]. Wang et al. [76] fabricated composite PVA membranes containing delaminated microporous aluminophosphate and showed distinct improvement on flux and separation factor. Guo et al. [77] incorporated γ-glycidyloxypropyltrimethoxysilane (GPTMS) into PVA by an in situ sol-gel method for ethylene glycol (EG) dehydration. The PVA-silica nanocomposite membranes effectively suppressed the swelling of PVA and exhibited desirable stability in aqueous EG solution. Adoor et al. [37] attempted to fabricate mixed matrix membranes (MMMs) containing sodium alginate (NaAlg), PVA and hydrophobic zeolite, i.e., silicalite-1. The incorporation of hydrophobic zeolite particles reduced swelling and led to increased selectivity but decreased permeability.

Natural polymeric materials including chitosan, alginate and agarose have advantages such as their abundant and renewable resources, low cost, non-toxic and biodegradable characteristics. This group of materials is hydrophilic in nature; but its swelling and instability are the major problems for dehydration applications. Intensive studies have been given to chitosan, produced from the *N*-deacetylation of chitin, for alcohol dehydration. Various modifications have been carried out to make the chitosan membranes more stable in water and to have better water permselectivity. Cross-linking with hexamethylene diisocyanate (HMDI) [78], glutaraldehyde [79], and sulfuric acid [80] have been investigated. Other modifications include blending with other polymers [81, 82] and incorporation of zeolite particles [83].

Prominent separation performance has been obtained from novel PBI hollow fiber pervaporation membranes [50, 51]. Synthesized from aromatic bis-*o*-diamines and dicarboxylates, PBI has superior hydrophilic nature and excellent solvent-resistance with robust thermal stability (Tg of 420°C). The brittleness of PBI was successfully overcome in a dual-layer composite form. The as-spun fibers without further cross-linking or heat treatment exhibit good separation performance for dehydration of tetrafluoropropanol (TFP) and isopropanol, as shown in table 5.3.

Table 5.3. Pervaporation performance of dual-layer hollow fibers with PBI as selective layer

Membrane	Feed & concentration	Operation T (°C)	Total flux (g/m^2hr)	α (water/ alcohol)	Membrane structure	Ref.
PBI/P84	85wt% TFP/water	60	332	1990	Dual-layer hollow fiber	[50]
PBI/PEI	85wt% IPA/water	60	867	2163	Dual-layer hollow fiber	[51]

Aromatic Polyimides

Recently, the development of pervaporation dehydration membranes based on aromatic polyimide has achieved promising results. Aromatic polyimides possess very attractive properties such as superior thermal stability, chemical resistance and mechanical strength. Although polyimides may exhibit instability at high temperatures and high humidity due to the hydrolysis of the imide rings, most polyimides are suitable for the dehydration of organic solvents under moderate conditions [84]. Conventionally, polyimides are synthesized by two-stage polycondensation of aromatic dianhydrides with diamines to form a soluble poly (amic acid), followed by imidization via thermal treatment [85]. The interactions between water molecules and the functional groups of polyimides are through hydrogen bonding (Figure 5.3). The small free volume and rigid polymer backbone contribute to the high water selectivity of polyimide membranes. As a result, without strong hydrophilicity, a heat-treated P84 co-polyimide asymmetric membrane may exhibit very limited swelling even in a high water content solution [18].

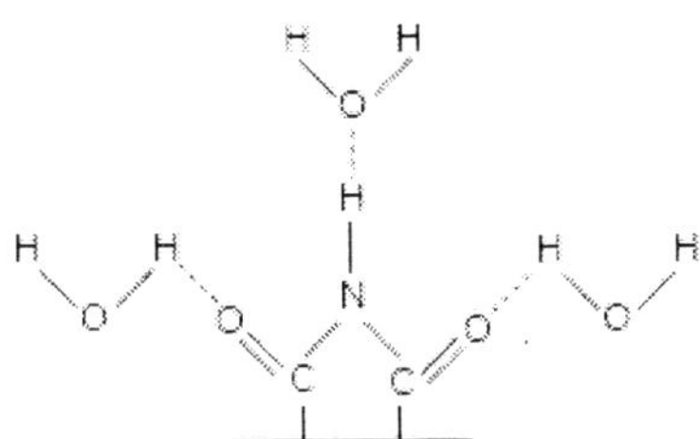

Figure 5.3. Interaction of water molecules with imide groups through hydrogen bonding [85].

The separation performance of polyimide membranes varies with the chemical composition and molecular structure of polymer chains, as well as the preparation conditions, and the operating conditions [28, 86, 87, 88, 89, 90]. Table 5.4 lists the pervaporation performance of recently developed polyimide membranes for pervaporation dehydration of alcohols. Although investigation of inherent membrane properties through dense membranes is essential, it is clear that research has moved from dense films to composite or asymmetric membranes because they have more commercial values.

Table 5.4. Summary of pervaporation performance of different polyimide membranes in aqueous alcohol systems

Membrane material	Feed & concentration	Operation T (°C)	Total flux (g/m²hr)	α (water/alcohol)	Membrane structure	Ref.
PEI	87.6wt% IPA/water,	25	36	173	Asymmetric	[13]
Polyetherimide (PEI) (Ultem 1000)	91wt% Ethanol/water	37	200	8	Dense	[91]
Kapton-H	90wt% Ethanol/water	75	12.12	850	Dense	[92]
PMDA-ODA	90wt% Ethanol/water	75	81.90	230		
BPDA-ODA/DABA (8/2)	90wt% Ethanol/water	75	18.49	1800		
BPDA-ODA/DABA (6/4)	90wt% Ethanol/water	75	23.23	1600		
BPDA-ODA/DABA (4/6)	90wt% Ethanol/water	75	68.2	150		
Polyimide PI2080	95vol% (~93wt%) Ethanol/water	60	1000	~900	Asymmetric	[86]
Polyimide prepared from BTDA/ODA	95vol% (~93wt%) Ethanol/water	25	37	1300	Composite	[88]
Polyetherimide (Ultem 1000) by cataphoretic electrodeposition	85wt% Ethanol/water	40	65	50	Composite	[94]
Polyimide from interfacial polymerization	90wt% Ethanol/water	40	1700	240	Composite	[93]
Polyimide prepared by dip coating	95vol% (~93wt%) Ethanol/water	60	600	550	Composite	[89]
BHTDA-DBAPB	90wt% Ethanol/water	55	650	50	Asymmetric	[90]

Matrimid ®	85wt% tert-butanol/water	100	3300	110	Asymmetric	[28]
P84 Co-polyimide	85wt% IPA/water	60	432.3	3866	Asymmetric	[18]
	85wt% IPA/water	60	883.5	10585	Hollow fiber	[57]

IPA: isopropanol

To develop composite membranes with polyimide as the selective layer, various preparation methods have been attempted, such as chemical vapor deposition and polymerization (CVDP), dip coating, and cataphoretic electrodeposition. These methods are able to produce a thin polyimide selective layer; however, the complexity of the membrane fabrication process is increased. On the other hand, the as-fabricated asymmetric polyimide membranes often show high flux but low separation factor due to defects in the selective skin layer [18, 57].

Post-treatments such as heat treatment and crosslinking have been conducted to reduce defects and enhance the separation property of polyimide membranes [18, 95, 96]. In general, heat treatment is easy to operate and effective for many materials such as polyimide, PBI, polysulfone and polyacrylonitrile (PAN). By inducing molecule relaxation and polymer chain repacking, heat treatment reduces membrane pore size and improves membrane selectivity [47, 97]. For example, Yanagishita et al. [86] found that heat treatment of polyimide membranes at 300°C for 3hr increased mechanical strength and separation factor for ethanol dehydration. Qiao et al. [18] observed noticeably smoothed surface roughness, reduced d-space and densified skin layer structure of P84 co-polyimide flat asymmetric membranes at heat treatment temperatures above 200°C. Liu et al. [57] demonstrated the application of heat treatment to P84 hollow fibers and obtained separation performance much superior to those asymmetric flat sheet membranes. The significant performance enhancement at a heat treatment temperature lower than the polymer's Tg, i.e., around 200°C, may be attributed to the local or segmental motions of polymer chains at β transition, as suggested by Zhou and Koros [98]. In addition, the segmental motions of polymer chains at a temperature above β transition enhance the formation of charge transfer complexes (CTCs) through their inherent electron donor (the diamine moiety) and electron acceptor (the dianhydride moiety) elements. The CTCs formation strongly depends upon heat-treatment temperature, i.e., the higher the heat treatment temperature, the more CTCs can be formed [99]. The intra- and inter-chain CTCs restrict the polymer chain mobility and act as crosslinking. The formation of CTCs can be characterized by both fluorescence and UV-vis spectrophotometer [100, 101, 102].

Adopted from gas separation membranes made from polyimide [102, 103, 104], chemical crosslinking has been proved as another economical and effective tool which can tune the pervaporation performance of polyimide membranes with or without the aid of thermal treatment. The modification of P84 polyimide asymmetric membranes with diamines for isopropanol dehydration was firstly investigated by Qiao and Chung [105]. With the introduction of amide groups after modification, P84 co-polyimide membranes exhibited higher hydrophilicity and apparently denser skin structure. There existed an optimum degree of crosslinking where separation factor achieved a maximum point then degraded; this was attributed to the increased hydrophilicity which caused excessive swelling. It was found that thermal treatment after crosslinking also affected membranes' property and performance. A low-temperature heat treatment facilitated the crosslinking reaction, while a high-temperature heat treatment caused the reaction reversed. The separation factor was further enhanced after heat treatment with a loss in flux.

Liu et al. [40] employed the crosslinking method to modify P84/PES dual-layer hollow fibers. Because the inner PES layer was inert to diamine compounds, the transport property of the inner layer was well maintained. As a result, the dual-layer hollow fibers after chemical modification showed a better performance enhancement than fibers after thermal treatment. Jiang et al. [106] demonstrated that chemical crosslinking by 1,3-propane diamine (PDA) for Matrimid® hollow fibers apparently improved membrane selectivity in water/isopropanol separation. In addition, a thermal pretreatment followed by chemical crosslinking was found effective in revitalizing and enhancing the membrane performance regardless the initial status of the hollow fiber (e.g. defective or defective free). However, extensive experimental data have revealed that the effectiveness of diamine modification varies significantly with diamines chemistry and structure, polyimide moieties and chain structure, and the pre- or post-heat treatment conditions. Therefore, one must take these factors into consideration when conducting the modification.

Other modification methods, e.g. blending with highly hydrophilic materials [96], incorporation of zeolite molecular sieves [107] and inorganic nanoparticles [108], have also shown effective in performance enhancement of polyimide membranes for the dehydration of organic solvents. Interestingly, the swell-up of polymer chains in the feed solution makes the adverse effect of interstitial defects between the polymer matrix and the inorganic particles much less significant compared to that in gas separation membranes [107]. However, so far these mixed matrix membranes are only demonstrated in dense films; it will be more interesting and challenging if the currently developed knowledge can be extended to fabricate membranes with a composite or asymmetric structure. In addition, the high temperature required (usually above T_g) to achieve a good interaction between the filler particles and the polymer matrix in the fabrication process probably can be reduced by the introduction of crosslinking agent and the modification of surface properties of inorganic particles, which may bring down the processing cost.

Hydrophobic Materials

Hydrophobic materials have higher stability in aqueous solutions. If the hydrophobic nature of the material can be changed to hydrophilic and the degree of modification can be controlled, this type material may become a good candidate for pervaporation dehydration. For example, poly(ether ether ketone) (PEEK) had been modified by sulfonation reaction for pervaporation separation of water/isopropanol mixtures and the hydrophilicity-hydrophobicity balance was controlled by different degrees of sulfonation [30]. Tu et al. [109] developed hydrophilic surface-grafted poly(tetrafluoroethylene) (PTFE) membranes with good performance and wide applications in pervaporation dehydration processes.

Inorganic Materials

Inorganic membranes are able to overcome the problems of instability and swelling of hydrophilic polymeric membranes. They show better structural stability and chemical resistance at harsh environments and high temperature operations [69, 110].

Table 5.5. Summary of pervaporation performance of some inorganic membranes in aqueous alcohol systems

Membrane Material	Feed & concentration (wt%)	Operation T (°C)	Total flux (g/m^2hr)	α (water/alcohol)	Membrane structure	Ref.
Zeolite NaA on Carbosep tube support	90% IPA	70	300	2000	Composite	[114]
Zeolite NaA from Mitsui Engineering and Shipbuilding	70% Ethanol	60	2100	2140	Composite	[110]
Ceramic commercial membrane	95% IPA	70	2100	600	Tubular	[112]
Pervap SMS® Silca membrane	90% IPA	70	3000	60	Porous	[111]
Zeolite NaA prepared by hydrothermal synthesis on tubular α-Al_2O_3 support	90% Ethanol	75	5600	10^3-10^4	Composite	[113]
Zeolite NaA prepared by microvave heating and conventional heating on α-Al_2O_3 support	95% IPA	70	1440	10000	Composite	[115]

Zeolite membranes have the advantages of high selectivity and high permeability due to their unique molecular sieving property and selective adsorption. The recently developed zeolite NaA, X and Y membranes exhibit impressive separation performance that is far superior to traditional polymeric membranes. The high separation factor is achieved because of the precise micropore structure of zeolite pores and the preferential sorption of water molecules. Microporous silica membranes are water selective and exhibit a much higher flux and less swelling but lower selectivity compared to polymeric membranes [111]. Ceramic membranes are resistant to microbes; they can be easily sterilized by steam or autoclave. Ceramic membranes show high water permeation flux and relatively high separation factor for alcohol dehydration [112]. The major drawbacks of inorganic membranes are (1) the higher cost of fabrication process compared to that of polymeric membranes and (2) the brittleness. However, the superior stability and higher separation performance may level off the initial fabrication and installation cost of inorganic membranes. The performance of recently developed inorganic membranes is summarized in table 5.5. It is obvious that the fabrication procedure also plays an important role on membrane performance. By lowering the transport resistance of the support layer and minimize the selective layer thickness, Sato and Nakane [113] developed NaA zeolite membranes with very high flux and promising water/alcohol separation factor.

5.5.2. Organophilic Membranes

In organo-selective membranes for the separation of small amount organics from water, the difference in solubility determines the membrane selectivity. This is because diffusivity always favors the smaller molecule, i.e., water. Membranes made from rubbery polymers such as poly(dimethyl siloxane) (PDMS) [116, 117], polyurethane [118], polybutadiene [119], polyamide-polyether block copolymers (PEBA™) [120] and poly[1-(trimethylsilyl)-1-propyne] (PTMSP) [121, 122], and hydrophobic inorganic materials such as zeolite silicalite-1 and ZSM-5, have been intensively investigated for the separation of organics from aqueous streams. PDMS are currently the benchmark material for this application because of its high affinity and low transport resistance for organics, as well as its stability in organic solutions [5].

The most important factor to advance organophilic pervaporation is to have breakthroughs in membrane materials and structure in addition to minimizing concentration polarization, optimizing the process, and improving energy efficiency [5, 42, 123]. The following approaches have been adopted: (1) modification of currently available membranes by crosslinking, grafting or incorporation of adsorbent fillers, (2) development of novel membrane structures, and (3) development of new polymeric materials. For example, Uragami et al. [124] crosslinked PDMS membranes with divinyl compound and found both permeability and benzene permselectivity of the membranes were improved. A novel polymeric-inorganic composite membrane made by coating cellulose acetate upon a tubular ceramic support was firstly developed by Song and Hong [125] for the dehydration of ethanol and isopropanol. Later this approach was adapted to coat a PDMS layer on top of a ceramic tubular support to extract ethanol from water [117, 126].

Using crosslinked PDMS as the selective layer and tubular non-symmetric ZrO_2/Al_2O_3 membranes as the support layer, Xiangli et al. [126] developed composite membranes with remarkably high flux (i.e., flux of $12300 g/m^2hr$ and separation factor of 6 for a feed ethanol concentration of 4.3wt% and temperature at 40°C). This performance is superior to the performance of PDMS composite membranes with a polymeric support, owing to the significantly reduced transport resistance of the ceramic support layer. Recently, Nagase et al. [127] synthesized siloxane-grafted poly(amide-imide) and polyamide with a new reactive diamino-terminated PDMS macromonomer. The newly developed material exhibited durability and good permselectivity toward several organic solvents with high permeation rates and reasonable separation factors (i.e., flux of $37.4 g/m^2hr$ and separation factor of 9.78 for a feed ethanol concentration of 9.24wt% and temperature at 50°C). Table 5.6 summarizes the recent development of PDMS membranes. Except for ethyl acetate and benzene removal, the separation factors of PDMS membranes for alcohol removal are still low. However, inorganic membranes (i.e. silicalite-1) and silicone rubber mixed matrix membranes have exhibited better separation performance [42].

Table 5.6. Summary of pervaporation performance of recently developed PDMS membranes for organic removal

Membrane material	Feed & concentration (wt%)	Operation T (°C)	Total flux (g/m²hr)	α (organic/ water)	Dense layer thickness (μm)	Membrane structure	Ref.
PDMS on ceramic	10% IPA	50	200	8	25	Composite	[117]
PDMS (Pervap1060)	1% Ethyl Acetate	30	400	300	8	Composite	[128]
	5% Methanol	30	230	6.5			
Cross-linked PDMSDMMA	0.05% Benzene	40	51.4	1853	270	Dense	[124]
PDMS coated on CA	5% Ethanol	40	1300	8.5	8	Composite	[129]
PDMS-copolymer coated on PVDF	10% Ethanol	~25	900	31	104	Composite	[130]
PDMS on ceramic support	4.3% Ethanol	70	12300	6	5-10	Composite	[126]
PDMS-grafted polyamide	7.11% Methanol	50	32.7	7.57	100-200	Composite	[127]
	9.24% Ethanol	50	37.4	9.78			
	6.25% Acetone	50	125	41.8			

PDMSDMMA: poly(dimethylsiloxane) dimethyl methacrylate macromonomer.
CA: cellulose acetate
PVDF: poly(vinylidene fluoride)

5.5.3. Organoselective Membranes

Albeit of the great potential in chemical and petrochemical industries, the separation of organic/organic mixtures using pervaporation is the least developed area. Pervaporation membranes have been applied to separate wide streams of organic/organic mixtures and basically these mixtures can be categorized into five major groups: polar/non-polar mixtures, e.g. methanol/methyl tert-butyl ether (MTBE) [131, 132]; aromatic/aliphatics, e.g. cyclohexane/benzene [67, 133]; aliphatic hydrocarbons, e.g. hexane/heptane [134]; isomers, e.g. C8 isomers (o-xylene, m-xylene, p-xylene and ethyl benzene) [135, 136, 137]; and enatioseparation, e.g. linalool racemic mixture separation [138]. Smitha et al. [39] have given a good literature summary on membrane materials and their performance for the above four categories, while Villaluenga and Tabe-Mohammadi [67] gave a deeper insight on the membranes developed for benzene and cyclohexane separation. Membrane materials are selected based on the solubility differences of organic components in membrane. By improving the interaction between membrane material and one permeating component, the separation performance can be enhanced.

Among the diversified applications in organic/organic separation by pervaporation, the separation of benzene/cyclohexane represents one of the most important but most difficult and complicated separation in petrochemical industry. The double bonds of benzene molecule have strong affinity to polar groups in a membrane; therefore hydrophilic membranes which possess polar groups such as PVA and benzoylchitosan membranes show selectivity to benzene [139, 140]. Benefit from the conjugated π bonds, graphite, carbon molecular sieve and carbon nanotube show preference to aromatics with effective π-π stacking interaction [36]. These inorganic materials have been used to enhance PVA membrane performance. Crystalline flake graphite was firstly incorporated into PVA membranes and resulted in significant increase of permeation flux and selectivity [34]. Later carbon nanotubes with or without wrapped with chitosan were introduced to the PVA matrix [36, 141]. The improvement in permeation flux and separation factor were attributed to the preferential affinity of carbon nanotubes towards benzene and the increased free volume by altering PVA polymer chain packing. Nam and Dorgan [38] attempted modification of the solubility selectivity of glassy polymer polyvinylchloride by physical blend with crosslinked rubbery materials; and the resultant membranes showed permselectivity toward benzene. Table 5.7 summarizes the recently developed membranes for benzene/cyclohexane separation.

5.6. Industrial Applications and Commercial Aspects

Corresponding to the three types of pervaporation membranes, the applications of pervaporation processes are mainly divided into three areas: (1) dehydration of alcohols or other aqueous organic mixtures; (2) removal of volatile organics from water; (3) organic/organic separation.

Dehydration of organic solvents such as alcohols, esters, ethers, and acids has become the most important application of pervaporation due to the high demand in industries and the difficulties to obtain the anhydrous form of these chemicals by traditional distillation technology. Both diffusion and sorption selectivity of water over organic solvents can be simultaneously sought by hydrophilic pervaporation membranes because water has smaller molecular size and stronger affinity to hydrophilic materials than the organic solvents. The first commercial membrane which consisted of a dense cross-linked PVA as the selective layer, an ultrafiltration poly(acrylonitrile) (PAN) and a fabric non-woven as the support layer was developed by GFT in 1980s for the dehydration of ethanol. Since then, 38 solvent dehydration plants for ethanol and isopropanol, 8 units for other solvents dehydration (i.e. ester) have been installed world widely [148].

Table 5.7. Summary of pervaporation performance of recently developed membranes for separation of benzene/cyclohexane mixture

Membrane material	Feed concentration benzene:cyclohexane (wt%:wt%)	Operation T (°C)	Flux (g/m^2hr)	α (benzene/cyclohexane)	Dense layer Thickness (μm)	Membrane structure	Ref.
Butyrylcellulose (BuCel)	0.5 :99.5	40	50	9.4	30	Dense film	[142]
	5 :95	40	180	7.5	30		
Polyimide	50:50	30	100	40	0.2	Composite	[143]
Co-polyimide	60:40	70	64	28	25	Dense film	[144]
Dinitrophenyl contained CA	50:50	70	10	101	40	Dense film	[145]
Modified polyurethanes	54 :46	25	3515	2.5	50-100	Dense film	[146]
Crosslinked hyperbranched-polyester on ethyl cellulose	10:90	30	1888	6.8	20-25	Dense film	[147]
NBR/SBR/PVC blend	50:50	60	100	13	50	Dense film	[38]
PVA–carbon nanotube /chitosan blend	50:50	50	65.9	53	80	Dense film	[34, 35, 36, 141]
Carbon molecular sieve filled PVA	50:50	50	59.25	23	80	Dense film	
Crystalline flake graphite filled PVA	50:50	50	90.7	100	80	Dense film	
	10:90	50	40.2	344	80	Dense film	

CA: cellulose acetate

NBR/SBR/PVC: acrylonitrile butadiene rubber (NBR)/styrene butadiene rubber (SBR)/ polyvinylchloride (PVC).

There are also numbers of attempts to employ pervaporation for organics removal from water which aim on water purification, pollution control, solvents/aroma compounds recovery and biofuel production from fermentation broth. Applications in this area include removal of trace amount of volatile organic compounds (VOCs) from aqueous streams. The emission of VOCs from industrial and municipal wastewater streams are of great concern due to the toxic and carcinogenic effects of VOCs. VOCs include solvents from petroleum industry, such as benzene, toluene and xylenes, and substances which contain chlorine, such as chloroform, 1,1,2-trichloroethane (TCA), trichloroethylene (TCE), perchloroethylene, and chlorobenzene. Due to the low solubilities of these compounds in water, the amount of these compounds dissolved in wastewater is very small; therefore treatment by distillation is not economically viable [68]. Traditionally, carbon adsorption and air stripping were employed as treatment processes; however, these treatments merely transfer the contaminant from water phase to another phase and further treatment is necessary. In addition, the regeneration of activate carbon is costly. Pervaporation is promising for VOCs removal or recovery by achieving the separation through preferential sorption of one component in the membrane without disruption of the process. If the concentration of the organic is sufficiently high in an aqueous stream, the recovery of organics is valuable. It has been demonstrated that a stream containing 2% ethyl acetate was concentrated to 96.7%, which was reused in the feed stream [149].

Combining with a fermentation process, pervaporation is applied to extract inhibitory products such as ethanol or butanol from a fermentation broth in order to increase the conversion rate [150]. As crude oil price reaches new highs, pervaporation become promising for biofuel (e.g., bio-ethanol and bio-butanol) recovery from fermentation broth [42, 151]. However, the flux and separation factor of pervaporation membranes for separating organics from water are still low, which seriously restrict the industrial-scale application of organophilic pervaporation [69]. Till 2002, only a few pervaporation systems for VOC removal were commercialized by Membrane Technology and Research, Inc [148].

Organic/organic separation by pervaporation has large potential applications in chemical, petrochemical and pharmaceutical industries. Research in this area is extremely challenging; nevertheless, the application of pervaporation in organic/organic separation has not acquired industrial acceptance because of the lack of advanced performance and the instability in organic solvents of currently available membranes [39]. The first and only pervaporation plant using organoselective membranes was built by Air Products in 1991 for removal of methanol from MTBE.

Nowadays, main global pervaporation membrane manufacturers and suppliers are: Sulzer Chemtech (Swiss), CM-Celfa Membrantrenntechnik (Swiss), GKSS (Germany), UBE Industries & Mitsui Engineering and Shipbuilding Co. Ltd (Japan), Membrane Technology and Research, Inc. (USA), and MegaVision Membrane Technology and Engineering Co Ltd, (China).

5.7. Conclusions and Future Perspective

Pervaporation is an important membrane separation technology that has shown promising results for organic separation. The various kinds of organic mixtures and the complicated interactions among permeant to permeant and permeant to membrane make the breakthroughs of developing novel membranes for pervaporation difficult. However, the recent proposal using permeance/permeability and selectivity to analyze pervaporation performance may provide better insight for future membrane design because the new parameters are able to reflect the intrinsic properties of a membrane material by excluding the effect of operating conditions. In addition, the advances in molecular dynamic simulation and modeling are helpful to predict a membrane performance for a specific application and therefore reduce the tedious experimental testing during membrane development.

Through modifications by crosslinking, blending or incorporation of inorganic particles, controlled fabrication process parameters, exploring new membrane materials and novel membrane structures, encouraging results have been obtained from post-GFT dehydration membranes. However, there is still space for future development. The potential candidate must surpass the trade-off line and be applied in composite or asymmetric configuration. The recent developments of dual-layer membrane configuration and hybrid polymer/inorganic membrane structure may lead to a brighter future for pervaporation membranes because they combine and synergize the properties of two components for practical applications.

Energy is a major concern globally due to resource depletion and recorded high prices for oil. Among many energy alternatives, biofuel (i.e., bio-ethanol and bio-butanol) is one of the environmentally friendly energy sources because of the environmental concerns of green house gases emission from petroleum-derived fuels. Therefore, breakthroughs in organophilic membranes especially for biofuel recovery from fermentation broths may have higher priority than others in the foreseeable future. Organoselective membranes for the separation of biofuel mixtures and butanol isomers may soon or later get attention. Advances in the fundamentals of transport mechanism and molecular engineering of pervaporation membranes for each particular organic-organic separation are urgently needed.

Acknowledgments

The authors would like to thank NUS and A*Star for funding this work with the grant numbers of R-398-000-001-018, R-398-000-029-305, R-398-000-044-305 and R-279-000-249-646.

References

[1] Baker, R. W.; Cussler, E. L.; Eykamp, W.; Koros, W. J.; Riley, R. L.; Strathmann, H. Membrane Separation Systems: Recent Developments and Future Directions. Noyes Data Corporation, New Jersey (1991).

[2] Feng, X.; Huang, R. Y. M. Liquid separation by membrane pervaporation: a review. *Ind. Eng. Chem. Res.* 36 (1997) 1048-1066.

[3] Park, H. C. Separation of alcohols from organic liquid mixtures by pervaporation. PhD thesis, University of Twente, Twente (1993).

[4] Bravo, J. L.; Fair, J. R.; Humphrey, J. L.; Martin, C. L.; Seibert, A. F.; Joshi, S. Fluid Mixture Separation Technologies for Cost Reduction and Process Improvement. Noyes Publications, Park Ridge (1986).

[5] Shao, P.; Huang, R. Y. M. Polymeric membrane pervaporation. *J. Membr. Sci.* 287 (2007) 162-179.

[6] Pearce, G. K. Pervaporation vs. distillation. A comparative costing for alcohol dehydration. *Proc. Int. Conf. Pervaporation Processes Chem. Ind.*, 4th (1989) 278-296.

[7] Wijmans, J. G.; Baker, R. W. The solution-diffusion model: a review. *J. Membr. Sci.* 107 (1995) 1-21.

[8] Mulder, M. H. V. Basic Principles of Membrane Technology. Kluwer Academic, Boston (1996).

[9] Crespo, J. G.; Böddeker, K. W. Membrane Process in Separation and Purification, Kluwer Academic Publishers (1995).

[10] Yeom, C. K.; Lee, S. H.; Lee, J. M. Pervaporative permeations of homologous series of alcohol aqueous mixtures through a hydrophilic membrane. *J. Appl. Polym. Sci.* 79 (2001) 703-713.

[11] Carretier, E.; Moulin, P.; Beaujean, M.; Charbit, F. Purification and dehydration of methylal by pervaporation. *J. Membr. Sci.* 217 (2003) 159-171.

[12] Huang, R. Y. M.; Yeom, C. K. Pervaporation Separation of aqueous mixtures using crosslinked poly(vinyl alcohol) (PVA). II. Permeation of ethanol-water mixtures. *J. Membr. Sci.* 51 (1990) 273-292.

[13] Huang, R. Y. M.; Feng, X. Dehydration of isopropanol by pervaporation using aromatic polyetherimide membranes. *Sep. Sci. Technol.* 28 (1993) 2035-2048.

[14] Wijmans, J. G.; Baker, R. W. A simple predictive treatment of the permeation process in pervaporation. *J. Membr. Sci.* 79 (1993) 101-113.

[15] Wijmans, J. G. Process performance = membrane properties + operating conditions. *J. Membr. Sci.* 220 (2003) 1-3.

[16] Guo, W. F.; Chung, T. S.; Matsuura, T. Pervaporation study on the dehydration of aqueous butanol solutions: a comparison of flux vs. permeance, separation factor vs. selectivity. *J. Membr. Sci.* 245 (2004) 199-210.

[17] Qiao, X. Y.; Chung, T. S.; Guo, W. F.; Matsuura, T.; Teoh, M. M. Dehydration of isopropanol and its comparison with dehydration of butanol isomers from thermodynamic and molecular aspects. *J. Membr. Sci.* 252 (2005) 37-49.

[18] Qiao, X. Y.; Chung, T. S.; Pramoda, K. P. Fabrication and characterization of BTDA-TDI/MDI (P84) co-polyimide membranes for the pervaporation dehydration of isopropanol. *J. Membr. Sci.* 264 (2005) 176-189.

[19] Mulder, M. H. V. Thermodynamic principles of pervaporation. in R. Y. M. Huang edited, Pervaporation Membrane Separation Processes, Elsevier Science Publishers, Amsterdam (1991).

[20] Matsuura, T. Synthetic Membranes and Membrane Separation Processes. CRC Press, Boca Raton, Florida (1993).

[21] Barton, A. F. M. Handbook of Solubility Parameters and Other Cohesion Parameters. CRC Press, Boca Raton, Florida (1990).

[22] Flory, P. J. Principles of Polymer Chemistry. Cornell University Press, New York (1953).

[23] Jonquieres, A.; Perrin, L.; Durand, A.; Arnold, S.; Lochon, P. Modelling of vapour sorption in polar materials: Comparison of Flory-Huggins and related models with the ENSIC mechanistic approach. *J. Membr. Sci.* 147 (1998) 59-71.

[24] Yoshikawa, M.; Yokoi, H.; Sanui, K.; Ogata, N.; Shimidzu, T. Polymer membrane as a reaction field. II. Effect of membrane environment on permselectivity for water/ethanol binary mixtures. *Polymer J.* 16 (1984) 653-656.

[25] Jonquieres, A.; Roizard, D.; Lochon, P. Use of empirical polarity parameters to describe polymer-liquid interactions: correlation of polymer swelling with solvent polarity in binary and ternary systems. *J. Appl. Polym. Sci.* 54 (1994) 1673-1684.

[26] Jonquieres, A.; Roizard, D.; Cuny, J.; Lochon, P. Solubility and polarity parameters for assessing pervaporation and sorption properties. A critical comparison for ternary systems alcohol/ether/polyurethaneimide. *J. Membr. Sci.* 121 (1996) 117-133.

[27] Reichardt, C. Solvents and Solvent Effects in Organic Chemistry, 2nd ed., Wiley-VCH, Weinheim (1998).

[28] Guo, W. F.; Chung, T. S. Study and characterization of the hysteresis behavior of polyimide membranes in the thermal cycle process of pervaporation separation. *J. Membr. Sci.* 253 (2005) 13-22.

[29] Bode, E.; Hoempler, C. Transport resistances during pervaporation through a composite membrane: experiments and model calculations. *J. Membr. Sci.* 113 (1996) 43-56.

[30] Huang, R. Y. M.; Shao, P.; Feng, X.; Burns, C. M. Pervaporation separation of water/isopropanol mixture using sulfonated poly(ether ether ketone) (SPEEK) membranes: transport mechanism and separation performance. *J. Membr. Sci.* 192 (2001) 115-127.

[31] Drioli, E.; Zhang, S.; Basile, A. On the coupling effect in pervaporation. *J. Membr. Sci.* 81 (1993) 43-55.

[32] Schaetzel, P.; Vauclair, C.; Luo, G.; Nguyen, Q. T. The solution-diffusion model order of magnitude calculation of coupling between the fluxes in pervaporation. *J. Membr. Sci.* 191 (2001) 103-108.

[33] Smit, E.; Mulder, M. H. V.; Smolders, C. A.; Karrenbeld, H.; Eerden, J. van; Feil, D. Modeling of the diffusion of carbon dioxide in polyimide matrices by computer simulation. *J. Membr. Sci.* 73 (1992) 247-257.

[34] Peng, F. B.; Lu, L. Y.; Hu, C. L.; Wu, H.; Jiang, Z. Y. Significant increase of permeation flux and selectivity of poly(vinyl alcohol) membranes by incorporation of crystalline flake graphite. *J. Membr. Sci.* 259 (2005) 65-73.

[35] Peng, F. B.; Lu, L. Y.; Sun, H. L.; Wang, Y. Q.; Liu, J. Q.; Jiang, Z. Y. Hybrid organic-inorganic membranes: solving the trade-off between permeability and selectivity. *Chem. Mater.* 17 (2005) 6790-6796.

[36] Peng, F. B.; Pan, F. S.; Sun, H. L.; Lu, L. Y.; Jiang, Z. Y. Novel nanocomposite pervaporation membranes composed of poly(vinyl alcohol) and chitosan-wrapped carbon nanotube. *J. Membr. Sci.* 300 (2007) 13-19.
[37] Adoor, S. G.; Prathab, B.; Manjeshwar, L. S.; Aminabhavi, T. M. Mixed matrix membranes of sodium alginate and poly(vinyl alcohol) for pervaporation dehydration of isopropanol at different temperatures. *Polymer.* 48 (2007) 5417-5430.
[38] Nam, S. Y.; Dorgan, J. R. Non-equilibrium nanoblends via forced assembly for pervaporation separation of benzene from cyclohexane: UNIFAQ-FV group contribution calculations. *J. Membr. Sci.* 306 (2007) 186-195.
[39] Smitha, B.; Suhanya, D.; Sridhar, S.; Ramakrishna, M. Separation of organic-organic mixtures by pervaporation-a review. *J. Membr. Sci.* 241 (2004) 1-21.
[40] Liu, R. X.; Qiao, X. Y.; Chung, T. S. Dual-layer P84/polyethersulfone hollow fibers for pervaporation dehydration of isopropanol. *J. Membr. Sci.* 294 (2007) 103-114.
[41] Koros, W. J.; Fleming, G. K.; Jordan, S. M.; Kim, T. H.; Hoehn, H. H. Polymeric membrane materials for solution-diffusion based permeation separations. *Prog. Polym. Sci.* 13 (1988) 339-401.
[42] Vane, L. M. A review of pervaporation for product recovery from biomass fermentation processes. *J. Chem. Technol. Biotechnol.* 80 (2005) 603-629.
[43] Pinnau, I.; Koros, W. J. Relationship between substructure resistance and gas separation properties of defect-free integrally-skinned asymmetric membranes. *Ind. Eng. Chem. Res.* 20 (1991) 1837-1840.
[44] Huang, R. Y. M.; Feng, X. Resistance model approach to asymmetric polyetherimide membranes for pervaporation of isopropanol/water mixtures. *J. Membr. Sci.* 84 (1993) 15-27.
[45] Ekiner, O. M.; Hayes, R. A.; Manos, P. Novel multicomponent fluid separation membranes. U.S. patent 5085676 (1992).
[46] Li, D. F.; Chung, T. S.; Wang, R.; Liu, Y. Fabrication of fluoropolyimide / polyethersulfone (PES) dual-layer asymmetric hollow fiber membranes for gas Separation. *J. Membr. Sci.* 198 (2002) 211-223.
[47] Chung, T. S. A review of microporous composite polymeric membrane technology for air-separation. *Polym. Polym. Comp.* 4 (1996) 269-283.
[48] Jiang, L. Y.; Chung, T. S.; Li, D. F.; Cao, C. Kulprathipanja S. Fabrication of Matrimid/ polyethersulfone dual-layer hollow fiber membranes for gas separation. *J. Membr. Sci.* 240 (2004) 91-103.
[49] Widjojo, N.; Zhang, S. D.; Chung, T. S.; Liu, Y. Enhanced gas separation performance of dual-layer hollow fiber membranes via substructure resistance reduction using mixed matrix materials. *J. Membr. Sci.* 306 (2007) 147-158.
[50] Wang, K. Y.; Chung, T. S.; Rajagopalan, R. Dehydration of tetrafluoropropanol (TFP) by pervaporation via novel PBI/BTDA-TDI/MDI co-polyimide (P84) dual-layer hollow fiber membranes. *J. Membr. Sci.* 287 (2007) 60-66.
[51] Liu, R. X.; Wang, K. Y.; Qiao, X. Y.; Chung, T. S. PBI/PEI dual-layer hollow fiber membranes for isopropanol dehydration through pervaporation. In progress (2008).
[52] Kesting, R. E. Synthetic Polymeric Membranes: A Structure Perspective. John Wiley & Sons, Inc., New York (1985).

[53] Pinnau, I.; Koros, W. J. Influence of quench medium on the structures and gas permeation properties of polysulfone membranes made by wet and dry/wet phase inversion. *J. Membr. Sci.* 71 (1992) 81-96.

[54] Paulsen, F. G.; Shojaie, S. S.; Krantz, W. B. Effect of evaporation step on macrovoid formation in wet-cast polymeric membranes. *J. Membr. Sci.* 91 (1994) 265-282.

[55] Wang, Y.; Jiang, L. Y.; Matsuura, T.; Chung, T. S.; Goh, S. H. Investigation of the fundamental differences between polyamide-imide (PAI) and polyetherimide (PEI) membranes for isopropanol dehydration via pervaporation. *J. Membr. Sci.* 318 (2008) 217-226.

[56] Niwa, M.; Kawakami, H.; Nagaoka, S.; Kanamori, T.; Shinbo, T. Fabrication of an asymmetric polyimide hollow fiber with a defect-free surface skin layer. *J. Membr. Sci.* 171 (2000) 253-261.

[57] Liu, R. X.; Qiao, X. Y.; Chung, T. S. The development of high performance P84 co-polyimide hollow fibers for pervaporation dehydration of isopropanol. *Chem. Eng. Sci.* 60 (2005) 6674-6686.

[58] Peng, N.; Chung, T. S. The effects of spinneret dimension and hollow fiber dimension on gas separation performance of ultra-thin defect-free Torlon® hollow fiber membranes. *J. Membr. Sci.* 310 (2008) 455-465.

[59] Leob, S.; Sourirajan, S. Sea water demineralization by means of an osmotic membrane. *Adv. Chem. Ser.* 38 (1963) 117-132.

[60] Pinnau, I.; Koros, W. J. Defect-free ultrahigh flux asymmetric membranes. U.S. Patent 4,902,422, (1990).

[61] Chung, T. S.; Kafchinski, E. R.; Vora, R. Development of a defect-free 6FDA-durene asymmetric hollow fiber and its composite hollow fibers. *J. Membr. Sci.* 88 (1994) 21-36.

[62] Clausi, D. T.; Koros, W. J. Formation of defect-free polyimide hollow fiber membranes for gas separations. *J. Membr. Sci.* 167 (2000) 79-89.

[63] Chung, T. S.; Teoh, S. K.; Hu, X. D. Formation of ultrathin high-performance polyethersulfone hollow-fiber membranes. *J. Membr. Sci.* 133 (1997) 161-175.

[64] Chung, T. S.; Kafchinski, E. R.; Foley, P. Development of asymmetric hollow fibers from polyimide for air separation. *J. Membr. Sci.* 75 (1992) 181-195.

[65] Ji, W. C.; Sikdar, S. K. Pervaporation using adsorbent-filled membranes. *Ind. Eng. Chem. Res.* 35 (1996) 1124-1132.

[66] Lipnizki, F.; Field, R. W.; Ten, P. K. Pervaporation-based hybrid process: a review of process design, applications and economics. *J. Membr. Sci.* 153 (1999) 183-210.

[67] Villaluenga, J. P. G.; Tabe-Mohammadi, A. A review on the separation of benzene/cyclohexane mixtures by pervaporation processes. *J. Membr. Sci.* 169 (2000) 159-174.

[68] Peng, M.; Vane, L. M.; Liu, S. X. Recent advances in VOCs removal from water by pervaporation. *J. Hazard. Mater.* B98 (2003) 69-90.

[69] Bowen, T. C.; Noble, R. D.; Falconer, J. L. Fundamentals and applications of pervaporation through zeolite membranes. *J. Membr. Sci.* 245 (2004) 1-33.

[70] Yu, J., Lee, C. H., Hong, W. H., Performances of crosslinked asymmetric poly(vinyl alcohol) membranes for isopropanol dehydration by pervaporation. *Chem. Eng. Process*. 41 (2002) 693-698.

[71] Guan, H. M.; Chung, T. S.; Huang, Z.; Chng, M. L.; Kulprathipanja S. Poly(vinyl alcohol) multilayer mixed matrix membranes for the dehydration of ethanol-water mixture. *J. Membr. Sci.* 268 (2006) 113-122.

[72] Namboodiri, V. V.; Ponangi, R.; Vane, L. M. A novel hydrophilic polymer membrane for the dehydration of organic solvents. *Europ. Polym. J.* 42 (2006) 3390-3393.

[73] Namboodiri, V. V.; Vane, L. M. High permeability membranes for the dehydration of low water content ethanol by pervaporation. *J. Membr. Sci.* 306 (2007) 209-215.

[74] Gao, Y.; Yue, Y.; Li, W. Application of zeolite filled pervaporation membrane. *Zeolite.* 16 (1996) 70-74.

[75] Chung, T. S.; Jiang, L. Y.; Li, Y.; Kulprathipanja, S. Mixed matrix membranes (MMMs) comprising organic polymers with dispersed inorganic fillers for gas separation. *Prog. Polym. Sci.* 32 (2007) 483-507.

[76] Wang, C.; Hua, W. M.; Yue, Y. H.; Gao, Z. Delaminated microporous aluminophosphate-filled polyvinyl alcohol membrane for pervaporation of aqueous alcohol solutions. *Micropor. Mesopor. Mater.* 105 (2007) 149-155.

[77] Guo, R. L.; Hu, C.; Pan, F. S.; Wu, H.; Jiang, Z. Y. PVA-GPTMS/TEOS hybrid pervaporation membrane for dehydration of ethylene glycol aqueous solution. *J. Membr. Sci.* 281 (2006) 454-462.

[78] Nawawi, M. G. M.; Huang, R. Y. M. Pervaporation dehydration of isopropanol with chitosan membranes. *J. Membr. Sci.* 124 (1997) 53-62.

[79] Huang, R. Y. M.; Pal, R.; Moon, G. Y. Crosslinked chitosan composite membrane for the pervaporation dehydration of alcohol mixtures and enhancement of structural stability of chitosan/polysulfone composite membranes. *J. Membr. Sci.* 160 (1999) 17-30.

[80] Lee, Y. M.; Nam, S. Y.; Woo, D. J. Pervaporation of ionically surface cross-linked chitosan composite membranes for water-alcohol mixtures. *J. Membr. Sci.* 133 (1997) 103-110.

[81] Chanachai, A.; Jiraratananon, R.; Uttapap, D.; Moon, G. Y.; Anderson, W. A.; Huang, R. Y. M. Pervaporation with chitosan/hydroxyethyl cellulose (CS/HEC) blended membranes. *J. Membr. Sci.* 166 (2000) 271-280.

[82] Sridhar, S.; Ganga, D.; Smitha, B.; Ramakrishna, M. Dehydration of 2-butanol by pervaporation through blend membranes of chitosan and hydroxy ethyl cellulose. *Sep. Sci. Technol.* 40 (2005) 2889-2908.

[83] Ahmad, A. L.; Nawawi, M. G. M.; So, L. K. Characterization and performance evaluations of sodium zeolite-Y filled chitosan polymeric membrane: Effect of sodium zeolite-Y concentration. *J. Appl. Polym. Sci.* 99 (2006) 1740-1751.

[84] Tanaka, K.; Okamoto, K. I. Structure and transport properties of polyimides as materials for gas and vapor membrane separation in Y. Yampolskii, I. Pinnau and B. D. Freeman edited, Materials Science of Membranes for Gas and Vapor Separation. John Wiley & Sons Ltd, Chichester (2007).

[85] Ohya, H.; Kudryavtsev, V. V.; Semenova, S. I. Polyimide membranes-applications, fabrications, and properties. Gordon and Breach Publishers, Amsterdam (1996).

[86] Yanagishita, H.; Maejima, C.; Kitamoto, D.; Nakane, T. Preparation of asymmetric polyimide membrane for water/ethanol separation in pervaporation by phase inversion process. *J. Membr. Sci.* 86 (1994) 231-240.

[87] Feng, X.; Huang, R. Y. M. Preparation and performance of asymmetric polyetherimide membranes for isopropanol dehydration by pervaporation. *J. Membr. Sci.* 109 (1996) 165-172.

[88] Yanagishita, H.; Kitamoto, D.; Haraya, K.; Nakane, T.; Tsuchiya, T.; Koura, N. Preparation and pervaporation performance of polyimide composite membrane by vapor deposition and polymerization (VDP). *J. Membr. Sci.* 136 (1997) 121-126.

[89] Yanagishita, H.; Kitamoto, D.; Haraya, K.; Nakane, T.; Okada, T.; Mastyda, H.; Idemoto, Y.; Koura, N.; Separation performance of polyimide composite membrane prepared by dip coating process. *J. Membr. Sci.* 188 (2001) 165-172.

[90] Fan, S. C. A study of polyamide and polyimide membranes for pervaporation and vapor permeation. PhD thesis, Chun Yuan University, Taiwan (2002).

[91] Huang, R. Y. M.; Feng, X. Pervaporation of water/ethanol mixtures by an aromatic polyetherimide membrane. *Sep. Sci. Technol.* 27 (1992) 1583-1597.

[92] Okamoto, K.; Tanihara, N.; Watanabe, H.; Tanaka, K.; Kita, H.; Nakamura, A.; Kusuki, Y.; Nakagawa, K. Vapor permeation and pervaporation separation of water-ethanol mixtures through polyimide membranes. *J. Membr. Sci.* 68 (1992) 52-63.

[93] Kim, J. H.; Lee, K. H.; Kim, S. Y. Pervaporation separation of water from ethanol through polyimide composite membranes. *J. Membr. Sci.* 169 (2000) 81-93.

[94] Qariouh, H.; Schué, R.; Schué, F.; Bailly, C. Sorption, diffusion and pervaporation of water/ethanol mixtures in polyetherimide membranes. *Polym. Int.* 48 (1999) 171-180.

[95] Yoshikawa, M.; Yokoi, H.; Sanui, K.; Ogata, N. Selective separation of water-alcohol binary mixture through poly(maleimide-co-acrylonitrile) membrane. *J. Polym. Sci.: Polym. Chem. Edition.* 22 (1984) 2159-2168.

[96] Chung, T. S.; Guo, W. F.; Liu, Y. Enhanced Matrimid membranes for pervaporation by homogeneous blends with polybenzimidazole (PBI). *J. Membr. Sci.* 271 (2006) 221-231.

[97] Tsai, H. A.; Ciou, Y. S.; Hu, C. C.; Lee, K. R.; Yu, D. G.; Lai, J. Y. Heat-treatment effect on the morphology and pervaporation performances of asymmetric PAN hollow fiber membranes. *J. Membr. Sci.* 255 (2005) 33-47.

[98] Zhou, F. B.; Koros, W. J. Study of thermal annealing on Matrimid fiber performance in pervaporation of acetic acid and water mixtures. *Polymer.* 47 (2006) 280-288.

[99] Hasegawa, M.; Horie, K. Photophysics, photochemistry, and optical properties of polyimides. *Prog. Polym. Sci.* 26 (2001) 259-335.

[100] Kawakami, H.; Mikawa, M.; Nagaoka, S. Gas transport properties in thermally cured aromatic polyimide membranes. *J. Membr. Sci.* 118 (1996) 223-230.

[101] Bos, A.; Punt, I. G. M.; Wessling, M.; Strathmann, H. Plasticization-resistant glassy polyimide membranes for CO_2/CH_4 separations. *Sep. Purif. Technol.* 14 (1998) 27-39.

[102] Shao, L.; Chung, T. S.; Goh, S. H.; Pramoda, K. P. Polyimide modification by a linear aliphatic diamine to enhance transport performance and plasticization resistance. *J. Membr. Sci.* 256 (2005) 46-56.

[103] Liu, Y.; Wang, R.; Chung, T. S. Chemical cross-linking modification of polyimide membranes for gas separation. *J. Membr. Sci.* 189 (2001) 231-239.

[104] Cao, C.; Chung, T. S.; Liu, Y.; Wang, R.; Pramoda, K. P. Chemical cross-linking modification of 6FDA-2,6-DAT hollow fiber membranes for natural gas separation. *J. Membr. Sci.* 216 (2003) 257-268.

[105] Qiao, X. Y.; Chung, T. S. Diamine modification of P84 polyimide membranes for pervaporation dehydration of isopropanol. *AIChE J.* 52 (2006) 3462-3472.

[106] Jiang, L. Y.; Chung, T. S.; Rajagopalan, R. Dehydration of alcohols by pervaporation through polyimide Matrimid® asymmetric hollow fibers with various modifications. *Chem. Eng. Sci.* 63 (2008) 204-216.

[107] Qiao, X. Y.; Chung, T. S.; Rajagopalan, R. Zeolite filled P84 co-polyimide membranes for dehydration of isopropanol through pervaporation process. *Chem. Eng. Sci.* 61 (2006) 6816-6825.

[108] Jiang, L. Y.; Chung, T. S.; Rajagopalan, R. Matrimid®/MgO mixed matrix membranes for pervaporation. *AIChE J.* 53 (2007) 1745-1757.

[109] Tu, C. Y.; Liu, Y. L.; Lee, K. R.; Lai, J. Y. Hydrophilic surface-grafted poly(tetrafluoroethylene) membranes using in pervaporation dehydration processes. *J. Membr. Sci.* 274 (2006) 47-55.

[110] Shah, D.; Kissich, K.; Ghorpade, A.; Hannah, R:; Bhattacharyya, D. Pervaporation of alcohol-water and dimethylformamVeride-water mixtures using hydrophilic zeolite NaA membranes: mechanisms and experimental results. *J. Membr. Sci.* 179 (2000) 185-205.

[111] Gallego-Lizon, T.; Ho, Y. S.; Santos, L. F. d. Comparative study of commercially available polymeric and microporous silica membranes for the dehydration of IPA/water mixtures by pervaporation/vapour permeation. *Desalination.* 149 (2002) 3-8.

[112] Verkerk, A. W.; Male, P. v.; Vorstman, M. A. G.; Keurentjes, J. T. F. Properties of high flux ceramic pervaporation membranes for dehydration of alcohol/water mixtures. *Sep. Purif. Technol.* 22 (2001) 689-695.

[113] Sato, K.; Nakane, T. A high reproducible fabrication method for industrial production of high flux NaA zeolite membrane. *J. Membr. Sci.* 301 (2007) 151-161.

[114] Jafar, J. J.; Budd, P. M. Separation of alcohol/water mixtures by pervaporation through zeolite A membranes. *Microporous Mater.* 12 (1997) 305-311.

[115] Huang, A. S.; Yang, W. S. Hydrothermal synthesis of NaA zeolite membrane together with microwave heating and conventional heating. *Mater. Lett.* 61 (2007) 5129-5132.

[116] Feng, X.; Huang, R. Y. M. Separation of isopropanol from water by pervaporation using silicone-based membranes. *J. Membr. Sci.* 74 (1992) 171-181.

[117] Hong, Y. K.; Hong, W. H. Influence of ceramic support on pervaporation characteristics of IPA/water mixtures using PDMS/ceramic composite membrane. *J. Membr. Sci.* 159 (1999) 29-39.

[118] Gupta, T.; Pradhan, N. C.; Adhikari, B. Separation of phenol from aqueous solution by pervaporation using HTPB-based polyurethaneurea membrane. *J. Membr. Sci.* 217 (2003) 43-53.

[119] Yoshikawa, M.; Wano, T.; Kitao, T. Specialty polymeric membranes. 1. Modified polybutadiene membranes for alcohol separation. *J. Membr. Sci.* 76 (1993) 255-259.

[120] Ji, W. C.; Sikdar, S. K.; Hwang, S. T. Modeling of multicomponent pervaporation for removal of volatile organic compounds from water. *J. Membr. Sci.* 93 (1994) 1-19.

[121] Fadeev, A. G.; Selinskaya, Y. A.; Kelley, S. S.; Meagher, M. M.; Litvinova, E. G.; Khotimsky, V. S.; Volkov, V. V. Extraction of butanol from aqueous solutions by pervaporation through poly(1-trimethylsilyl-1-propyne). *J. Membr. Sci.* 186 (2001) 205-217.

[122] Gonzalez-Velasco, J. R.; Gonzalez-Marcos, J. A.; Lopez-Dehesa, C. Pervaporation of ethanol-water mixtures through poly(1-trimethylsilyl-1-propyne) (PTMSP) membranes. *Desalination.* 149 (2002) 61-65.

[123] Lipnizki, F.; Hausmanns, S.; Ten, P. K.; Field, R. W.; Laufenberg, G. Organophilic pervaporation: prospects and performance. *Chem. Eng. J.* 73 (1999) 113-129.

[124] Uragami, T.; Ohshima, T.; Miyata, T. Removal of benzene from an aqueous solution of dilute benzene by various cross-linked poly(dimethylsiloxane) membranes during pervaporation. *Macromolecules.* 36 (2003) 9430-9436.

[125] Song, K. M.; Hong, W. H. Dehydration of ethanol and isopropanol using tubular type cellulose acetate membrane with ceramic support in pervaporation process. *J. Membr. Sci.* 123 (1997) 27-33.

[126] Xiangli, F. J.; Chen, Y. W.; Jin, W. Q.; Xu, N. P. Polydimethylsiloxane (PDMS)/ceramic composite membrane with high flux for pervaporation of ethanol-water mixtures. *Ind. Eng. Chem. Res.* 46 (2007) 2224-2230.

[127] Nagase, Y.; Ando, T.; Yun, C. M. Syntheses of siloxane-grafted aromatic polymers and the application to pervaporation membrane. *Reactive Funct. Polym.* 67 (2007) 1252-1263.

[128] Kujawski, W. Pervaporative removal of organics from water using hydrophobic membranes. Binary mixtures. *Sep. Sci. Technol.* 35 (2000) 89-108.

[129] Li, L.; Xiao, Z. Y.; Tan, S. J.; Pu, L. Z.; Zhang, B. Composite PDMS membrane with high flux for the separation of organics from water by pervaporation. *J. Membr. Sci.* 243 (2004) 177-187.

[130] Chang, C. L.; Chang, M. S. Preparation of multi-layer silicone/PVDF composite membranes for pervaporation of ethanol aqueous solutions. *J. Membr. Sci.* 238 (2004) 117-122.

[131] Nam, S. Y.; Lee, Y. M. Pervaporation separation of methanol/methyl t-butyl ether through chitosan composite membrane modified with surfactants. *J. Membr. Sci.* 157 (1999) 63-71.

[132] Yoshikawa, M.; Yoshioka, T.; Fujime, J.; Murakami, A.; Pervaporation separation of MeOH/MTBE through agarose membranes. *J. Membr. Sci.* 178 (2000) 75-78.

[133] Wang, Y. C.; Li, C. L.; Huang, J.; Lin, C.; Lee, K. R.; Liaw, D. J.; Lai, J. Y. Pervaporation of benzene/cyclohexane mixtures through aromatic polyamide membranes, *J. Membr. Sci.* 185 (2001) 193-200.

[134] Knight, K. F.; Duggal, A.; Shelden, R. A.; Thompson, E. V. Dependence of diffusive permeation rates and selectivities on upstream and downstream pressures: V. Experimental results for the hexane/heptane (ideal) and toluene/ethanol (nonideal) systems. *J. Membr. Sci.* 26 (1986) 31-50.

[135] Wessling, M.; Werner, U.; Hwang, S. T. Pervaporation of aromatic C8-isomers. *J. Membr. Sci.* 57 (1991) 257-270.

[136] Miyata, T.; Iwanoto, T.; Uragami, T. Characteristics of permeation and separation for propanol isomers through poly(vinyl alcohol) membranes containing cyclodextrin. *J. Appl. Polym. Sci.* 51 (1994) 2007-2014.

[137] Kusumocahyo, S. P.; Kanamori, T.; Sumaru, K.; Iwatsubo, T.; Shinbo, T. Pervaporation of xylene isomer mixture through cyclodextrins containing polyacrylic acid membranes. *J. Membr. Sci.* 231 (2004) 127-132.

[138] Paris, J.; Molina-Jouve, C.; Nuel, D.; Moulin, P.; Charbit, F. Enantioenrichment by pervaporation. *J. Membr. Sci.* 237 (2004) 9-14.

[139] Yamasaki, A.; Shinbo, T.; Mizoguchi, K. Pervaporation of benzene/cyclohexane and benzene/n-hexane mixtures through PVA membrane. *J. Appl. Polym. Sci.* 64 (1997) 1061-1065.

[140] Inui, K.; Tsukamoto, K.; Miyata, T. Permeation and separation of a benzene/cyclohexane mixture through benzoylchitosan membranes. *J. Membr. Sci.* 138 (1998) 67-75.

[141] Peng, F. B.; Hu, C. L.; Jiang, Z. Y. Novel poly(vinyl alcohol)/carbon nanotube hybrid membranes for pervaporation separation of benzene/cyclohexane mixtures. *J. Membr. Sci.* 297 (2007) 236-242.

[142] Uragami, T.; Tsukamoto, K.; Miyata, T. Performance of butyrylcellulose membranes for benzene/cyclohexane mixtures containing a low benzene concentration by pervaporation. *Biomacromolecules.* 5 (2004) 2116-2121.

[143] Yanagishita, H.; Arai, J.; Sandoh, T.; Negishi, H.; Kitamoto, D.; Ikegami, T.; Haraya, K.; Idemoto, Y.; Koura, N. Preparation of polyimide composite membranes grafted by electron beam irradiation. *J. Membr. Sci.* 232 (2004) 93-98.

[144] Yang, L. M.; Kang, Y. R.; Wang, Y. L.; Xu, L. W.; Kita, H.; Okamoto, K. I. Synthesis of crown ether-containing copolyimides and their pervaporation properties to benzene / cyclohexane mixtures. *J. Membr. Sci.* 249 (2005) 33-39.

[145] Kusumocahyo, S. P.; Ichikawa, T.; Shinbo, T.; Iwatsubo, T.; Kameda, M.; Ohi, K.; Yoshimi, Y.; Kanamori, T. Pervaporative separation of organic mixtures using dinitrophenyl group-containing cellulose acetate membrane. *J. Membr. Sci.* 253 (2005) 43-48.

[146] Wolińska-Grabczyk, A. Effect of the hard segment domains on the permeation and separation ability of the polyurethane-based membranes in benzene/cyclohexane separation by pervaporation. *J. Membr. Sci.* 282 (2006) 225-236.

[147] Luo, Y. J. Pervaporation properties of EC membrane crosslinked by hyperbranched-polyester acrylate. *J. Membr. Sci.* 303 (2007) 183-193.

[148] Jonquieres, A.; Clement, R.; Lochon, P.; Neel, J.; Dresch, M.; Chretien, B. Industrial state-of-the-art of pervaporation and vapour permeation in the western countries. *J. Membr. Sci.* 206 (2002) 87-117.

[149] Blume, I.; Wijmans, J. G.; Baker, R. W. The separation of dissolved organics from water by pervaporation. *J. Membr. Sci.* 49 (1990) 253-286.

[150] Strathmann, H.; Gudernatsch, W. Pervaporation in biotechnology in R. Y. M. Huang edited Pervaporation Membrane Separation Processes. Elsevier Science Publisher B.V., New York (1991).

[151] O'Brien, D. J.; Roth, L. H.; McAloon, A. J. Ethanol production by continuous fermentation-pervaporation: a preliminary economic analysis. *J. Membr. Sci.* 166 (2000) 105-111.

In: Advances in Membrane Science and Technology
Editor: Tongwen Xu
ISBN: 978-1-60692-883-7

Chapter VI

Membrane Bioreactor: Theory and Practice

In S. Kim*[*1], Namjung Jang[2] and Xianghao Ren[1]

[1] Dep. of Env. Sci. and Eng., Gwangju Institute of Sci. and Tech. (GIST), Republic of Korea

[2] Environmental/Regional Planning Team, Jeonbuk Development Institute, Republic of Korea

Abstract

The characteristics of a membrane bioreactor (MBR) process, using different types of MBR, were compared with a conventional activated sludge process. During membrane filtration using the MBR, the fouling factors were divided into suspended solids and soluble materials in relation to the characteristics of Extracellular Polymeric Substances (EPS) and Soluble Microbial Products (SMP), respectively. The relationship between EPS and SMP was defined, with their effects on membrane fouling also investigated. A modified fouling index and fouling reduction methods were established for the monitoring and control of fouling of the MBR. Simple flux decline models were introduced for the MBR. Finally, the current MBR practices have been summarized, which indicated that the submerged MBR was the most promising biological process for the wastewater treatment and water reuse.

Keywords: MBR; Fouling; EPS; SMP; Molecular weight distributions; Hydrophobicity.

[*] Correspondence concerning this chapter should be addressed to In S. Kim at iskim@gist.ac.kr

1. Membrane Bioreactor (MBR)

1.1. What Is a MBR?

A membrane bioreactor (MBR) is a combination of biological activated sludge and physical membrane separation processes. Conventional activated sludge (CAS) is the most commonly using process in wastewater treatment facilities, where aerobic bacteria decompose various organic contaminants into water and carbon dioxide. In the CAS process, aerobic bacteria are produced in the aeration tank by the decomposition of BOD components using air bubbles. The excess bacteria (sludge) are then separated in a clarifier by difference in their specific gravity, resulting in the settling of sludge, with the subsequent disposal problems associated in the CAS operation. In the MBR process, the membrane modules take the place of the clarifier to minimize the sludge problems associated with CAS.

Stephenson et al. [1] cited Smith et al. [2] as first attempting to combine membrane technology and CAS who; in 1969, made use of ultrafiltration membranes for solid-liquid separation for CAS. In another early report, in 1970, Hardt et al. [3] used a 10 L aerobic bioreactor for treating a synthetic sewage using a dead-end ultrafiltration membrane for sludge separation. The first full-scale commercial aerobic MBR process was applied in the late 1970s in North America, and then in Japan in the early 1980s [1]. The introduction of an aerobic MBR into Europe did not occur until the middle of 1990s [1].

The MBR process incurs high maintenance cost due to the replacement of the membrane system, which has blocked theist extended application. However, advances in membrane technology have resulted in lower cost, higher quality membranes for application to MBR. Nowadays, MBR has gained considerable attention in wastewater treatment, as well as for water reclamation and reuse, and has become the most promising process in biological water treatment systems.

1.2. Advantages and Disadvantages of the MBR

The MBR process shows several advantages over CAS, which are outlined, as followings:

A. Better Effluent Quality

Theoretically, the size of aerobic bacteria is larger than that of membrane pores; therefore, all bacteria cells will remain in the bioreactor, which creates good conditions for bacteria as the have a much lower relative growth rate than aerobic microorganisms, such as nitrifying bacteria [4]. The enzymes that would otherwise be lost in the conventional sedimentation step are also retained. A membrane can retain high molecular weight soluble compounds; thereby, improve the opportunity for their oxidation in biological reactors. The membrane in the MBR plays a role not only in solid-liquid separations, but also in organics removal due to adsorption.

B. High Resistance to Shock Loads

The MBR absorbs various hydraulic and organic loads within the system, which also deals with the sludge problem, resulting in a stable effluent quality. Ueda and Hata [5] reported that the MBR was also able to cope with up to three fold short-term inflow fluctuations.

C. Low Sludge Production

The MBR produces less sludge than CAS due to the relatively low F/M ratio and high sludge age [6, 7] reported that the yield coefficient for an aerobic membrane separation process for treating municipal wastewater was 0.23 kg·SS/kg·COD removed, which was below the 0.3 to 0.5 kg·SS/kg·COD removed by a conventional activated sludge process.

D. Systems Are Simple and Compact

They can be operated with biomass concentrations around 10,000 ~ 20,000 mg/L, and up to 50,000 mg/L, with zero-sludge wastage [8], compared with 1,500-3,500 mg/L for CAS. If the reaction rate is assumed to be directly proportional to the biomass, the shorter hydraulic retention times also means the volume of the aeration tank can be smaller. The MBR also requires no post-treatment for additional solids removal.

E. Easy Automatic Control and Maintain, Less Labor Requirement

Since the MBR are simple and compact, as described in item "D", their operation within a treatment process can be easily automated and controlled with an automatic system, resulting in the need for less labor.

However, there are several disadvantages associated with the application of MBR for greater economic benefits and better efficiency.

F. Membrane Fouling

Membrane fouling caused by various factors in the bioreactor can reduce the plant efficiency, shorten the membrane life and lead to frequent regeneration of the membrane. Reducing membrane fouling is the most important research area for the MBR. This will be discussed further in section 2.

G. Decreased Biomass Viability and Accumulation of Non-Reactive Compounds

A long solids retention time (SRT) results in decreased biomass viability. Retention and accumulation of non-reactive compounds within the bioreactor as a result of the membrane can lead to microbial inhibition or toxicity, and fouling of the membrane surface. Therefore, adequate SRT controls are required for the activity of microorganisms and to prevent fouling in the MBR.

H. High Oxygen Consumption

The MBR requires high oxygen consumption and minimum maintenance energy compared with CAS due to the high biomass concentrations.

I. Difficulties in Disposing of Chemical Waste from Cleaning

The disposal of chemical waste from membrane cleaning also presents difficulties, which can also lead to secondary pollution.

1.3. Types of MBR

The application of MBR has dramatically increased, with examples of MBR in practice shown in section 4. First, the aerobic MBR started from a cross-flow type of reactor, with a normal membrane filtration system. The membrane module is submerged in the bioreactor, which is currently the most common type, but the type is not fixed in research. New types of MBR module have been developed for low energy consumption and high resistance to membrane fouling, i.e. the trial of transverse-flow type module introduced in this section.

A. Cross-Flow System

In this configuration, also called side-stream MBR, the retentate is recycled back to the bioreactor (figure 1 (a)). In the cross-flow MBR; however, the shear stress produced by cross-flow velocity breakdown bacterial cells and induces the discharge of granulated matter, such as glycogen and cell wall fragments [9]. The shear hampers the microbial activity of microorganisms, and the smaller particles contribute to the increased mean specific filtration resistance [10]. Membranes in an external loop also have a higher energy requirement [11].

B. Submerged System

A membrane module consists of a strong and flexible hollow fiber directly submerged into a bioreactor (figure 1 (b)), which was first introduced by Yamamoto et al. [12]. Permeate flux is drawn from the outside of the fiber to the module ends. Bubble aerators are equipped at the bottom of the membrane modules to agitate the membrane fiber to reduce membrane fouling and provide a portion of the biological process oxygen required. Hollow fiber membranes are set in motion, and then come into contact with each other, leading to a further mechanical cleaning effect. However, immersed membrane pore spaces may be prone to clogging. Even if intermittent suction and low-pressure operation are applied to the submerged MBR, membrane fouling is inevitable. This limited permeate flux requires a greater membrane area and increases the complexity of the membrane filtration systems.

C. Transverse-Flow System

Hollow fiber membrane modules can be of tube-side fed or shell-fed types. The shell-side fed module can be operated with either a longitudinal (feed flow along the fiber) or transversal flow (feed flow perpendicular to the fiber). In a transversal flow membrane module the fibers play a role as turbulence promoters, without the need to apply higher feed flow velocities or auxiliary turbulence promoters (figure 1 (c)). Therefore, compared with longitudinal flow modules, transverse-flow modules require less membrane area and less energy consumption to produce a given amount of permeate [13]. However, producing methods with this kind of transversal membrane module is rather complicated. Although much effort has been put into making transverse membrane modules for commercial use, using a centrifugal casting and other complex techniques [14, 15], transversal flow microfiltration modules can be produced in the laboratory, but no information of the production of transverse-flow modules have been reported until now.

As well as the application of membrane filtration with CAS, there has been much research on the combination of biological process with membrane filtration. The anaerobic MBR (anaerobic process + membrane filtration) is another potential process, even though the membrane maintenance system can be complex for application to a real plant compared to the aerobic MBR. Also, the combination of membrane filtration and a biological wastewater treatment process has been the focus of researchers, such as A_2O, sequencing batch reactor (SBR), rotating biological contactor (RBC), aerobic granulation reactor and biological media processes, and so on.

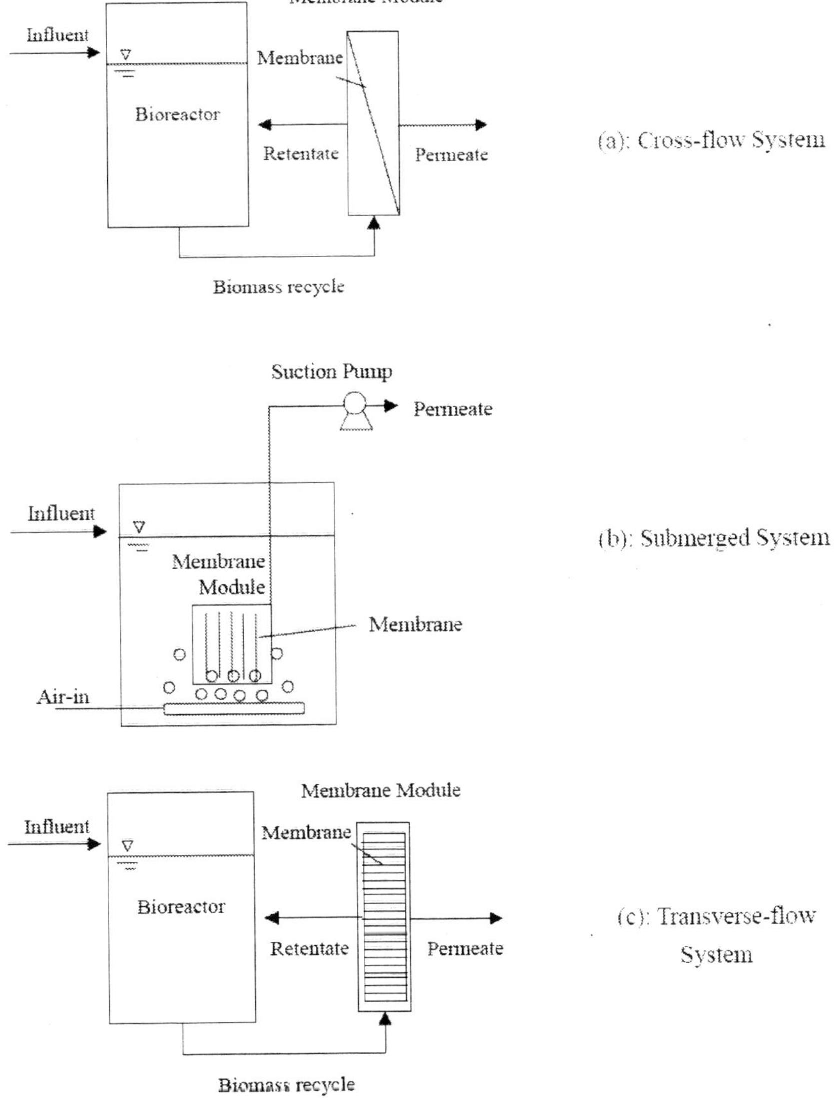

Figure 1. Schematic diagrams of different types of MBR.

2. Membrane Fouling in MBR

2.1. Fouling Factors of MBR

Membrane fouling in the MBR is attributed to the physicochemical interactions between the biofluid (soluble materials and suspended solids) and membrane, and the most serious problem affecting the membrane separation system. The fouling leads to permeate flux decline, resulting in the need for greater membrane cleaning and replacement, which increase the operating costs.

Chang et al. [16] summarized three categories of MBR fouling: membrane characteristics (configuration, mineral, hydrophobicity, porosity, pore size), biomass characteristics (mixed liquor suspended solids; MLSS, extracellular polymeric substances; EPS, floc structure, dissolved matter, floc size) and operating conditions (configuration, cross flow velocity, aeration, hydraulic retention time, solids retention time, trans-membrane pressure). Fouling factors relating to the MBR are very complex, with their relationships being extremely hard to find, as all the factors are linked and depend on the characteristics of the microorganism communities within the bioreactor.

In this section, to simplify the complex matrix of components, factors affecting bio-fouling were divided into two categories; suspended solids and soluble materials in the mixed liquor (ML).

The rheological and physiological characteristics of suspended solids (sludge or biological floc) influence the filterability of sludge and the formation of a cake layer on the membrane surface. Wu et al. [17] reported that the capillary suction time (CST; sludge filterability test) had the greatest impact on membrane fouling, followed by SCOD, carbohydrate, EPS, proteins, apparent viscosity and MLSS. The MLSS concentration is not always proportional to membrane fouling. Some researchers have suggested a higher sludge concentration resulted in less fouling, implying that membrane fouling is related not only to the sludge quantity, but to its characteristics also [18, 19]. Filamentous bacteria can also affect the hydrophobicity of sludge due to their hydrophobic cell walls. The presence of filamentous bacteria is not always associated with poor settleability (high sludge volume index (SVI)), which to some extent supports the idea of their role as a backbone in the flocs [20]. However, their overgrowth is always associated with settling problems, which could increase membrane fouling.

Soluble materials are able to block the membrane pores, and these rejected solutes build up in the region adjacent to membrane, increasing their concentrations compared to the bulk solution. The accumulated solutes at the membrane surface result in a concentration gradient of solutes between the membrane surface and the bulk solution. This gradient results in a diffusive flow of solutes or particles from the membrane surface back into the bulk solution. The creation of this 'concentration polarization' boundary layer and the back flow of solutes from the membrane surface to the bulk create a hydrodynamic resistance, which must be overcome by ΔP. Also, the mechanism of fouling by soluble materials related with the characteristics of solute, such as concentration, electrochemical character and components etc. Soluble microbial products (SMP) especially are the focus of main soluble materials in the biological treatment process.

Figure 2 shows the biological fouling components, key variables and parameters of fouling in a MBR. The main components of suspended solids and soluble materials are biological floc and SMP, respectively. The influence of size/structure and hydrophobicity of EPS are key variables characteristics of a biological floc. Also, the size/structure of a biological floc formed by an EPS matrix is affected by the hydrophobicity. The molecular weight distribution (MWD) and hydrophobicity are key variables characteristics of the SMP. The MWD is linked with the hydrophobicity. Also, EPS, via their release, can affect the characteristics of SMP. These key variables are affected by the microorganism community, operation conditions and influent characteristics. This simplified fouling structure is described further in the next sections.

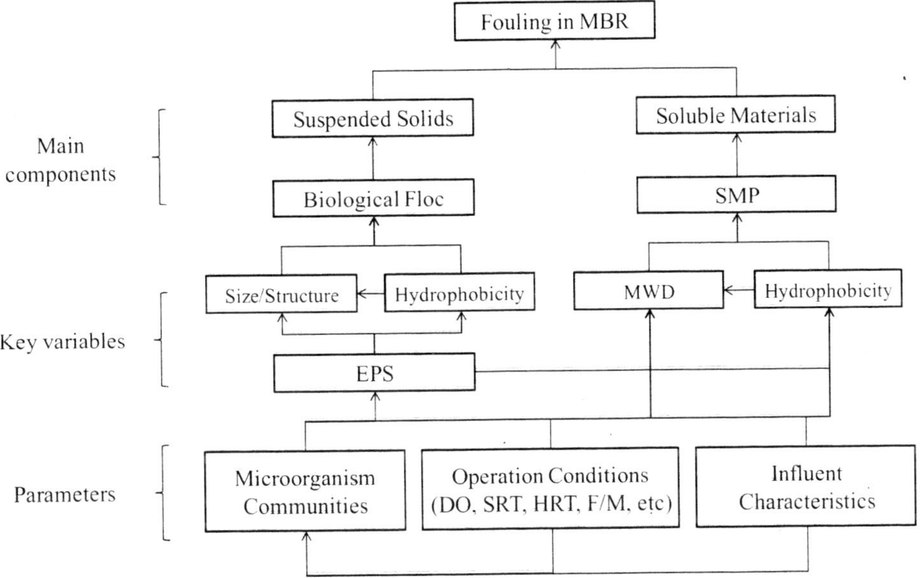

Figure 2. Factors influencing membrane fouling in a MBR.

2.2. Suspended Solids Verses Soluble Materials in the Fouling

There has been research on the fractionation effect of ML on the fouling caused by suspended solids and soluble materials. Bouhabila et al. [21] separated sludge samples into suspended solids and soluble materials (colloids and solutes), and reported the relative role of different fractions in membrane fouling compared to other researchers [22, 23]. Also, some researchers have shown other fractionation results [24, 25]. Table 1 shows the results of the fractionation of suspended solids and soluble materials. The differences in the results might come from the substrate, biomass concentration, methods for fractionation and aeration intensity.

Despite these differences, it appears that suspended solids are only partially responsible for the fouling, with colloids being the main contributor. If the aeration intensity was sufficient to prevent the attachment of suspended solids, the portion of cake resistance should theoretically decrease. However, Meng et al. [26] reported that a large aeration intensity of

800 L/h resulted in severe breakup of biological flocs, with the release of soluble materials into the bulk solution, causing soluble materials to become the main foulants in their MBR operation. Also, aeration is essential for the bio-flocculation and oxidation of organic materials by biomass. Jin et al. [27] reported that the rate of membrane fouling in a low DO reactor was 7.5 times faster than that in a high DO reactor. They indicated that the biofilm porosity on the membrane surface in the low DO (0.65) was smaller than that in the high DO (0.85), which resulted in lower filterability under low DO conditions (increase specific cake resistance). Therefore, it is necessary to apply the optimum aeration intensity, with fine air bubbles, to the MBR system in relation to efficient oxygen transfer for biomass growth, and high intensity of aeration, with course air bubbles, near the membrane for the control of fouling.

Table 1. Relative role of different sludge fractions in membrane fouling

Reference	Membrane (%)	Solutes (%)	Colloids (%)	Suspended solids (%)
Bouhabila et al. [21]	-	26	50	24
Defrance [22]	-	5	30	65
Wisniewski and Grasmick [23]	-	52	25	23
Chang and Lee [24] (YM3 non-foaming)	38	7		55
Trussell [25] SRT 2d	21	64		15
Trussell [25] SRT 10d	9	88		3

2.3. Extracellular Polymeric Substances (EPS) and Soluble Microbial Products (SMP)

A. What Are EPS?

The production of EPS is a general property of microorganisms in natural environments, and has been shown to occur both in prokaryotic (bacteria, Archaea) and in eukaryotic (algae, fungi) microorganisms. Wingender et al. [28] reported that EPS, defined as "EPS of biological origin, participate in the formation of microbial aggregates", with another definition being "Organic polymers of microbial origin, which; in biofilm systems, are frequently responsible for the binding of cells and other particulate materials (cohesion) and to the substratum (adhesion)". EPS can be classified into bound (sheaths, capsular polymers, condensed gel, loosely bound polymers, and attached organic material) and soluble EPS (soluble macro-molecules, colloids, and slimes) [28]. However, no standard words are use to define EPS. Some researchers used EPS or ECP as the sum of bound EPS and soluble EPS; whereas, others only focused on SMP themselves in the effluent of a biological treatment process. In this section, EPS refers to bound EPS and SMP, which are equal to soluble EPS.

Early biofilm research often assumed that polysaccharides were the most abundant components of EPS. However, Nielsen et al. [29] reported that proteins and nucleic acids also appeared in significant amounts or even predominated in EPS. Dignac et al. [30] reported that the organic analyses of sludge and EPS showed that EPS were predominantly composed of proteins (Carbohydrate and proteins were major components of EPS to an extent of 70-80 %). Even though some research has found humic acids, uronic acids and DNA as components of EPS, Liu and Fang [31] summarized that carbohydrates and proteins were the most commonly analytes found in an EPS study.

B. Production of EPS

Wingender et al. [28] reported that the localization and composition of EPS may be the result of different processes: active secretion, shedding of cell surface material, cell lysis and adsorption from the environment. They also reported that the relationship of EPS production and biomass growth were unclear. Some researchers have reported less EPS production with rapidly consuming substrate and bacteria growth, and others have shown controversial results. Laspidou and Ritmann [32] reported that the relationship between EPS production and the rate of biomass growth seems to depend on the kind of microorganisms involved and the system conditions.

C. What Are SMP?

Soluble microbial products (SMP) are defined as soluble cellular components released during cell lysis, which diffuse through the cell membrane, and are lost during synthesis or usually form the majority of the effluent chemical oxygen demand (COD) and biochemical oxygen demand (BOD) from a biological treatment process. SMP can be divided into two categories; 1) substrate-utilization-associated products (UAP), which are produced directly during substrate metabolism, and 2) biomass-associated products (BAP), which are the biomass formed presumably as part of decay [33].

D. Production of SMP

Rittmann and McCarty [33] showed the formation of SMP via the following equation:

$$r_{SMP} = k_{UAP} k X_a + k_{BAP} X_a \qquad (1)$$

where, k =specific substrate-utilization rate (mg Substrate/mg Active Biomass • d)

X_a =active biomass concentration (mg/L)
k_{UAP}=UAP-formation coefficient (mg UAP/mg Substrate)
k_{BAP}=BAP-formation coefficient (mg BAP/mg Active biomass • d)

E. Unified Theory for EPS and SMP

Laspidou and Ritmann [32] suggested a unified theory of how active and inert biomass, EPS, UAP and BAP relate to each other. The schematic representation of the unified model is shown in figure 3. The summary of their hypotheses is shown below.

1) SMP are the sum of the BAP and UAP.
2) EPS are hydrolyzed to form BAP.

3) UAP are formed directly from, and in proportion to, substrate utilization.
4) EPS are formed directly during, and in proportion to, substrate utilization.
5) Active biomass is composed of EPS and active cells, while a portion of EPS is included with the inert biomass.
6) True dead cell residue is produced as part of the endogenous decay of biomass, which comprises part of the inert biomass.
7) BAP and UAP cycle back to become electron-donor substrate cells, since they are biodegradable.

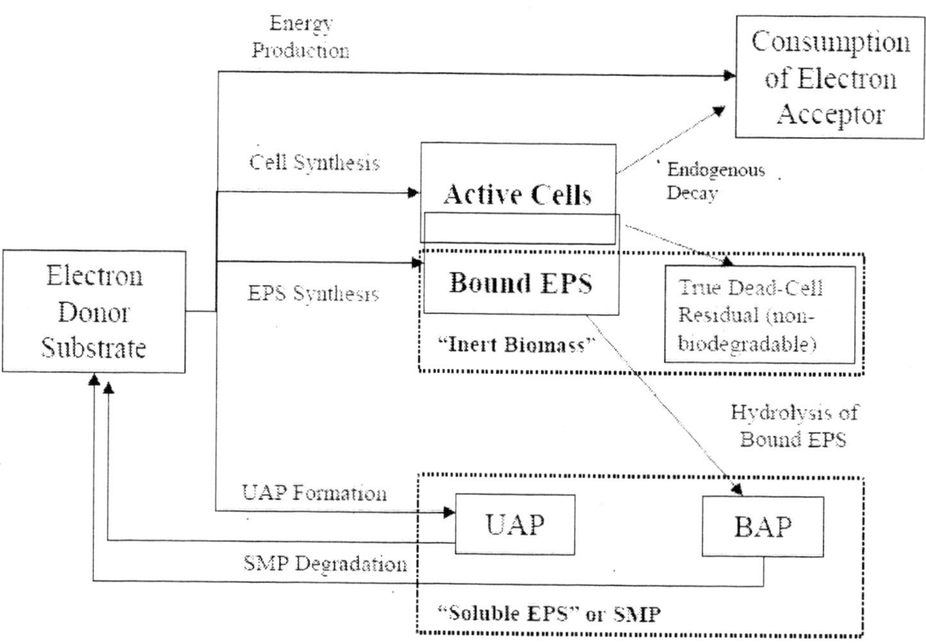

Figure 3. Schematic representation of the unified model for active biomass, EPS, SMP and inert biomass (adopted from Laspidou and Ritmann [32]).

F. Extraction of EPS

Various methods are used for EPS extraction, such as heat, sonication, EDTA, formaldehyde, sodium hydroxide, sodium chloride and so on. The controversial results obtained from EPS experiments may be the result of the different EPS extraction methods employed. Therefore, it is necessary to set standard protocol for the extraction of EPS. In this section; of the various extraction methods, the cation exchange resin (CER) method is introduced due to its simple procedure, without cell rupturing or damage. Also, the CER is easily separated from extraction materials, without the need for chemical or biological reactions, which causes no inhibition during the analyses of the extractants.

Frølund et al. [34] reported the extraction of EPS using a cation exchange resin (CER), which they evaluated and found to be superior to other commonly used methods in terms of yield and minimal disruption of the exopolymers. The protein-carbohydrate ratio in the extract varied significantly with extraction time. Irrespective of the stirring intensity and amount of CER, the results showed that no or only very little lysis occurred within the first 2 hours of extraction. They suggested a mild extraction method, with minimum risk of inducing cell lysis, a short time (0.5 to 1h), 600 rpm stirring intensity and 70 g CER/g VS.

The mechanism of EPS extraction by CER is described in figure 4. To simplify the EPS matrix, the EPS are assumed to be mainly composed of proteins and carbohydrates. Cells, cations, sodium ions, proteins and carbohydrates are shown as circles. Divalent cation ions, such as calcium and magnesium, play a bridging role between the cell and EPS. EPS aggregate the cells and form a microbial floc (figure 4 (a)). CER has sodium ions (figure 4 (b)), which exchange divalent cations with sodium ions (figure 4 (d)) during the physicochemical reaction. Monovalent ions have weaker bonding strengths than divalent ions; therefore, EPS are extracted to the bulk solution (figure 4 (c)). The total extraction mixture contains both of EPS and SMP. The EPS concentration can be obtained by subtraction of the SMP from the total mixture.

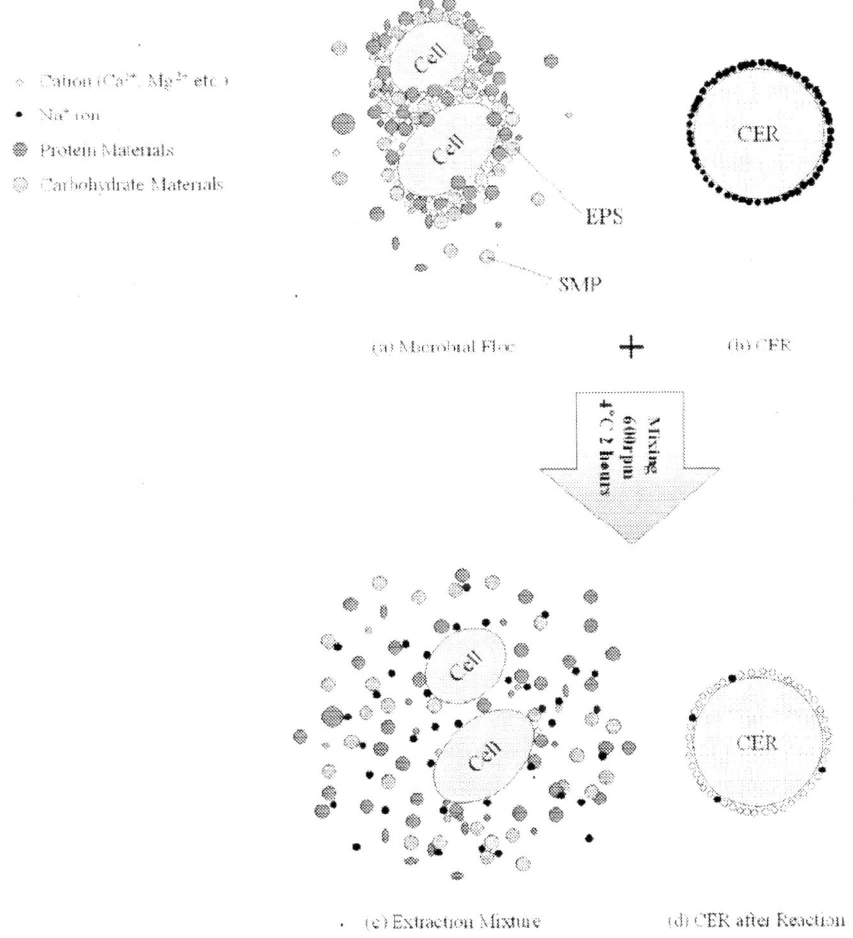

Figure 4. Mechanism of EPS extraction by CER (not to scale).

2.4. EPS and Bio-Flocculation

A. Mechanism of Bio-Flocculation

Biological flocs have been found to have a net negative charge, which is the result of functional groups on the EPS. Sobeck and Higgins [35] reported that the portion of carbohydrates in EPS was made up of uronic acids, which contain a carboxyl group. Carboxyl groups contribute to the negative charge of biological flocs due to their unprotonated characteristics within typical pH ranges. Also, proteins are rich in amino acids, which also contain carboxyl groups, such as glutamic and aspartic acids. They reported the mechanism of bio-flocculation, compared three theories, and concluded that the divalent cation bridging (DCB) theory best described the major role of cations in bio-flocculation. Their works are summarized below.

DLVO theory

The DLVO theory, named after its developers, Derjaguin, Landau, Verwey and Overbeek, is a classical colloidal theory that describes charged particles as having a double layer of counter ions surrounding the particle. The double layer, or cloud of ions, surrounding the particle results in repulsion of adjacent particles and inhibits aggregation. From the DLVO theory, as the ionic strength increases, the size of the double layer decreases, which decreases the repulsion between particles, causing short-range attractive forces that promotes aggregation. However, the limitation of DLVO theory is that the result showed floc deterioration under steady state conditions even though an increase of ionic strength occurs due to the addition of sodium to the system. Interestingly, the result of a batch test of short-term interaction showed that DLVO interactions improved the flocculation of existing particles.

Alginate theory

Alginate is a polysaccharide produced by bacteria, which is typically made up of repeated mannuronic and guluronic acids. The unique composition of this polysaccharide results in the formation of alginate gels in the presence of calcium ions. This gel is typically referred to as the "egg-box model". However, the limitation of alginate theory is that the addition of magnesium results in the promotion of floc properties. It is unlikely that calcium alone is the key to the role of cations in bioflocculation. Also, the role of proteins in bio-flocculation is as important as that of polysaccharides, such as alginate.

Divalent cation bridge theory

The notion of DCB theory was incorporated by McKinney [36] in 1952 and Tezuka [37] in 1969. According to the DCB theory, divalent cations bridge negatively charged functional groups within the EPS, which helps to aggregate and stabilize the matrix of biopolymer and microbes and; therefore, promote bioflocculation. The DCB theory has been supported by other work, which has demonstrated that the addition of sodium caused deterioration in the floc properties due to displacement of divalent cations from the binding sites within the floc [38]. Higgins and Novak [38] reported that when the sum of the monovalent cation concentrations (Na^+, $NH_4{}^+$, and K^+) divided from the sum of the divalent cation concentrations (Ca^{2+} and Mg^{2+}) was greater than 2, this could cause deterioration in the floc property. However, bio-flocculation does not seem to depend only on electrochemical reactions. Urbain et al. [20] described the importance of hydrophobic reactions to bio-flocculation. They reported that internal hydrophobic bonding was involved in the flocculation mechanism, which balanced with hydrophilic bonding, determines the sludge settling properties in a highly hydrated system, such as biological sludges.

B. Relationship of EPS and Bio-Flocculation

Liu and Fang [31] reported that the relationship between EPS and bio-flocculation, and summarized that this might conflict with another theory for the following reasons:

1) Activated sludge retention is a complex process of variables, such as type of wastewater, nutrient level, sludge retention time (SRT), dissolved oxygen and reactor configuration.
2) The amount and composition of EPS for a given sludge are strongly dependent on the extraction methods. The lack of a standardized EPS extraction procedure makes comparison and interpretation of published results difficult.
3) There is no standard characterization protocol for EPS.

The characteristics of an EPS matrix should affect the size and structure of bio-flocculation. However, it is difficult to find standard relationships between bio-flocculation and EPS, as the characteristics of an EPS matrix are determined by the microorganism communities, which depend on various operation conditions and influents.

2.5. Hydrophobicity and Bio-Flocculation

A. Hydrophobicity of Sludge

With an EPS matrix due to the role cations, the hydrophobicity is another important key to bio-flocculation. The hydrophobicity of *Corynebacterium glutamicum* is higher when grown under phosphate-saturated conditions compared with cells grown under phosphate-depleted conditions [39]. Phosphate limitation induces the synthesis of teichuronic acids rather than teichoic acids in gram positive bacteria. However, it has been reported that additional changes might occur in the cell wall structure, because these two polymers are mainly hydrophilic. Therefore, proteic constituents of bacteria, such as fimbrae, have been reported to probably mediate the hydrophobic interactions between cell surfaces.

Jorand et al. [40] reported that the quantity of EPS released by sonication of a hydrophilic strain cultured in the laboratory was greater than that of a hydrophobic strain. Some strains in activated sludge flocs maintain their surface properties (hydrophobicity/hydrophilicity), irrespective of the growth phase, but conversely, others show variations in their course of growth (hydrophobicity is different regarding the type of microorganism and the nutrient conditions).

Jorand et al. [41] found six pieces of evidence for the role of hydrophobic interactions in bio-flocculation.

1) The presence of hydrophobic cells in activated sludge.
2) The preferential adhesion of hydrophobic bacteria to the flocs.
3) Spectroscopic evidence showing the presence of highly pronounced hydrophobic zones inside activated sludge flocs.
4) That activated sludge flocs trapped low water soluble organic compounds.
5) The existence of a positive link between sludge settleability and floc hydrophobicity.
6) A significant increase in flocculation when short alkyl chains were grafted onto a synthetic polymer.

B. Methods for Hydrophobicity Analysis

Relative hydrophobicity (RH) is often measured by the contact angle, bacterial adhesion to hydrocarbons (BATH), salt aggregation and acid precipitation, as well as from adsorption by hydrophobic resin. Liu and Fang [31] have reported methods to test for hydrophobicity, such as BATH and XAD resin. However, there is no standard measurement of hydrophobicity. The hydrophobicity of flocculated sludge may be measured for the sludge as is [41, 42] or after disintegration by ultrasonication [20]. Urbain et al. [20] reported that the measurement of disintegrated sludge was crucial to bio-flocculation.

The BATH method can be applied to find the hydrophobicities of EPS and SMP, as well as sludges [25]. The procedures are as follows: A 50 ml sample is uniformly agitated in 50 ml of hydrocarbon in a separating funnel for 10 min. During the mixing, the hydrophobic materials react with the hydrocarbon, with the hydrophilic materials remaining in the water phase. After 10 min, when the two phases have completely separated, the 40 ml aqueous phase is transferred to other glassware to exclude the error from the separation line. The relative hydrophobicity is expressed as the ratio of the aqueous phase concentration after emulsification (S_e) to that of the initial sample concentration (S_i), as shown below.

$$\text{Relative Hydrophobicity (\%)} = 100 \times \left(1 - \frac{S_e}{S_i}\right) \qquad (2)$$

2.6. SMP and Membrane Fouling

A. Molecular Weight Distribution of SMP

Barker and Stuckey [43] summarized the molecular weight distribution (MWD) of SMP in their review paper, and showed GPC (gel permeation chromatography) and stirred cell as methods for the analysis of SMP MWD. Kuo and Parkin [44] found the distribution of SMP showed the same distribution as aerobic systems: bimodal, with the majority of SMP having a MW of less than 1kDa or greater than 10 kDa, while very little SMP had MWs between 1 and 10 kDa. Confer and Logan [45, 46] showed that small MW compounds (<1kDa) accumulated in solution during both carbohydrate and protein degradations.

However, the release of EPS could increase the concentration and MWD of SMP. Jang et al. [47] reported that floc deterioration caused an increase of the SMP concentration under oxygen deficient conditions (the protein SMP concentration, carbohydrate SMP portion of the high molecular weight components were increased), which was assumed to cause severe membrane pore blocking. It is important to monitor the MWD of soluble materials in the bioreactor to control fouling. Soluble materials at a high concentration could cause less membrane fouling than a low concentration containing high MW materials.

B. Filtration of Protein and Carbohydrate

Kim and Jang [48] reported, from a Fourier transform infrared (FT-IR) analysis, that proteins and carbohydrates were the dominant foulants on the surface of fouled membrane from a MBR. Also, Kimura et al. [49] reported that carbohydrates were the dominant membrane foulants, with a high F/M making the foulants more proteinaceous from their FI-IR and nuclear magnetic resonance (NMR) analyses. It is important to understand the filtration mechanism or characteristics of proteins and carbohydrates. Many researchers have focused on macromolecule protein and carbohydrate fouling during membrane filtration [25, 50-54].

Kelly and Zydney [51] reported that the rate of BSA aggregation during protein storage increased with increasing pH. Also, the increased hydrophobicity affected the protein aggregation due to the protein-protein interactions. Protein aggregation plays an important role in protein fouling, and the mechanism has been well described in the literature [50]. They reported that the initial flux decline was due to the deposition of large BSA aggregates onto the membrane surface, which then served as attachment sites for the deposition of the bulk protein.

Trussell [25] reported the carbohydrate SMP and total SMP were well correlated with the rates of steady-state membrane fouling (R^2=0.72 and R^2=0.77) comparing to protein (R^2=0.36). Also, Cicek et al. [55] showed the correlation between the soluble carbohydrate concentration was better (R^2=0.87) than that of SCOD (<0.1 μm filtration), with better membrane permeability (R^2=0.80), implying further support for the role of carbohydrate SMP in membrane fouling.

2.7. Solids Retention Time (SRT) and Food to Microorganism Ratio (F/M)

Shin and Kang [56] reported that SMP accumulated in the MBR under long SRT conditions (infinite SRT > 20 day of SRT). Chang and Lee [24] reported that the tendency for membrane fouling decreased with increasing SRT. Nuengjamnong et al. [57] reported that specific cake resistances decreased with increasing SRT, whereas, the EPS concentration decreased with the CER extraction method. It seems that the tendency for fouling can not be determined by the SRT. However, a too long or infinite SRT in the operation of MBR causes the accumulation of non-reactive compounds, with toxicity towards microbial activity. Therefore, the optimum conditions of SRT are required in the MBR as the influent.

Some research has focused on the F/M, which is an important operating parameter. Nagaoka et al. [58] reported that a reactor with a higher loading rate showed a sudden increase in the trans-membrane pressure, while a reactor with a lower loading rate showed a delayed pressure increase. Kimura et al. [49] reported that F/M and membrane flux were important operating parameters. Neither the concentration of DOC nor the viscosity of mixed liquor showed any clear relationship with membrane fouling. They also showed that a high

F/M would cause the foulant to be more proteinaceous in their FTIR and NMR (^{13}C nuclear magnetic resonance) analyses. Trussell [25] reported that the rate of steady state fouling increased with increasing F/M in his pilot SMBR.

2.8. Modified Fouling Index (MFI) for MBR

Membrane fouling can be monitored using the trans-membrane pressure or permeate flux in the MBR. However, coping with severe membrane fouling after certain changes to the ML characteristics is too late. Therefore, monitoring the changes in the filterability of ML is constantly required. In this section, a modified fouling index is suggested as a possible monitoring method.

Carman [59] developed a standard model for incompressible cakes based on a modified form of Darcy's law for flow through porous media.

Flow through the cake,

$$J = \frac{1}{A}\frac{dV}{dt} = \frac{P_c}{\mu r^* L} \tag{3}$$

Flow through the support media,

$$J = \frac{1}{A}\frac{dV}{dt} = \frac{P_m}{\mu R_m} \tag{4}$$

where,

J = filtrate velocity (m/s)
V = filtrate volume (m^3)
L = cake thickness (m)
A = filtration area (m^2)
P_m = pressure differences across the support medium (Pa)
P_c = pressure differences across the cake (Pa)
μ = absolute viscosity (Pa·s)
r^* = inverse of the permeability coefficient ($1/m^2$)
R_m = support medium resistance (1/m)

Let $P_t = P_c + P_m$,

$$J = \frac{1}{A}\frac{dV}{dt} = \frac{P_t}{\mu(r^* L + R_m)} \tag{5}$$

Because the cake thickness is difficult to measure in dynamic filtration tests, the expression r^* is replaced by $\alpha\omega$, giving,

$$\frac{dV}{dt} = \frac{P_t A}{\mu(\alpha\omega\frac{V}{A} + R_m)} = \frac{P_t A^2}{\mu(\alpha\omega V + R_m A)} \qquad (6)$$

where, α = specific resistance (m/kg)

ω = mass of cake deposited per unit volume of filtrate (kg/m^3)

The integrated form of equation (6),

$$\frac{t}{V} = \frac{\mu\alpha\omega}{2P_t A^2}V + \frac{\mu R_m}{P_t A} \qquad (7)$$

The modified fouling index (MFI) can be obtained from the slope of equation (7), using the viscosity and pressure corrections, as shown below [60].

$$MFI = \frac{\mu_{20}}{\mu} \cdot \frac{\Delta P}{210} \cdot \tan a \;\; [\tan a = \frac{\mu\alpha\omega}{2P_t A^2}] \qquad (8)$$

Table 2. MFI values in the MBR operation

References	Min. MFI (10^3 s/L^2)	Max. MFI (10^3 s/L^2)
Kim and Jang [48]	0.5	9.3
Jang et al. [47]	4.1	36.4
Jang et al. [61]	16.8	47.6
Jang [62]	2.0	39.1

*A stirred batch cell (8400, Amicon, USA) was used to measure the permeate volume, with an ultrafiltration membrane (nominal molecular weight limit 300 kDa, polyethersulfone, 41.8cm2, Amicon, USA), under constant pressure (10 psi).

This equation may be a useful index for sludge conditions in predicting the fouling level, assuming that the pressure and filtration membrane are standardized. Table 2 shows a summa ry of the MFI values. A higher MFI signifies a higher fouling rate. These results could be use d as a simple index to determine the sludge conditions required for fouling, even if standard t est apparatus, further research of the sludge level specification (such as very good, good, nor mal, bad and danger) and the building of database of influents are required. For example, the index could imply that the sludge conditions are "good" in a MBR below a MFI of 5.0×10^3 s/L^2.

2.9. Fouling Reductions for MBR

Until now, only the factors affecting fouling of MBR have been introduced. To simplify the complex relationships between these factors, biological floc and SMP were set as the main components of fouling. In terms of the introduced fouling structure (figure 2), methods for the reduction of fouling in the MBR have been developed.

With respect to suspended solids, the maintenance of "good" operating conditions is required for an adequate biological floc. The "good" could refer to good filterability or a strong, big structure, without too many filamentous bacteria. For example, Kim and Jang [48] attempted to promote bio-flocculation with the addition of calcium ions, and concluded that the cake layer resistance was reduced due to the better flocculation caused by the calcium bridges and the increased hydrophobicity of the EPS under optimum calcium operating conditions. Hwang et al. [63] and Lee et al. [64] added a membrane fouling reducer (cationic polymeric materials) to cause bio-flocculation, which also enhanced the membrane filterability.

With respect to soluble materials, the maintenance of a low SMP concentration and minimization of high MWD materials are required in the bioreactor. Initially, it is important to prevent the accumulation of SMP and the release of EPS caused by floc deterioration, which usually stems from oxygen depletion, a low divalent cation concentration or the toxicity of materials. For example, Lesage et al. [65] added powdered activated carbon (PAC) to the bioreactor, which showed that PAC could structure the biomass, then deposit at the membrane surface, decrease the protein and carbohydrate concentrations in supernatant, and cause a reduction in membrane fouling. Also, the addition of PAC allowed adsorption of toxic compounds and SCOD materials, which enhanced the effluent quality [66, 67].

There has also been much research on the reduction of fouling in a MBR via membrane modification (new material, pore geometry, charge properties, anti-foulant coating and so on), advanced operation applications (sub-critical flux, air flow control or sparging, module configuration and so on) and membrane chemical cleaning strategies (application of various cleaning agents, cleaning under operation and so on).

3. Membrane Flux Decline Models

Usually, ultrafiltration (UF) and microfiltration (MF) membranes are applied to a MBR process, even if some research has attempted to use nanofiltration (NF) membranes [68, 69]. Transport in UF and MF membranes is fundamentally different compared to that in nanofiltration (NF) and reverse osmosis (RO) membranes. In the case of UF and MF, specific path ways (pores) exist for the passage of the solvent and solute; whereas, NF and RO are governed by partitioning and diffusion for the transport of material. Therefore, UF and MF may be considered porous membranes, while NF and RO are non-porous membranes [70]. There are many models of flux decline for the design of membrane processes. However, there is no universal model for a MBR as a result of the various fouling factors mentioned in section 2. Several basic flux decline models have been introduced for the MBR.

3.1. Resistance-in-Series Model

Darcy's Law states that the flux is directly proportional to the potential pressure drop and inversely proportional to the resistance. The total resistance can be divided into simplified resistances, which originate from various causes of fouling (figure 5). For porous membrane systems, the steady state flux can be expressed by the resistance-in-series model given below.

$$J = \frac{\Delta P}{R_t \cdot \mu} \tag{9}$$

$$R_t = R_m + R_c + R_p + R_a + R_g \tag{10}$$

where,

J: permeate flux,
P: transmembrane pressure,
η: viscosity of the permeate,
R_t: total resistance,
R_m: intrinsic membrane resistance,
R_c: cake resistance formed by cake layer deposited over membrane surface,
R_p: resistance caused by pore plugging onto the membrane pore or membrane surface,
R_a: resistance caused by solute adsorption onto the membrane pore,
R_g: gel layer resistance formed by gel layer attached over membrane surface.

Even though each resistance can be obtained by empirical methods (not predictive model), this model is very useful for separating the causes of fouling. The cake layer (R_c) and pore blocking (R_p) resistances are known as the main fouling mechanisms in the MBR [24, 71]. R_c can be defined as the resistance to filtration due to solids and particles that are able to be removed from the membrane surface by mechanical means (mainly caused by biological floc). R_p can be the resistance to filtration due to adsorbed organics and micro-particles that cannot be mechanically removed from the membrane surface (mainly caused by SMP). The separation of the total resistance was fractionated in some research by stirred of a batch cell using sludge samples [21, 23, 24].

In a bioreactor, the fractionation of resistance can be conducted using the procedures used by Trussell [25]. Before removing the MBR from service and emptying the bioreactor, the membrane permeate was collected. The MBR was refilled with the collected permeate, without detachment of the cake from the membrane surface. The flux profile in relation to pressure was performed without aeration to determine R_t. Next, aeration was performed, and the membrane was allowed to relax for 4 hours without suction. The bioreactor was emptied, with the sludge removed from the membrane surface. The collected permeate was then refilled into the bioreactor. A second flux profile was performed, with aeration, to determine R_c. Chemical cleaning of the membrane was conducted with 1g/L NaOCl solution for 12 hours. After the chemical cleaning, the reactor was emptied and filled with tap water.

$Na_2S_2O_3$ was added for dechlorination, with a final flux profile performed with tap water to determine R_p.

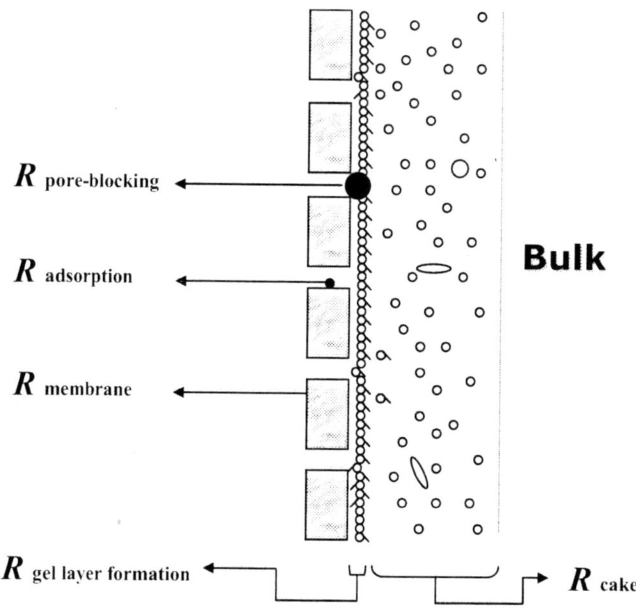

Figure 5. Various resistances in relation to permeate flux decline.

3.2. Hagen-Poiseuille Equation

The simplest representation is one where the membrane is considered as a number of parallel cylindrical pores, perpendicular or oblique to the membrane surface. Assuming that all the pores have the same radius, then we may write the steady state flux as below:

$$J = \frac{\varepsilon r^2 \Delta P}{8 \mu \tau \Delta x} \tag{11}$$

where,

ε = surface porosity (the ratio of pore area to membrane area multiplied by the number of pores),

r = pore radius,

τ = pore tortuosity factor (=1 for perpendicular pores),

x = membrane thickness (pore length).

From this model, increasing the number of pores on the membrane surface, ε, increasing the pore radius, r; or increasing the driving pressure, ΔP, should increase the flux. While, increasing the viscosity (decreasing in the water temperature), μ, increasing the tortuosity, τ, or membrane thickness, x, will decrease the flux. This equation generally holds for pure water, but when suspended solids and soluble materials are introduced to the system the membrane performance will deviate from equation 11.

3.3. Concentration Polarization Model

Concentration polarization describes the tendency of the solute to build up at the membrane solution interfacial region. Film theory can be used to model the concentration polarization effect. A control volume is drawn to the membrane surface, with the mass balance equation applied to the solute. The mass balance consists of the mass of solute carried to the membrane, which equals the mass of solute diffusing from the membrane back to the bulk and the mass of solute passing through the membrane. The mass balance and boundary conditions are as follows:

$$J_C = -D\frac{dC}{dt} + JC_P \quad (12)$$

$$C = C_m \text{ at } x=0 \quad (13)$$

$$C = C_B \text{ at } x=\delta \quad (14)$$

where,

C: solute concentration,
D: diffusion coefficient of the solute,
x: distance from the feed side of the membrane,
C_p: solute concentration in permeate,
δ : thickness of the concentration boundary layer,
C_m: concentration at the membrane surface,
C_B: concentration in bulk.

If the solute is completely retained ($C_p=0$), the concentration polarization under steady state can be defined as shown below:

$$J = k \ln \frac{C_m}{C_B} \quad (15)$$

where,

$k = \frac{D}{\delta}$: mass transfer coefficient.

This model can be modified to account for the system conditions, such as laminar flow and Brownian diffusive transport, fully developed flow, shear induced diffusion and so on.

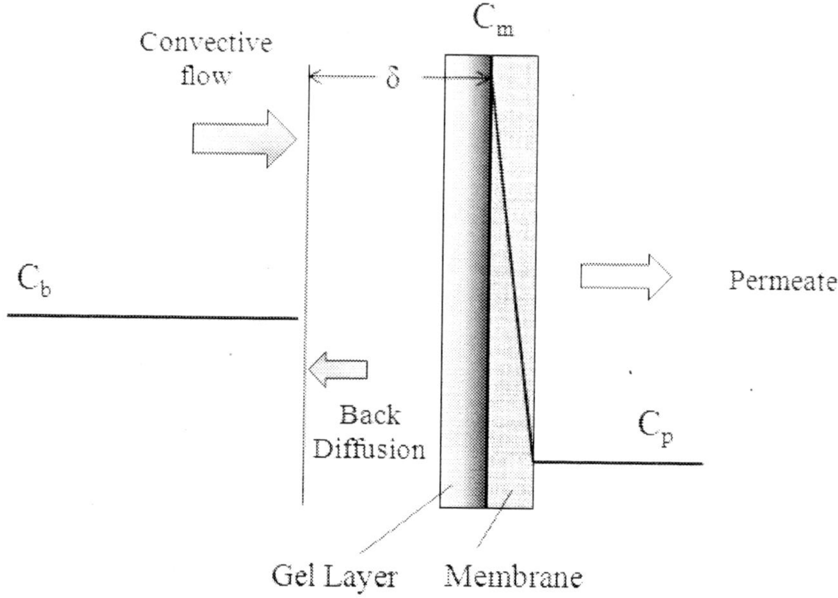

Figure 6. Concentration polarization.

3.4. Happel's Cell Model

Particle accumulation in the cake layer provides an additional resistance to the flow of filtrate, which also reduces the flux. The pressure drop in the cake layer is associated with the frictional drag resulting in the flow of filtrate through the dense layer of accumulated particles [72]. The steady state flux can be expressed as shown below:

$$J = \frac{\Delta P_c}{\frac{kT}{D} A_s(\theta) M_c} \tag{16}$$

$$A_s = \frac{1 + \frac{2}{3}\theta^5}{1 - \frac{3}{2}\theta + \frac{3}{2}\theta^5 - \theta^6}, \quad \theta = (1-\varepsilon)^{1/3} \tag{17}$$

where,

kT/D $(=6\pi\mu a_p)$ = frictional drag coefficient,

k = Boltzmann constant,

T = absolute temperature,

μ = solvent viscosity,

a_p = particle radius,

M_c = total number of particles accumulated in the cake layer,

$A_s(\theta)$ = correction function accounting for the effect of retained neighboring particles,

ε = porosity of cake layer of accumulated particles.

3.5. Constant Pressure Filtration Laws

Hermia [73] proposed different filtration models, which can be applied to non-Newtonian fluids and for constant pressure filtration. The characteristic form is shown as:

$$\frac{d^2t}{dV^2} = k\left(\frac{dt}{dV}\right)^n \tag{18}$$

where,

n = 2.0 for complete blocking
n = 1.5 for standard blocking
n = 1.0 for intermediate blocking
n = 0 for cake filtration
k = filtration constants

These equations can be expressed as simple linear equations relating the filtrate flow (Q), volume (V) and time (t) (Kim [74]). The standard pore blocking and cake filtration models are summarized below.

A. Standard Pore Blocking Model (SPBM)

When the particle diameter is much less than the membrane pore diameter, the particles will pass through the pores and be deposit on the walls, which will subsequently reduce the pore diameter and; thus, the pore volume.

$$\frac{t}{V} = \frac{K_s}{2}t + \frac{1}{Q_0} \tag{19}$$

where,

Q_0 = initial permeate flux
K_s = filtration constant

B. Cake Filtration Model (CFM)

In the case of large particles that cannot enter through the pores, a deposit in the form of a cake will be formed on the membrane surface.

$$\frac{t}{V} = \frac{K_c}{2} V + \frac{1}{Q_0} \quad (20)$$

where,

Q_0 = initial permeate flux
K_c = filtration constant

4. Current Practice of Membrane Bioreactor (MBR)

4.1. Market Share

There are many international companies that manufacture and sell MBR, but two companies, GE water and Process Technologies and Kubota Corporation, dominate more than 90% of the MBR market. As shown in table 3, GE water and Process Technologies and Kubota Corporation occupy about 60 and 35%, with other MBR companies only occupying less than 5% of the world MBR market.

Table 3. Information on the global MBR market; 2004

Company	MBR sales ($ Millions E)	Number of installed MBRs (E)	Installed MBR capacity (mgd E)	Market share (%)
GE	130.2	334.0	198.3	60.0
Kubota	75.5	654.0	67.6	35.0
Other	10.9	524	15.7	5.0

Source: Business Communications Co., Inc.

4.2.Company Profile

- Aqua Tech (Korea)

The BIOSUF (Biological treatment with submerged ultra-filtration) process is a typical MBR system produced by Aqua Tech. The main processes of the BIOSUF are anoxic, aerobic and submerged ultra-filtration zones, where a submerged tubular U/F module zone is separated from the aerobic zone, as shown in figure 7. The HRT of the flux control zone is 8-24 hr, and the SRT of anoxic and aerobic zones is 50-365 days. The BOD loading rate and F/M ratio of anoxic and aerobic zones are 2.5 kg BOD_5/m^3/day and 0.10-0.15 kg BOD_5/kg MLSS/day, respectively. The process is applied to nightsoil and animal wastewater, municipal water, middle water and industrial water treatment. The capacity of the process is more than 0.79 MGD (3,000 m^3/d). The performance of the BIOSUF process obtained from Aqua Tech for municipal water treatment is shown in table 4.

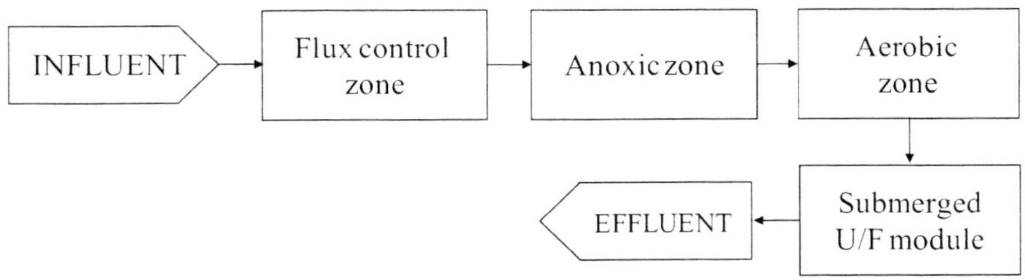

Figure 7. The schematic diagram of the BIOSUF process used for municipal water treatment, as modified from Aqua Tech.

Table 4. The results obtained with the use of the BIOSUF process for municipal water treatment

Parameter	Influent	Effluent	Removal rate
BOD (mg/L)	200	< 1	99.5 %
SS (mg/L)	200	None	100 %
E. coli.	-	None	100 %
TN (mg/L)	50	2.4	94.4 %
TP (mg/L)	10	0.9	70.0 %

- Enviroquip, Inc. (USA)

The SymBio process was developed for simultaneous nitrification and denitrification using MBR technology of Enviroquip, Inc. The wastewater flows into an anoxic zone, passes through a screen, and then goes into a pre-aeration zone where the simultaneous nitrification and denitrification occur, with a level of dissolved oxygen (DO) below 1.0 mg/L (figure 8). The capacity of this system is more than 3.0 MGD (11,400 m^3/d). Table 5 shows the water quality obtained using the SymBio process in Big Bear's municipal wastewater treatment plant.

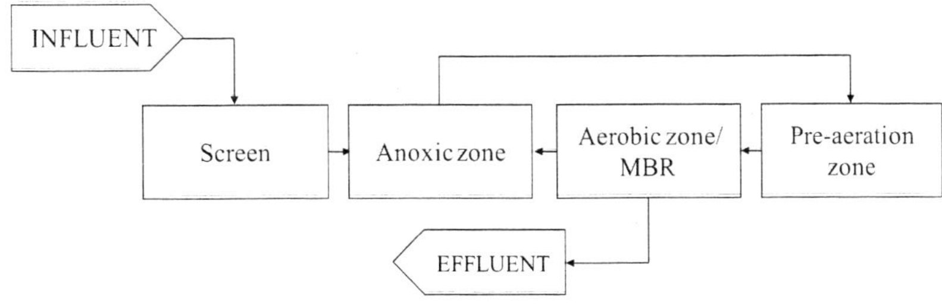

Figure 8. Schematic diagram of the SymBio process. The process is modified from Enviroquip, Inc.

Table 5. The results obtained from the use of the SymBio system at a municipal wastewater plant

Parameter	Influent	Effluent
BOD (mg/L)	281	7.0
TSS (mg/L)	329	7.0
NH4-N (mg/L)		1.0
TKN (mg/L)	35	3.1 (TIN)
Nitrate-N (mg/L)		1.9

Source: Enviroquip, Inc.

- GE water and Process Technologies (Canada)

The GE Company is the world's largest manufacturer and system integrator of hollow fiber membrane technologies and complimentary products. The ZeeWeed membrane bioreactor (MBR) process is a GE technology, which consists of a suspended growth biological reactor integrated with an ultrafiltration membrane system, using the ZeeWeed hollow fiber membrane. The treatment capacity of the membrane bioreactor installations is over 450 million gallon per day (MGD) (1,709,996 m^3/d). Two applications of this system are a 11 MGD (41,800 m^3/d) at the Brescia wastewater treatment plant in Italy and a 40 MGD (151,000 m^3/d) at the Gwinnett county tertiary municipal wastewater treatment plant in the USA. Figure 9 and table 6, sourced and modified from ZENON Environmental Inc., show a schematic diagram of the ZeeWeed process and the results of treated water from the Brescia wastewater treatment plant.

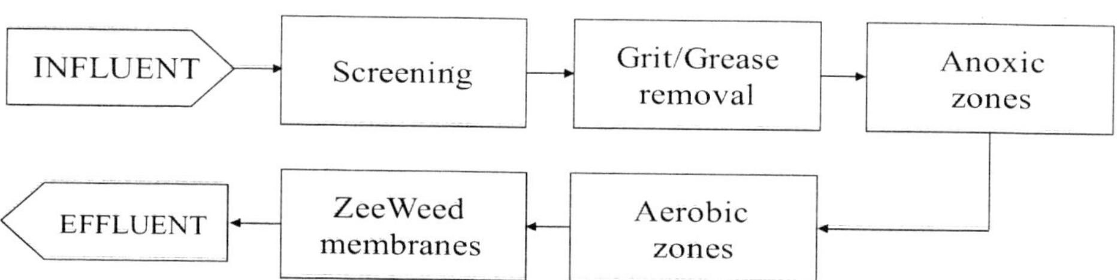

Figure 9. Schematic diagram of the ZeeWeed process.

Table 6. The results obtained using the ZeeWeed membrane bioreactor

Parameter	Influent	Effluent
BOD (mg/L)	255	10
COD (mg/L)	505	20
TSS (mg/L)	290	N.D.
TKN (mg/L)	50	2
TN (mg/L)	> 50	< 10

*N.D. = Not Detectable.

- Hans Huber AG (Germany)

The Huber UF membrane bioreactor has two different plate-type modules; a vacuum upstream membrane (VUM) and a vacuum rotating membrane (VRM). The VUM module consists of several vertically arranged 0.04 μm pore size stationary membrane plates; whereas, the VRM has the advantages of hollow fiber systems, and incorporates trapezoidal membrane plates. The capacities of the Huber MBR at the Buchloe industrial, Scwagalp municipal, Knautnaundorf municipal and Veurne industrial wastewater treatment plants are 0.013 (49.4 m^3/d), 0.06 (228 m^3/d), 0.08 (304 m^3/d) and 0.125 (475 m^3/d) MGD, respectively. A schematic diagram and the removal efficiency of the Huber MBR sourced from Hans Huber AG are shown in figure 10 and table 7, respectively.

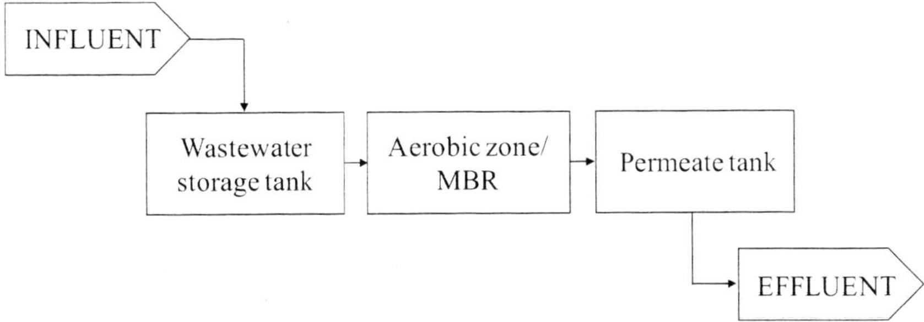

Figure 10. Schematic diagram of the Huber membrane bioreactor process for industrial starch production.

Table 7. Removal efficiencies of the Huber membrane bioreactor applied to wastewater treatment at the Abu Dhabi slaughterhouse

Parameter	Unit	Removal efficiency
Grease	(mg/L)	90-98 %
Solids	(mg/L)	90-96 %
BOD_5/COD	(mg/L)	85-90 %

Source: Hans Huber AG.

- Hyundai Engineering Co., Ltd and the Korea Express (Korea)

The HANT® process is a typical MBR technology of the two companies, and an effective treatment system with a hollow fiber membrane. The HRT, SRT and MLSS of the aerobic tank are 3.1 hr, 15-30 days and 7,000-15,000 mg/L, respectively. The treatment capacity of the HANT® process is 40 MGD (150,000 m^3/d). A schematic of the HANT® process and the water quality, sourced and modified from Hyundai Engineering Co., LTD and The Korea Express, are shown in figure 11 and table 8.

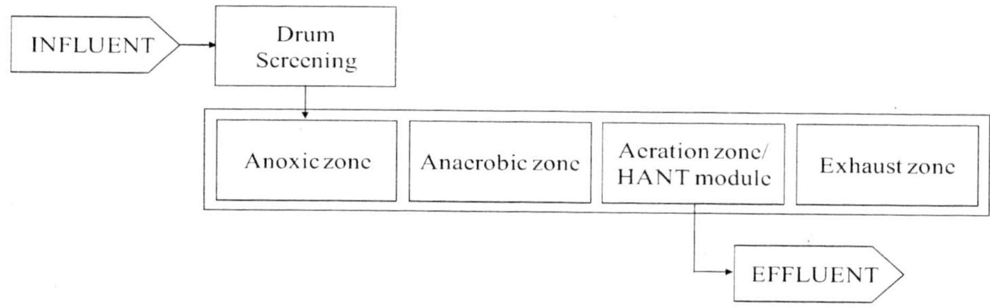

Figure 11. Schematic diagram of the HANT® process.

Table 8. Performance of the HANT® process

Items	Unit	Results			Remark error range (%)
		Influent	Effluent	Removal efficiency (%)	
pH	-	7.4	6.9	-	±10
COD_{Cr}	mg/L	241.3	12.0	95.0	±10
COD_{Mn}*	mg/L	64.2	6.0	90.7	±10
BOD_5	mg/L	96.5	1.0	99.0	±5
TN	mg/L	31.7	8.0	74.8	±10
TP	mg/L	3.2	0.5	84.4	±10
SS	mg/L	85.9	0.2	99.8	±1
E. Coli.	MPN/100ml	4300	None	-	±1

* CODMn from Korea analytical method
** The other items are from standard method (U.S. EPA).

- ITT Industries (USA)

The dual stage MBR is an advanced treatment plant from ITT Industries, and has a hollow-fiber UF membrane module with a 0.2 μm of membrane pore size. The hollow-fiber UF module consists of a single ended, free floating membrane fiber design. Figure 12 and table 9, obtained from ITT Industries, show a schematic diagram of the dual stage MBR process and the general water quality obtained from the use of the DF-MBR process for wastewater treatment.

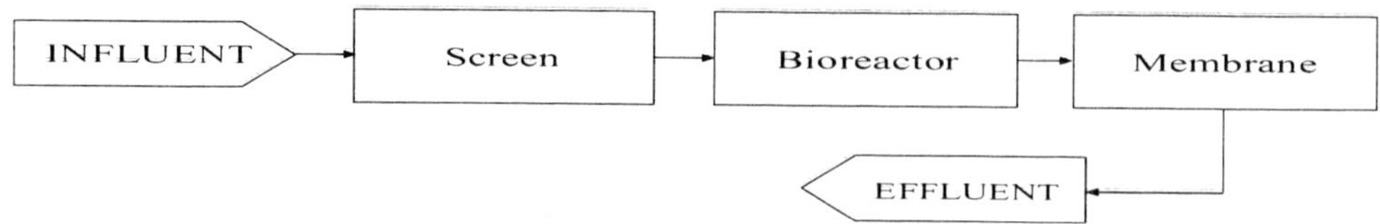

Figure 12. Schematic diagram of the dual stage MBR process.

Table 9. Water quality obtained using the DF-MBR process

Items	Effluent
BOD_5 (mg/L)	< 2
TSS (mg/L)	< 2
NH_3-N (mg/L)	< 1
TN (mg/L)	< 3
TP (mg/L)	< 1
Turbidity (NTU)	< 0.5

- Korea Membrane Separation (KMS) (Korea)

KMS is very active in the application of MBR processes and have developed a new wastewater treatment technique (KSMBR process) in co-operation with Korea Water Resources Corporation and Ssangyong Engineering Construction Co., Ltd. The KSMBR module is a hollow fiber membrane, and the process combines the MBR process with a dynamic state bioreactor system using a tri-sectional aeration SBR process and flow alternation technology operation (figure 13). The operative flux and pressure of the process are 0.3 m^3/m^2/day and -5cmHg~-40cmHg, respectively. The capacities of the KSMBR applied in the industrial wastewater treatment plant in Daegu city and in municipal wastewater treatment plant of Daejeon city are 6.58 MGD (25,000 m^3/d) and 4.74 MGD (18,000 m^3/d), respectively. The performance of the KSMBR process obtained from KMS for sewage treatment is shown in table 10.

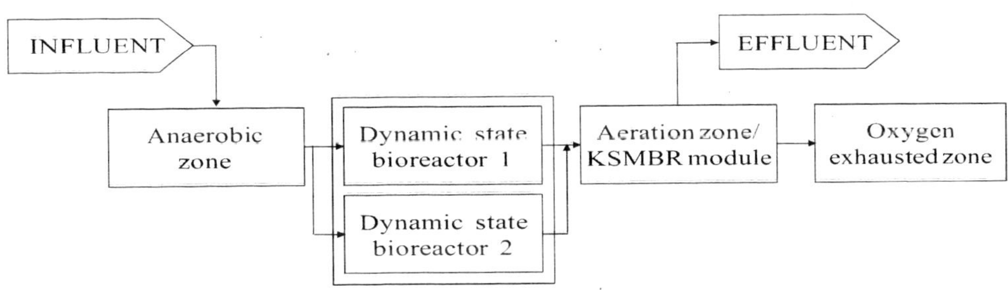

Figure 13. Schematic diagram of the KSMBR process. The process is modified from Korea Membrane Separation (KMS).

The PURON system, a typical MBR product from KMS, consists of a long hollow fiber membrane with tiny pores approximately 0.05 μm in size. The design of the PURON module is different from other general MBR modules, in that there is no top header for trapping hair and other debris, which can cause clogging of the fibers. Instead, the upper ends of the hollow fiber membranes move freely, with a seaweed-like action. The PURON module is available in standard membrane area sizes of 1500, 500 and 235 square meters, with a treatment capacity of over 37 MGD (140,000 m^3/d). Two of applications of this process are a 0.5 MGD (2,000 m^3/d) industrial MBR at Sobelgra in Belgium and a 0.17 MGD (630 m^3/d) municipal MBR at Simmerath in Germany. A schematic of the PURON process and the operational data, sourced and modified from Koch Membrane Systems (KMS), Inc., are shown in figure 14 and table 11, respectively.

Table 10. Performance of the KSMBR process for sewage treatment

Items	Influent	Effluent	Removal rate
BOD (mg/L)	90.6 (31.6-331.0)	1.3 (0.4-2.3)	98.3 (95.9-99.6) %
COD_{Cr} (mg/L)	116.2 (55.1-428.0)	19.0 (9.5-30.8)	81.3 (63.7-92.8) %
COD_{Mn} (mg/L)	42.3 (17.2-154.0)	5.4 (3.1-8.5)	85.8 (69.8-95.8) %
SS (mg/L)	97.0 (7.0-1,244)	0.7 (0.0-2.0)	97.9 (92.3-100) %
TN (mg/L)	25.2 (17.4-62.9)	8.3 (2.6-14.1)	66.1 (44.2-89.7) %
TP (mg/L)	3.1 (1.4-10.6)	0.6 (0.1-1.2)	79.8 (57.7-93.8) %
E. Coli. (number/ml)	22,000 (1,200-59,000)	0 (0-0)	100 (100-100) %

- Koch Membrane Systems (KMS), Inc. (USA)

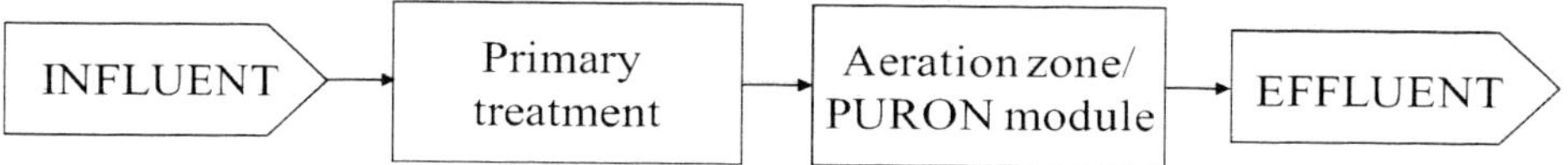

Figure 14. Schematic diagram of the PURON process.

Table 11. The results obtained with the use of the PURON membrane bioreactor*.

Parameter	Influent (mg/L)	Loads (kg/d)	Effluent (mg/L)
COD	1880-2100	4000	100-200
BOD	700-930	2000	2-5
SS	330-460	800	0
TN	35-50	100	1-2
TP	13-15	30	< 1

*These data are the results of the PURON system applied at Sobelgra, Belgium.

- KOLON E and C (Korea)

The KIMAS process is a typical MBR process of KOLON E and C, which consists of a KIMAS module, a combination of a hollow fiber membrane and cassette. The nominal pore size and permeate area per module are 0.1 μm and 20 m^2, respectively. In the KIMAS process, the KIMAS module is not submerged in the aerobic zone, but set behind this zone (figure 15). The capacities of the municipal water treatment plant in Icheon city and the industrial wastewater treatment plant in Cheonan city are 0.29 MGD (1,100 m^3/d) and 0.24 MGD (900 m^3/d), respectively. The general water quality obtained using the KIMAS process, sourced from KOLON E and C, is shown in table 12.

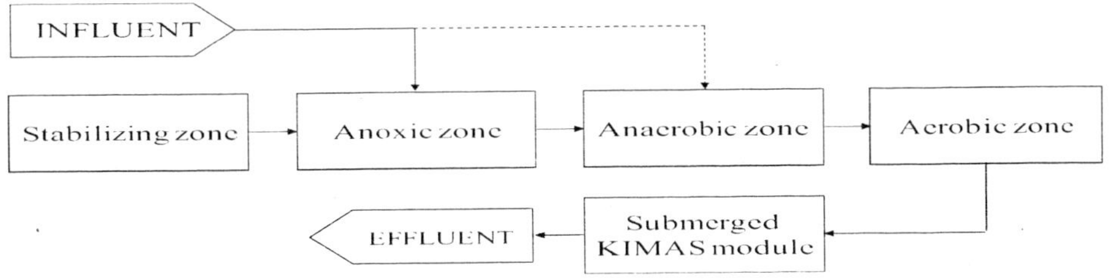

Figure 15. Schematic diagram of the KIMAS process. The process is modified from KOLON E and C.

Table 12. General performance of the KIMAS process

Items	Influent	Effluent	Removal rate
BOD (mg/L)	150.4	0.7	99.5 %
COD_{Cr} (mg/L)	68.6	6.5	90.3 %
COD_{Mn} (mg/L)	243	16.7	93.0 %
SS (mg/L)	92.7	0.05	99.9 %
TN (mg/L)	45.8	7.3	83.9 %
TKN (mg/L)	40.6	1.1	97.3 %
TP (mg/L)	6.3	2.3	65.4 %
E. Coli. (number/ml)	55,167	0.5	99.99 %

- Kubota Corporation (Japan)

The Kubota submerged MBR system uses plate and frame polyolefin membrane panels instead of hollow fiber membranes. The normal pore size of the membrane and MLSS of the Kubota reactor are 0.4 μm and 15,000-20,000 mg/L, respectively. The capacity of the MBR system ranges from 0.02 MGD (114 m^3/d) to 20 MGD (76,000 m^3/d). The system is applied at the Daldowie municipal wastewater treatment plant, with 3 MGD (11,400 m^3/d), at the Swanage municipal wastewater treatment plant, with 3.4 MGD (12,920 m^3/d) and at the Narraghmore housing development wastewater treatment plant, with 0.02 MGD (76 m^3/d). The capacity of the Kubota submerged MBR system at Al Ansab in Oman will be 20 MGD (76,000 m^3/d) when the plant is completed. A schematic diagram and the water quality of the Kubota submerged MBR system used at Kubota Hanshin Office are shown in figure 16 and table 13, respectively.

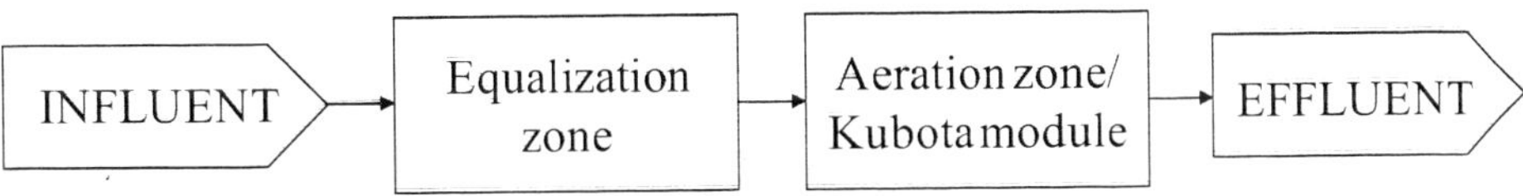

Figure 16. Schematic diagram of the Kubota submerged MBR system, modified from Kubota Corporation.

Table 13. Performance of the Kubota submerged MBR for a kitchen wastewater reuse system; 29-Jan, 2003

Items	Influent	Effluent
BOD (mg/L)	300	4
COD_{Mn} (mg/L)	200	3
SS (mg/L)	150	0.1
TN (mg/L)	20	3.9
TP (mg/L)	3	0.1

Source: Kubota Corporation.

- Mitsubishi Rayon Co., Ltd. (Japan)

The SteraporeSURTM and SteraporeSADFTM series are main modules in Mitsubishi's submerged MBR systems, which employ a hollow fiber polyethylene MF element. The material and pore size of the hollow-fiber membrane in SteraporeSURTM /SteraporeSADFTM are polyethylene/PVDF and 0.4 μm/0.4 μm, respectively. The capacities of the SteraporeSURTM MBR system applied for wastewater treatment at a semiconductor plant and a miscellaneous water treatment plant at an airport are 1.05 MGD (4,000 m^3/d) and 0.71 MGD (2,700 m^3/d), respectively. The capacities of the SteraporeSADFTM MBR system applied at a wastewater treatment facility of a chemical plant and at a domestic wastewater treatment plant at a prefecture are 0.32 MGD (1,200 m^3/d) and 7.89 MGD (30,000 m^3/d), respectively. A schematic diagram and the water quality of the Mitsubishi MBR system are shown in figure 17 and table 14, respectively.

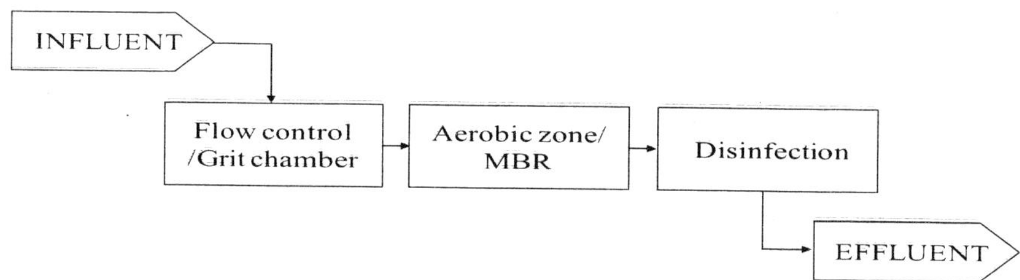

Figure 17. Schematic diagram of the Mitsubishi MBR system, modified from Mitsubishi Rayon Co., Ltd.

The X-Flow AirLift^{TM} MBR is a technology from Norit, which consists of an airlift tubular ultrafiltration membrane. The size of UF membrane and MLSS of the MBR system are 5 mm and 15,000 mg/L, respectively. The capacity of the system ranges from less than 0.063 MGD (240 m^3/day) to more than 0.95 MGD (3,600 m^3/day). The AirLift^{TM} MBR system has been applied at the Millsborough wastewater treatment plant, with 1.1 MGD (4,200 m^3/day), the Xuzhous cigarette factory, with 0.53 MGD (2000 m^3/day), and the Ootmarsum wastewater treatment plant, with 0.95 MGD (3,600 m^3/day). Figure 18 and table 15, modified from Membrane Bioreactors for Municipal Wastewater Treatment of STOWA report (2002), show a schematic diagram and the water quality obtained with the use of the X-flow pilot plant.

Table 14. Water quality obtained with the use of the Mitsubishi MBR system applied at a domestic wastewater treatment plant in China

Items	Influent	Effluent
BOD (mg/L)	200	4.8
COD_{Cr} (mg/L)	450	30.8
SS (mg/L)	200	< 5
NH_4-N (mg/L)	40	0.28
TP (mg/L)	7	0.06

Source: Mitsubishi Rayon Co., Ltd.

- Norit (The Netherlands)

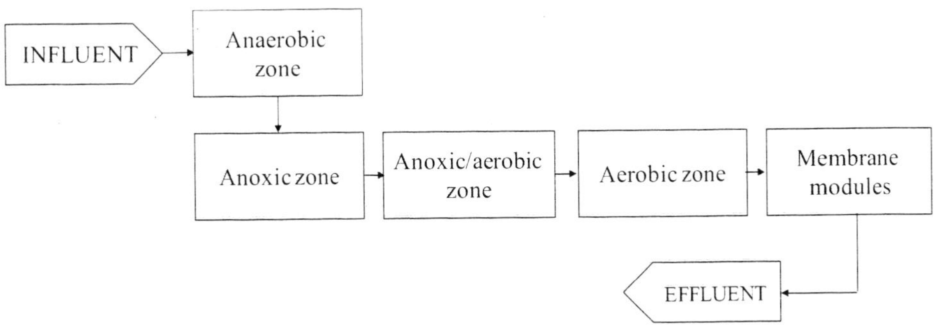

Figure 18. Schematic diagram of the X-Flow AirLift™ MBR pilot process.

Table 15. The results obtained using the X-Flow AirLift™ MBR pilot process

Parameter	Influent	Effluent	Removal rate
COD (mg/L)	569	36	94.0 %
TN (mg/L)	56	7.8	78.3 %
TP (mg/L)	11.3	1.4	88.0 %

- Pall Corporation (USA)

The Pall Aria™ MBR system was developed to treat wastewater at the Pall Corporation, which uses a Microza hollow fiber membrane module. The pore rating and fiber OD/ID of the module are 0.1 μm and 1.3 mm/0.7 mm, respectively. By immersing the membrane system in the aerobic biological zone, the Pall Aria™ MBR system is used for the removals of phosphorous and nitrogen, and for pre-RO reuse/discharge. The capacity of the MBR system is 0.3-4.2 MGD (840-16,080 m^3/d), with a removal efficiency, sourced from Pall Corporation, as shown in table 16.

Table 16. Microbial and particular removals using the Pall membrane system

Parameter	Microfiltration (MF)	Ultrafiltration (UF)
Giardia	> 6 (log)	> 6 (log)
Cryptosporidium	> 6 (log)	> 6 (log)
MS2 Coliphage or Bacteriophage	0.5-2.5 log*	4.5-6 log*
Turbidity	< 0.1 NTU	< 0.1 NTU

*Virus removal varies depending on the coagulation process upstream of the system.

- Pure Envitech (Korea)

The dynamic flow-membrane bioreactor (DF-MBR) is the main technology used in the Pure Envitech MBR process, which employs an ENVIS as the membrane module, consisting of a plate and frame membrane module. The effective membrane area, permeate flux and pressure of the ENVIS module are 0.98 m2, 0.3-0.5 m^3/d and 0~-40 cmHg, respectively. The DF-MBR mainly consists of anoxic, anaerobic and aerobic zones, with a submerged ENVIS module, where the influent alternately flows into the anoxic and aerobic zones (figure 19). The capacity of the DF-MBR applied at the Pyeongtaek wastewater treatment plant is 2.16 MGD (8,200 m^3/d). The general water quality of the DF-MBR process, obtained from Pure Envitech, is shown in table 17.

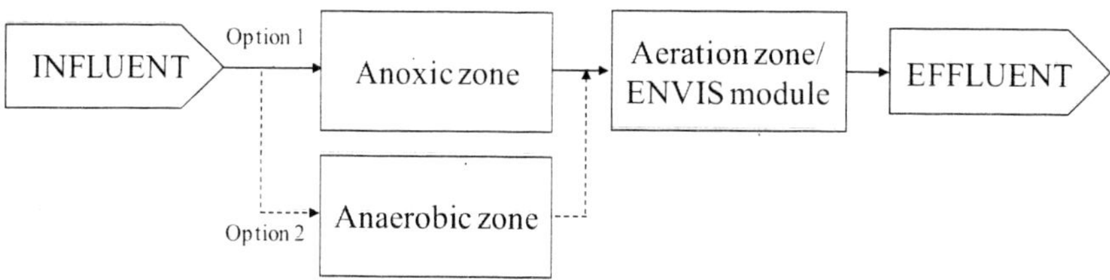

Figure 19. Schematic diagram of the DF-MBR system. This figure has been modified from the DF-MBR process of Pure Envitech.

Table 17. Water quality obtained with the use of the DF-MBR process

Items	Influent	Effluent	Removal rate
BOD (mg/L)	125.5	1.3	99.2 %
COD_{Cr} (mg/L)	224.2	10.1	95.2 %
SS (mg/L)	153.0	-	-
TN (mg/L)	38.9	9.8	75.2 %
NH_4-N (mg/L)	25.3	0.14	99.4 %
TP (mg/L)	4.1	0.8	74.9 %
E. Coli. (number/100ml)	$4.1 * 10^4$	N.D.	100 %

- SAEHAN Industries (Korea)

The SZ membrane system is the MBR technology of SAEHAN, which uses immersed hollow fiber membranes. The ZeeWeed membrane module, made by GE water and Process Technologies, is submerged in an aerobic zone (figure 20). The capacity of SZ membrane system is more than 13.2 MGD (50,000 m^3/d). Three applications of this system are for the treatments of surface water at Ki-Heung city (50,000 m^3/d), alkaline waste water at Ki-Heung city (1,000 m^3/d) and waste water at Yu-Soo city (250 m^3/d). The water quality of the wastewater treated using this system, as obtained from SAEHAN Industries, is shown in table 18.

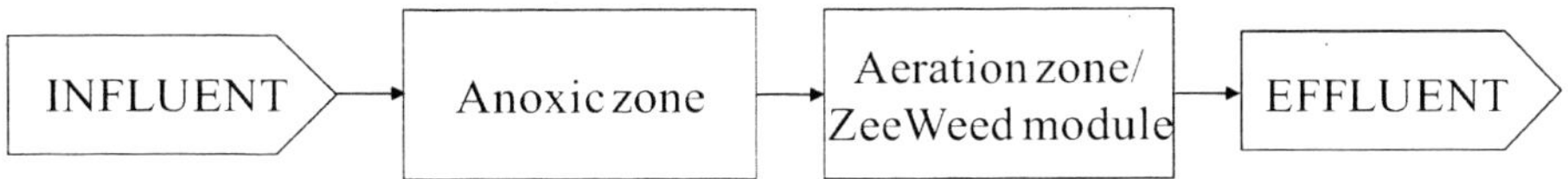

Figure 20. Schematic diagram of the SZ membrane process. The process is modified from SAEHAN Industries.

Table 18. The results obtained using the SZ membrane process for wastewater treatment

Parameter	Permeate
BOD (mg/L)	< 5
TSS (mg/L)	< 2
NH_3-N (mg/L)	< 1
TN (mg/L)	< 10
TP (mg/L)	< 0.1

- Siemens Water Technologies (Germany)

The MemJet submerged membrane bioreactor is a Siemens Water Technologies, which consists of a hollow fiber membrane module. The SRT is typically 10-50 days, with a normal MLSS concentration of 8,000-14,000 mg/L. The treatment capacities of two applications are 0.67 MGD (2,546 m^3/d) at the Calls Creek wastewater facility in Watkinsville, Georgia, and 0.16 (608 m^3/d) at the Park Place wastewater facility. The MemJet system mainly consists of a fine screen, and anoxic and aeration zones, with a MemJet module (figure 21). The water quality of the MemJet membrane bioreactor system applied at Watkinsville, Georgia, is shown in table 19, sourced from Siemens Water Technologies.

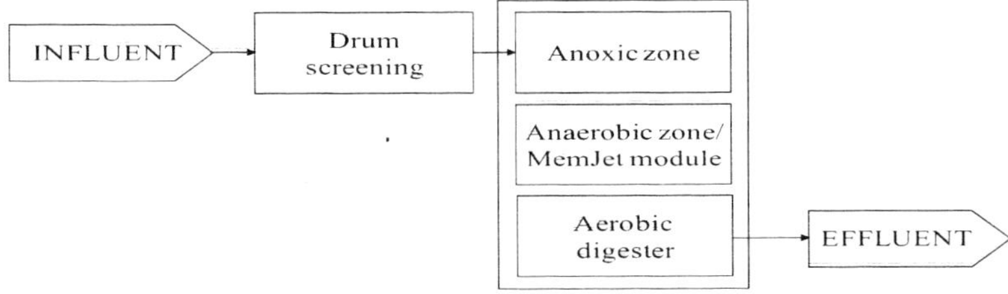

Figure 21. Schematic diagram of the MemJet process. The figure is modified from Siemens Water Technologies.

Table 19. The results obtained with the use of the MemJet membrane bioreactor

Parameter	Influent	Effluent
BOD (mg/L)	250	< 2
TSS (mg/L)	250	< 2
TKN (mg/L)	28	< 1
Turbidity (NTU)		< 2

- Veolia Water Solution and Technologies (France)

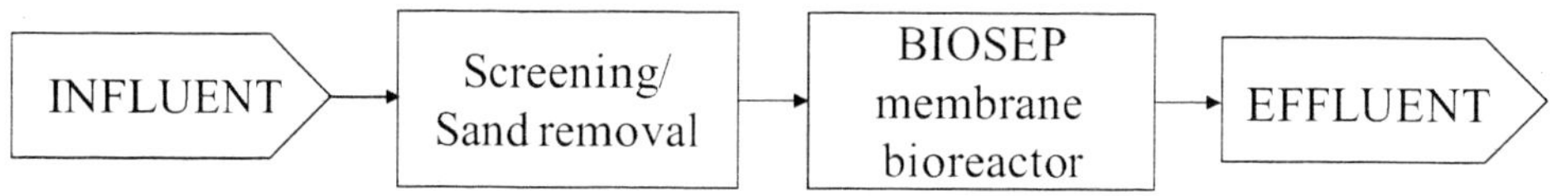

Figure 22. Schematic diagram of the BIOSEP process.

Table 20. Performance of the BIOSEP process for municipal and industrial effluents

Items	Unit	Effluent
COD	mg/L	< 30
BOD_5	mg/L	< 5
SS	mg/L	< detectable threshold
TN	mg/L	< 10-15
TP	mg/L	< 0.5-2

Source: Veolia Water Solution and Technologies.

- Other companies

Table 21. Other MBR related companies

Country	Company
Germany	ABFALL-ABWASSER-ANLAGENTECHNIK GMBH (A3)
USA	CeraMem Corporation
Australia	Envirogen, Inc.
Canada	IPEC Consultants Ltd.
Singapore	Keppel Corporation
Germany	Martin Systems AG
UK	Millenniumpore
France	Novasep Group
USA	Parkson Corporation
France	Polymem SA
China	Shanghai Yiming Filtration Tech
USA	Shaw Group
USA	TriSep Corporation
The Netherlands	Tyriqua BV

The BIOSEP is a MBR technology of Veolia Water, which employs either flat sheet or hollow fiber membranes, with an immersed system. The MLSS concentration and capacity of the MBR system range from 8,000 to 15,000 mg/L and 0.013 to 0.38 MGD, respectively. A schematic diagram and the general performance of the BIOSEP process for municipal and industrial effluents, sourced and modified from Veolia Water Solution and Technologies, are shown in figure 22 and table 20, respectively.

References

[1] T. Stephenson, S. Judd, B. Jefferson, K. Brindle, Membrane bioreactors for wastewater treatment, London, IWA publishing, 2000.

[2] C.V. Smith, D. Gregorio, R.M. Talcott, The use of ultrafiltration membrane for activated sludge separation, In: Proceeding of the 24th Annual Purdue Industrial Waste Conference, Purdue University, West Lafayette, Indiana, 1969.

[3] F.W. Hardt, L.S. Clesceri, N.L. Nemerow, D.R. Washington, Solids separation by ultrafiltration for concentrated activated sludge, *J. Wat. Pollut. Control Fed.* 42 (1970) 2135-2148.

[4] C. Chiemchaisri, K. Yamamoto, Biological nitrogen removal under low temperature in a membrane separation bioreactor, *Wat. Sci. Tech.* 28 (1993) 325-333.

[5] T. Ueda, K. Hata, Domestic wastewater treatment by a submerged membrane bioreactor with gravitational filtration, *Wat. Res.* 33 (1999) 2888-2892.

[6] K. Yamamoto, M. Hiasa, T. Mahmood, T. Matsuo, Direct solid-liquid separation using hollow fiber membrane in an activated sludge aeration tank, *Wat. Sci. Tech.* 21 (1989) 43-54.

[7] V. Urbain, R. Benoit, J. Manem, Membrane bioreactor: a new treatment tool, *J. AWWA* 88 (1996) 75-86.

[8] E.B. Müller, A.H. Stouthamber, H.W. Verseveld, D.H. Eikelboom, Aerobic domestic wastewater treatment in a pilot plant with complete sludge retention by crossflow filtration, *Wat. Res.* 29 (1995) 1179-1189.

[9] Y. Shimizu, Y.I. Okuno, K. Uryu, S. Ohtubo, A. Watanabe, Filtration characteristics of hollow fiber microfiltration membranes used in membrane bioreactor for domestic wastewater treatment, *Wat. Res.* 30 (1996) 2385-2392.

[10] J.S. Kim, J.H. Lee, I.S. Chang, Effect of pump shear on the performance of a crossflow membrane bioreactor, *Wat. Res.* 35 (2001) 2137-2144.

[11] T. Ueda, K. Hata, Y. Kikuoka, Treatment of domestic sewage from rural settlements by a membrane bioreactor, *Wat. Sci. Tech.* 34 (1996) 189-196.

[12] K. Yamamoto, M. Hiasa, T. Mahmood, T. Matsuo, Direct solid-liquid separation using hollow fiber membrane in an activated sludge aeration tank, *Wat. Sci. Tech.* 21 (1989) 43-54.

[13] F.N.M. Knops, H. Futselaar, I.G. Rácz, The transversal flow microfiltration module:theory, design, realization and experiments, *J. Membr. Sci.* 73 (1992) 153-161.

[14] Zenon Environmental Inc., Development of an advanced transverse flow nanofiltration membrane process for high performance desalination, *Water Treatment Report*. (9) USA, 1995.

[15] L. Van Dijk, G.C.G. Roncken. Membrane bioreactors for wastewater treatment: the state of the art and new developments, *Wat. Sci. Tech.* 35 (1997) 35-41.

[16] I.S. Chang, P. Le Clech, B. Jefferson, S. Judd, Membrane fouling in membrane bioreactors for wastewater treatment, *J. Environ. Eng.* 128 (2002) 1018-1029.

[17] Z. Wu, Z. Wang, Z. Zhou, G. Yu, G. Gu, Sludge rheological and physiological characteistics in a pilot-scale submerged membrane bioreactor, *Desalination*. 212 (2007) 152-164.

[18] W.T. Lee, S.T. Kang, H.S. Shin, Sludge characteristics and their contribution to microfiltration in submerged membrane bioreactors, *J. Membr. Sci.* 216 (2003) 217-227.

[19] S. Rosenberger, M. Kraume, Filterability of activated sludge in membrane bioreactors, *Desalination*. 146 (2002) 373.

[20] V. Urbain, J.C. Block, J. Manem, Bioflocculation in activated sludge: an analytic approach, *Wat. Res.* 27 (1993) 829-838.

[21] E.H. Bouhabila, R. Ben Aïm, H. Buisson, Fouling characterization on membrane bioreactors, *Sep. Purif. Technol.* 22-23 (2001) 123-132.

[22] L. Defrance, Bioréacteur à membrane pour le traitement des eaux résiduaires: etute du colmatage de membranes minerals et amélioration du flux de peréat par application de techniques hydrodynamics, Thesis Technology University of Compiégne, France, 1997.

[23] C. Wisniewski, A. Grasmick, Membrane bioreactor for treatment of polluted effluents: biological performance and advantages, Med. Fac, Landbouw. Univ. Gent. 1996.

[24] I.S. Chang, C.H. Lee, Membrane filtration characteristics in membrane-coupled activated sludge system-the effect of physiological states of activated sludge on membrane fouling, *Desalination*. 120 (1998) 221-233.

[25] R.S. Trussell, The effect of organic loading on process performance and membrane fouling in a submerged membrane bioreactor treating municipal wastewater, PhD Dissertation, Civil and Environmental Engineering. University of California, Berkeley, 2004.

[26] F. Meng, F. Yang, B. Shi, H. Zhang, A comprehensive study on membrane fouling in submerged membrane bioreactors operated under different aeration intensities, *Sep.Purif. Technol.* 2007, in press.

[27] Y.L. Jin, W.N. Lee, C.H. Lee, I.S. Chang, X. Huang, T. Swaminathan, Effect of DO concentration on biofilm structure and membrane filterability in submerged membrane bioreactor, *Wat. Res.* 40 (2006) 2829-2836.

[28] J. Wingender, T.R. Neu, H.C. Flemming (Eds.), Microbial extracellular polymeric substances characterization, Structure and Function, Springer-Verlag Berlin Heidelberg, 1999.

[29] P.H. Nielsen, A. Jahn, R. Palmgren, Conceptual model for production and composition of exopolymers in biofilms, *Wat. Sci. Tech.* 36 (1997) 11-19.

[30] M.F. Dignac, V. Urbain, D. Rybacki, A. Bruchet, D. Snidaro, P. Scribe, Chemical description of extracellular polymers: implication on activated sludge floc structure, *Wat. Sci. Tech.* 38 (1998) 45-53.

[31] Y. Liu, H.H.P. Fang, Influence of extracellular polymeric substances (EPS) on flocculation, settling, and dewatering of activated sludge, *Crit. Rev. Environ. Sci. Technol.* 33 (2003) 237-273.

[32] C.S. Laspidou, B.E. Rittmann, A unified theory for extracellular polymeric substances, soluble microbial products, and active and inert biomass, *Wat. Res.* 36 (2002) 2711-2720.

[33] B.E. Rittmann, P.L. McCarty, Environmental biotechnology: principles and application. New York, McGraw Hill, 2001.

[34] B. Frølund, R. Palmgren, K. Keiding, P.H. Nielsen, Extraction of extracellular polymers from activated sludge using a cation exchange resin, *Wat. Res.* 30 (1996) 1749-1758.

[35] D.C. Sobeck, M.J. Higgins, Examination of three theories for mechanisms of cation-induced bioflocculation, *Wat. Res.* 36 (2002) 527-538.

[36] R.E. McKinney, A fundamental approach to the activated sludge process II. A proposed theory of floc formation, *Sewage Ind. Wastes.* 24 (1992) 280-287.

[37] Y. Tezuka, Cation-dependent flocculation in a Flavobacterium species predominant in activated sludge, *Appl. Microbiol.* 17 (1969) 222-226.

[38] M.J. Higgins, J.T. Novak, The effect of cations on the settling and dewatering of activated sludges: laboratory experience, *Water Environ. Res.* 69 (1997) 215-224.

[39] J.N. Bűchs, N. Mozes, C. Wandrey, P.G. Rouxhet, Cell adsorption control by culture conditions, *Appl. Microbial. Biotechnol.* 40 (1998) 179-185.

[40] F. Jorand, P. Guicherd, V. Urbain, J. Manem, J.C. Block, Hydrophobicity of activated sludge flocs and laboratory-growth bacteria, *Wat. Sci. Tech.* 30 (1994) 211-218.

[41] F. Jorand, F. Boué-Bigne, J.C. Block, V. Urbain, Hydrophobic/hydrophilic properties of activated sludge exopolymeric substances, *Wat. Sci. Tech.* 37 (1998) 307-315.

[42] R. Bura, M. Cheung, B. Liao, J. Finlayson, B.C. Lee, I.G. Droppo, G.G. Leppard S.N. Liss, Composition of extracellular polymeric substances in the activated sludge floc matrix, *Wat. Sci. Tech.* 37 (1998) 325-333.

[43] D.J. Barker, D.C. Stuckey, A review of soluble microbial products (SMP) in wastewater treatment systems, *Wat. Res.* 33 (1999) 3063-3082.

[44] W.C. Kuo, G.F. Parkin, Characterization of soluble microbial products from anaerobic treatment by molecular weight distribution and nickel-chelating properties, *Wat. Res.* 30 (1996) 915-922.

[45] D.R. Confer, B.E. Logan, Molecular weight distribution of hydrolysis products during biodegradation of model macromolecules in suspended and biofilm cultures I. Bovine serum albumin, *Wat. Res.* 31 (1997) 2127-2136.

[46] D.R. Confer, B.E. Logan, Molecular weight distribution of hydrolysis products during biodegradation of model macromolecules in suspended and biofilm cultures II Dextran and dextrin, *Wat. Res.* 31 (1997) 2137-2145.

[47] N. Jang, X. Ren, K. Choi, I.S. Kim, Comparison of membrane biofouling in nitrification and denitrification for the membrane bioreactor (MBR)", *Wat. Sci. Tech.* 53 (2006) 43-49.

[48] Kim, N. Jang, The effect of calcium on the membrane biofouling in the membrane bioreactor (MBR), *Wat. Res.* 40 (2006) 2756-2764.

[49] K. Kimura, N. Yamato, H. Yamamura, Y. Watanabe, Membrane fouling in pilot-scale membrane bioreactors (MBRs) treating municipal wastewater, *Environ. Sci. Technol.* 39 (2005) 6293-6299.

[50] S.T. Kelly, A.L. Zydney, Mechanisms for BSA fouling during microfiltration, *J. Membr. Sci.* 107 (1995) 115-127.

[51] S.T. Kelly, A.L. Zydney, Protein fouling during microfiltration: Comparative behavior of different model proteins, *Biotech. Bioeng.* 55 (1997) 91-100.

[52] F. Meacle, A. Aunins, R. Thornton, A. Lee, Optimization of the membrane purification of a polysaccharide-protein conjugate vaccine using backpulsing, *J. Membr. Sci.* 161 (1999) 171-184.

[53] M. Manttari, L. Puro, J. Nuortila-Jokinen, M. Nystrom, Fouling effects of polysaccharides and humic acid in nanofiltration, *J. Membr. Sci.* 165 (2000) 1-17.

[54] M. Manttari, M. Nystrom, Critical flux in NF of high molar mass polysaccharides and effluents from the paper industry, *J. Membr. Sci.* 170(2000) 257-273.

[55] N. Cicek, M.T. Suidan, P. Ginestet, J. Audic, Role of soluble organic matter on filtration performance of a membrane bioreactor, WEFTEC, Chicago, IL. 2002.

[56] H.S. Shin, S.T. Kang, S.Y. Nam, W.T. Lee, S.R. Chae, Simultaneous removal of organics and nutrients in a single reactor coupled with membrane, *J. Korea Soc. Environ. Eng.* 23 (2001) 971-978.

[57] E. Namkung, B.E. Rittmann, Soluble microbial products (SMP) formation kinetics by biofilms, *Wat. Res.* 20 (1986) 795-806.

[58] H. Nagaoka, S. Yamanishi, A. Miya, Modeling of biofouling by extracellular polymers in a membrane separation activated sludge system, *Water Sci..Technol.* 38 (1998) 497-504.

[59] P.C. Carman, Fundamental principals of industrial filtration (A critical review of present knowledge), *Trans. Inst. Chem. Eng.* 16 (1938) 168-188.

[60] J.C. Schippers, J. Verdouw, The modified fouling index, a method of determining the fouling characteristics of water, *Desalination.* 32 (1980) 137-148.

[61] N. Jang, X. Ren, G. Kim, C. Ahn, J. Cho, I.S. Kim, Characteristics of soluble microbial products and extracellular polymeric substances in the membrane bioreactor for water reuse, *Desalination.* 202 (2007) 90-98.

[62] N. Jang, R.S. Trussell, R.P. Merlo, D. Jenkins, S.W. Hermanowicz, I.S. Kim, Fractionations of EPS (extracellular polymeric substances) molecular weight and filtration resistances as F/M (food/microorganism) in the submerged membrane bioreactor, *ICOM2005,* Korea, 2005.

[63] B.K. Hwang, W.N. Lee, P.K. Park, C.H. Lee, I.S. Chang, Effect of membrane fouling reducer on cake structure and membrane permeability in membrane bioreactor, *J, Membr. Sci.* 288 (2007) 149-156.

[64] W.N. Lee, I.S. Chang, B.K. Hwang, P.K. Park, C.H. Lee, X. Huang, Changes in biofilm architecture with addition of membrane fouling reducer in a membrane bioreactor, *Process Biochem.* 42 (2007) 655-661.

[65] N. Lesage, M. Sperandio, C. Cabassud, Study of a hybrid process: adsorption on activated carbon/membrane bioreactor for the treatment of an industrial wastewater, *Chem. Eng. Process.* 2007, in press

[66] G.T. Seo, H.I. Ahan, J.T. Kim, Y.J. Lee, I.S. Kim, Domestic wastewater reclamation by submerged membrane bioreactor with high concentration powdered activated carbon for stream restoration, *Wat. Sci. Tech.* 50 (2004) 173-178.

[67] G. Munz, R. Gori, G. Mori, C. Lubello, Powdered activated carbon and membrane bioreactors (MBR-PAC) for tannery wastewater treatment: long term effect on biological and filtration process performances, *Desalination.* 207(2007) 349-360.

[68] J.H. Choi, K. Fukushi, K. Yamamoto, A submerged nanofiltration membrane bioreactor for domestic wastewater treatment: the performance of cellulose acetate nanfiltration membranes for long-term operation, *Sep. Purif. Technol.* 52 (2007) 470-477

[69] J.H. Choi, S. Dockko, K. Fukushi, K. Yamamoto, A novel application of a submerged nanofilration membrane bioreactor (NF MBR) for wastewater treatment, *Desalination.* 146 (2002) 413–420.

[70] M. Mulder, Basic Principles of Membrane Technology, KLUER ACADEMIC PUBLISHERS, DORDRECHT, 1996.

[71] A.L. Lim, R. Bai, Membrane fouling and cleaning in microfiltration of activated sludge wastewater. *J. Membr. Sci.* 216 (2003) 279-290.

[72] L. Song, M. Elimelech, Theory of concentration polarization in crossflow filtration. *J. of Chem. Soc. Faraday Trans.* 91(1995) 3389-3398.

[73] J. Hermia, Constant pressure blocking filtration law application to power-law non-Newtonian fluid, *Trans. Inst. Chem. Eng.* 60 (1982) 183-187.

[74] S.H. Kim, Application of adsorption and membrane-adsorption hybrid system in water and wastewater treatment, PhD thesis, Dep. of Chem. Tech., Chonnam National University, Republic of Korea, 2002.

In: Advances in Membrane Science and Technology
Editor: Tongwen Xu
ISBN: 978-1-60692-883-7

Chapter VII

Membrane Integration Processes in Industrial Applications

Hong-Joo Lee [*] ***and Seung-Hyeon Moon***
Department of Environmental Science and Engineering,
Gwangju Institute of Science and Technology (GIST),
1 Oryong-dong, Buk-gu, Gwangju 500-712, Korea

Abstract

The industrial applications of membrane processes have been increasing for production of various chemicals. Often the membrane based separation and purification processes integrated with other physical and chemical processes have advantages in terms of environmental-friendliness and energy efficiency. The process integration consisting of non-membrane processes such as ion exchange, adsorption, chemical and biochemical reactions as well as other membrane processes has been employed in order to utilize the effectiveness of the membrane processes. Also a substantial number of industrial applications are found especially in biotechnology and chemical engineering as the integrated processes of eletrodialysis (ED), electrodialysis with bipolar membrane (EDBPM), reverse osmosis (RO), ultrafiltration (UF), and pervaporation (PV). Also the applications of integrated ion exchange membrane processes have been expanded with the competitiveness of the productivity and cost saving for production of ionic products and purification of water.

1. Introduction

A membrane is a permeable or semi-permeable phase acting as a barrier of certain species. The membrane controls the transport of various species through itself, one product being depleted in certain components and a second product concentrated. Various membranes

[*] Correspondence concerning this chapter should be addressed to Hong-Joo Lee at hjlee68@paran.com

have been developed for the separations of mixtures of gases and vapors, miscible liquids and solid/liquid and liquid/liquid dispersions and dissolved solids and solutes from liquids. The main uses of membrane separations in an industrial scale are in reverse osmosis (RO), nanofiltration (NF), ultrafiltration (UF), microfiltration (MF), pervaporation (PV), and electrodialysis (ED) [1]. The applications of membrane separation processes have grown in industrial applications such as food industries, biotechnology, chemical industries and membrane based energy devices in addition to membrane based separation and purification processes.

Transport of selected species through the membrane is achieved by applying a driving force across the membrane. The species in RO, NF, UF, and MF are separated by the pressure between two liquid phases. In addition, PV separates mixtures of miscible liquids. For the electrically charged species, the membranes fixed positive or negative charges, so called ion exchange membrane, separate by exclusion of ions of the same charge as carried in the membrane phase. The applications of ion exchange membrane processes can be classified into three representative types [1]: (i) mass separation processes involving the separation of components from electrolytes, such as ED, diffusion dialysis, Donnan dialysis, and EDBPM (electrodialysis with bipolar membrane), (ii) chemical synthesis processes involving an electrochemical reaction producing certain chemicals, such as chlorine-alkaline electrolysis and production of hydrogen and oxygen via water electrolysis, (iii) energy conversion and storage processes involving the conversion of chemicals into electrical energy and vice versa, such as fuel cells and electrical batteries.

A single membrane process is sometimes inefficient in solving given separation problems. Therefore, integrated membrane processes have been developed to overcome the limits of a single process. By integrating a membrane process with non-membrane or other membrane processes, the process can achieve an efficiency level which is not achievable by either process used alone. The applications of membranes processes have been expanded with the competitiveness over traditional, high energy intensive, environmentally undesirable and costly processes [2,3]. More comprehensive integration with pressure-driven membrane separation processes, adsorption, and chemical unit operation and electrochemical/biological reactors can be found in the integrated ion exchange membrane processes. This review mainly considers important industrial or pilot applications of membrane integration based on ion exchange membranes processes and additionally other integrated membrane processes such as RO and UF.

2. Membrane Technology Integration Based on Desalting Electrodialysis

The conventional desalting ED is the most commercially relevant ion exchange membrane separation process to produce potable water from brackish water and seawater. Currently water desalination and the pre-concentration of NaCl from seawater are still ones of the most important large scale applications of the conventional ED. Other applicable areas can be found in the food and chemical industries such as the demineralization of whey or deacidification of fruit juice [4-6]. Also, the process is applied to recover toxic or valuable components from effluents of galvanic and metal surface finishing processes. The production of high quality process water for power plants and the electronic industry by the integrated ED system with other processes such as ion exchange or reverse osmosis is considered as one of the important applications. Some of the more important large scale industrial applications of conventional ED as well as related applications [7,8]. Table 1 summarizes selected industrial applications of ED and related processes.

Table 1. Industrial applications of ED and related processes

Application	Process development	Problems related to applications
Brackish water desalination	Commercial	Concentration of feed and costs, Process reliability
Industrial process water	Commercial	Scale formation, Membrane stability, Product water quality and costs
Production of table salt	Commercial	Production cost, Membrane fouling
Industrial effluent	Commercial	Treatment cost, Membrane fouling
Food and biotechnology	Commercial or pilot phase	Membrane fouling, Product loss, Application experience and costs
Concentration of RO brine	Pilot phase	Concentration of feed and costs, Waster disposal

2.1. Desalination of Brackish Water and Seawater

The desalination of brackish water and seawater has become a reliable method for water supply all over the world [9]. It has already been practiced successfully for many decades and the technical and economical feasibility is obvious. However, the common processes for the thermal desalination such as multi-effect distillation (MED) and multi-stage flash (MSF) require large quantities of energy for the treatment of brackish water and seawater [10,11]. Recently, interests on the desalination using membrane processes such as RO and ED have significantly increased. It is considered that RO is one of the most versatile desalination methods for the treatment of water of any salinity from brackish water to high salinity seawater, which overtook MSF as the leading desalination technology [9].

Compared to RO, the advantages of ED are the following [12]: (i) high water recovery rates for raw water with high sulfate content, (ii) long operation life of membranes due to higher chemical and mechanical stability, (iii) less membrane fouling of scale, (iv) less raw water pretreatment, and (v) flexible process to various feed water quality. However, the desalination of brackish water by ED requires post-treatment procedure. Furthermore, the generation of chlorine gas at the anode can lead to corrosion problems in the surrounding of the plant [12].

In the desalination of seawater with salt contents in the range of 35,000-45,000 mg/L, the two-stage procedure with the combination of RO and ED was considered for the long term operation and the satisfactory product water quality. The salt concentration of the seawater could be reduced to about 2000 to 4000 mg/L in the first stage using RO and to less than 500 mg/L in the second stage using ED [13,14]. The two-stage combining ED with RO showed higher reliability and possibly lower costs compared to a one-stage system [2,15].

An example of the hybrid approach for economical zero liquid discharge is the installation of EDR (Electrodialysis reversal) plus RO along with evaporation and crystallization at the Doswell Combined Cycle Power Plant in Hanover County, Virginia. Use of membrane processes with the combined EDR and RO to pre-concentrate and reclaim plant wastewaters has resulting in a 62 % downsizing of the evaporator system and reduction of overall capital costs by $ 900,000 [16].

2.2. Pre-Concentration of Salt

NaCl is the major ionic constituent of seawater and a potential feedstock for the manufacture of food grade salt. The concentration of diluate salt solutions by ED is used in the treatment of certain industrial effluent to recover valuable or toxic constituents or to avoid excessive salt discharge into wastewater treatment plants. It is of importance to produce certain salts from various raw water sources [1].

Of special interest is the concentration of NaCl from seawater before evaporation. ED is widely used as a pre-concentration step in the production of salt before evaporation [17]. Large plants with a capacity of 20,000 to more than 200,000 tons of table salt per year are in operation in Japan. The success of ED in Japan is related to the development of ion exchange membranes with a high selectivity to monovalent ions at a low cost. The integrated ED systems for salt pre-concentration consist of three major parts: pretreatment of seawater, concentration by ED and salt production by evaporative crystallization. ED concentrates filtrated seawater (3 % salt content) to produce brine at approximately 20 % for the evaporative crystallization of salt [17].

2.3. Production of Industrial Waters

Certain quality standards in terms of dissolved solids and colloidal materials depending o n the water sources should be considered for the process selection to produce industrial water s. Traditionally, precipitation, filtration, and ion exchange are used in the production of indus trial waters [18,19]. Today, these processes are replaced or complimented more and more by i on exchange membrane processes. Major applications of ED in the production of industrial w aters include (i) predemineralization of boiler feed water, (ii) desalination of cooling water fo r reuse, (iii) desalination of process water for chemical production processes, and (vi) desalin ation of contaminated industrial water for reuse. ED has clear cost advantages over the compe ting processes, such as ion exchange and RO, for certain raw water compositions.

2.3.1. *Applications on the Preparation of Industrial Waters*

Van der Hoek *et al.* reported three different integrated membrane systems in the production of industrial waters for the extension of 13 million m^3/year [20]. For the treatment of Rhine River water, the raw water was treated sequentially by ozonation, biologically activated carbon filtration, slow sand filtration and RO in the integrated membrane system I. In the integrated membrane system II, RO was replaced to EDR. In addition, Rhine River water was treated sequentially by EDR, ozonation, biologically activated carbon filtration and slow sand filtration in the integrated membrane system III. The integrated EDR systems showed lower energy consumption and chemical consumption in the industrial water application.

For the preparation of boiler feed water, large quantities of raw water should be demineralized. And, multiple or redundant ion exchange techniques require depending on the level of deionization. The most significant costs are attributed to the regeneration of the ion exchange resin. The regeneration costs could be reduced drastically with the predemineralization of the feed water by ED [21]. Another benefit of the predemineralzation by ED is the significant reduction in waste disposal.

2.3.2. *Applications on the Recycling of Industrial Waters*

Recently, there has been a substantial decrease in the quality of raw water for industrial use with a corresponding increase in water costs. Also, the discharge of certain industrial wastewaters is often problematic and costly because of high salt concentrations. For recycling wastewater back into an industrial process it should undergo certain purification steps which usually include a desalination process. ED can be used very effectively and economically in recycling industrial process water due to high recovery rates and stable operation at wide pH range and at somewhat high temperature. A reduction of raw water intake and wastewater discharge have been achieved by increasing the cycles on concentration in cooling towers using acids, anti-precipitants, and side stream clarification and softening [22].

Another example of a successful application of ED is the recycling of produced water from crude oil production. In many oil wells, steam is injected into the ground to heat and liquefy the crude oil so that it can be pumped. The steam, which is mixed with the oil, is condensed and pumped out of the ground. The water is then separated from the oil by an oil/water separator. The recovered water usually also contains large quantities of salt and hardness extracted from the ground. Since the salt concentration in the produced water is relatively high, *i.e.*, in the order of 5,000 to 10,000 mg/L, large quantities of salt are used for the regeneration of the ion exchange resins. Substantial saving in regeneration costs can be achieved by the EDR as pre-softening desalination step. Before EDR unit the produced water is cooled down from ca. 80 °C to about 45 °C. In the EDR system, 80-90% of the total dissolved solids are removed depending on the feed water salt concentration. The partially demineralized water is then further treated in a softener and finally recycled to the steam generator [23].

2.3.3. Applications on the Treatment of Industrial Effluents

Very often toxic or valuable components should be removed from an industrial effluent to avoid pollution of the environment [24]. A major source of pollution are effluents of the metal processing industry including heavy metals such as nickel, zinc, cadmium, chromium, silver, mercury, and copper ions, which are toxic even in relatively low concentrations. The effluents are often difficult to separate from a mixture with other salts for safe disposal. Ion exchange membrane processes play increasingly important roles in separating, concentrating, and recycling these heavy metals in industrial effluents. In combination with other processes, such as ion exchange, UF and MF, the application of ED provides a solution to severe pollution problems and saves substantial production costs by recovering and recycling valuable components and water. One of the typical examples is the recovery of nickel, cadmium, and copper from rinse solutions of galvanic processes [25].

Another interesting application is the recovery and recycling of water and sulfuric acid from the rinse solution of a lead battery production line. The integrated membrane process consisting of several processes including ED, RO, NF, and MF was developed to recycle the water and sulfuric acid from the spent rinse water for the battery production line [1]. The first step in the process of the spent rinse water of the electrode production line is the precipitation of lead sludge in a settling tank. The supernatant solution containing sulfuric acid and various metal ions and particles is fed to MF. The retentate of the MF is recycled to the sedimentation tank. The filtrate is passed through NF unit, rejecting the metal ions. The retentate and sulfuric acid is fed to a neutralization tank. The NF filtrate is further processed in an ED unit. The diluate stream of the ED unit, which contains a small amount of sulfuric acid, is fed to RO unit. In the RO system, the filtrate which is free of sulfuric acid and metal ions is directly recycled to the rinsing tank. The RO concentrate is partially recycled into the ED unit and sulfuric acid is fed to the neutralization tank. Figure 1 shows the treatment and recycling of acid rinse water from a lead battery production line. The overall mass balance shows that 88 % of the sulfuric acid and 25 wt% of the rinse water are recovered in the integrated ED system. It is reported that the system reduced costs for water and sulfuric acid and substantial saving in neutralization chemicals and sludge disposal costs [22].

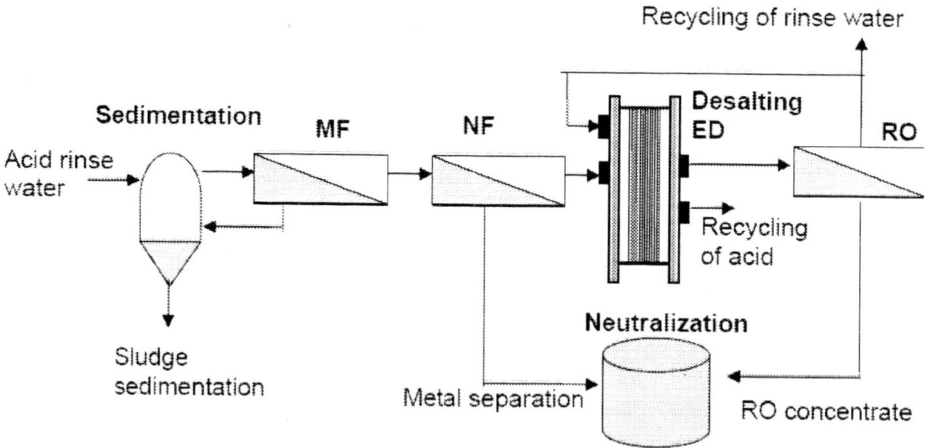

Figure 1. Integrated ED system for the treatment and recycling of spent rinse water from a acid battery production line.

The metal plating and the metal surface treatment containing zinc, cadmium, silver and copper provides a multitude of economically increasing application of ion exchange membrane processes. In the metal surface treatment, ED is combined with other processes to recover acids from a pickling bath [26-28]. The acids for the pickling, such as H_2SO_4, HCl, HNO_3, HF, etc., deteriorate with time and become ineffective in the treatment process. Since the spent pickling solution contains various heavy metals and substantial amount of acids, the obtained sludge should be deposited in an environmentally friendly method after the solution is neutralized. From a stainless steel pickling bath, the ED process was used to recover acids from the effluent of to recycle HNO_3 and HF. Here, the diluate from the ED unit containing HNO_3 to the neutralization process, while the concentrate is directly recycled into the pickling bath [27]

Many interesting applications of ED can be found in wastewater treatments including the paper and pulp industry. Here, ED has been successfully used to remove NaCl selectively from the chemical recovery cycle of Kraft pulp production. The accumulation of chloride ions gives series problems in the pulping process. Using ED, chloride is selectively removed with a high selectivity for monovalent ions [29].

2.4. Food Processing

Integration of ion exchange membrane processes is also applied to the production of organic acids from industrial feeds and the utilization of resources in the food and biotechnology. The process integration of ED can be found in pretreatment of feed solutions and combination of fermentation. In order to decrease fouling potentials and ensure stable operation performance, the feed solution is necessary to treat before the ED stack. Pressure-driven membrane processes are used to remove foulants (particles, bacteria, cellulose, proteins, or suspended solids) and multivalent ions. Also, adsorption and ion exchange can be employed to remove organics which foul anion exchange membranes during the ED process [30]. Many applications of conventional ion exchange in the food industries can be replaced by ion exchange membranes because of better economics, simpler operation, and less waste materials. Typical applications of ED in the food industry are summarized in table 2 [30].

Table 2. Examples for the ED application in the food industries

Food industry	Application of electrodialysis	Ref.
Diary industry	Demineralization of cheese whey Demineralization of non-fat milk	54,55
Sugar industry	Demineralization of molasses in sugar production Demineralization of waste molasses Demineralization of polysaccharide	59,60,61
Fermentation industry	Demineralization of soy sauce Desalination of amino acids Recovering of organics acids	35,40,43, 44, 68
Fruit juice and beverages	Deacidication of fruit juices Concentration	63

2.4.1. Applications on the Separation of Organic Acids

In the fermentation of organic acids, separation, concentration, and purification are necessary for the product preparation. The related conventional techniques include precipitation, acidification, extraction, crystallization, distillation, ion exchange, and adsorption. The conventional fermentation process produces calcium lactate salt precipitate, which should be reacidified by a strong acid, thus yielding calcium salt as a by-product which causes high chemical cost and waste generation [31,32]. One of the important ED applications in the food industry is the recovery of lactic acid from the fermentation broth [33-37].

Up to now, the more feasible strategy in the organic acid fermentation is to integrate ED with fermentation by external auxiliary treatments such as pressure-driven membrane processes, adsorption and ion exchange as an integrated ED system [33, 38-40]. In the ED fermentation, the broth is fed into ED in batch or continuously after pretreated, and the effluent is recycled to fermentation. ED has been used as an alternative for the precipitation for recovering lactic acid because the process can remove produced lactic acid continuously and enhance the production of lactic acid by removing ionic nutrients such as sulfate and phosphate ions, resulting in decrease in productivity of lactic acid [41,42]. In addition, the performance of the ED fermentation could be enhanced by keeping bacteria off membrane using immobilized cells and MF in the integrated ED system [33,43,44].

2.4.2. Applications on the Separation of Amino Acids

The amphoteric properties can be used for the separation of amino acids by ion exchange membranes when the pH value of the raw solution exists between the respective isoelectric points. The electromigration of amphoteric amino acids through ion exchange membranes occurs only in the presence of H^+ or OH^- ions [45-47].

Solutions of amphoteric amino acids can be desalted by ED. The operating conditions should be controlled because of the concentration depletion on the membrane surface which leads to water splitting and fouling at the surface of the membranes. Sandeaux *et al.* reported that the extraction degrees varied from 80% to 100% in the separation of different types of amino acids, such as anion types (aspartic acid, glutamic acid), cation types (histidine, arginine, lysine) and zwitterions types (serine, glycine, threonine, alanine, proline, valine, isoleucine, leucine, phenylalanine) [48]. In addition, the separation of several amino acids were studied using the desalting ED such as phenylalanine [49] hydroxyphenylglycine [50, 51], glycine [46], glutamic acid [52], taurine and glycine [47] arginine [47,53], and alanine [45].

2.4.3. Applications on the Demineralization in Food Industry

A large amount of whey can be obtained in the diary industry as in the production of various types of cheese. Whey provides an excellent source of protein, lactose, vitamins, and minerals. However, it is not suitable as food materials because of the high salt content. Demineralization of whey under conditions would be desirable without decomposition of the proteins and vitamins as valuable products. The application of ion exchange process to demineralize whey was not successful because of severe fouling of the exchange resins and a substantial pH shift causing denaturation of protein [54].

ED was successfully used for the demineralization of whey and skimmed milk due to the development of ion exchange membranes having the stablility in solutions with a good chemical stability [55-57]. The degree of demineralization, depending on the individual salts, shows the composition of cheese whey and skimmed milk demineralized in the plant. The total removal of salts is ca. 90 % of cheese whey and 70 % for the skimmed milk. In addition, the demineralization rate of monovalent ions is higher than that of the multivalent ions.

ED also can be applied in the food industry such as the demineralization of soy sauce and sugar malasses. The salt content should be reduced without losing fragrance, color, or taste for human consumption [58]. It is reported that the degree of desalination increased to 75.1 % in the desalination of soy sauce using an ED system [58]. In addition, various works have been carried out to remove melassigenic ions in a sugar industry including ion exchange process, adsorption, coagulation and membrane processes [59]. Using an ED system, melassigenic ions for beet sugar solutions could be removed and the quality of the three examined solutions (juice, syrup and mother liquor B) was improved without fouling effects [60,61].

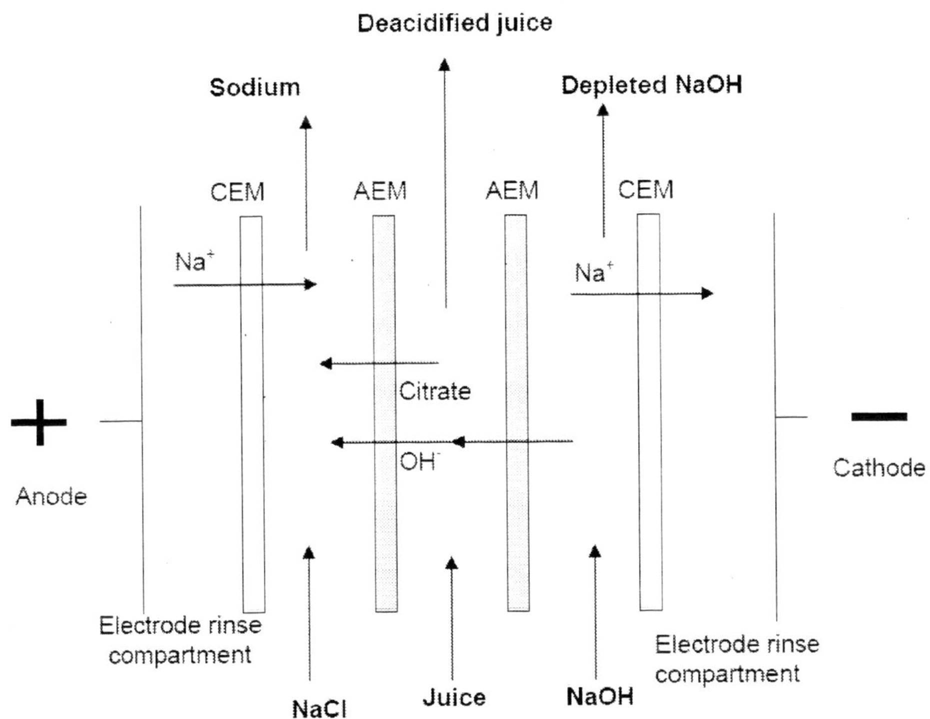

Figure 2. Illustration of deacidification from a fruit juice using ED(AEM: anion exchange membrane, CEM: cation exchange membrane, BPM: bipolar membrane).

2.4.4. Applications on the Fruit Juices and Beverages

Other applicable areas can be found in the deacidication of fruit juice [4-6]. Tartaric acid, its salts, potassium bitartrate and calcium tartrate, which form crystalline deposits during aging, are normal constituents of grape juice and wines [6,62]. ED can be used to prevent the precipitation of tartrates from grape musts, wines, and concentrates. Also, cations such as potassium and calcium are removed in the ED process.

Adjusting the ratio between sugar and acid in grape musts and wines is sometimes desirable to improve sensory quality. The high acidity of some fruit juices reduces their quality and usefulness in food preparation. Citric and malic acids are responsible of the high acidity of the fruit juices [6,63]. In the conventional process, ion exchange has been considered for the deacidification of citric and malic acids from the fruit juices. However, it gives a poor quality juice in terms of the sensory properties of a product, involving taste, color, and odor and the large amount of effluent during the regeneration phase of the resins. As a novel deacification process, citrate and malate anions from the juice are extracted in the ED process and replaced by OH^- ions provided by NaOH solution in the cell. Figure 2 shows illustration of deacidification from a fruit juice using ED. The deacidification by ED has some advantages in terms of physicochemical and sensorial analyses, *i.e.*, (i) no change in the cation concentration, (ii) a slight color change, and (iii) good flavor due to the partial elimination of organic anions.

3. Membrane Technology Integration Based on Electrodialysis with Bipolar Membranes

The conventional method for generating H^+ and OH^- ions from water is electrolysis, which generates gases (O_2 and H_2) and consumes large amount of energy. Nowadays, special ion exchange membranes, so called bipolar membrane, are used to produce H^+ and OH^- ions by the water-splitting directly without gas generation. The applicable potentials of ED have been greatly extended by the introduction of bipolar membranes consisting of an anion and a cation selective layer joined together. In the electric field across the membrane, transport of electric charge can be accomplished by H^+ and OH^- ions produced by the splitting of water [64,65]. The main features of the bipolar membranes are the follows: (i) the production of H^+ and OH^- ions by the water-splitting on the membrane surface, (ii) low energy consumption without gas generation during water splitting, and (iii) no oxidation and/or reduction on the electrode surface. Since bipolar membranes became available as commercial products, a very large number of potential applications of the EDBPM (Electrodialysis with bipolar membrane) have been studied extensively on a laboratory or pilot plant scale [66-68]. A great potentials have been found in the areas of chemical production and separation, biochemical engineering, environmental conservation, etc. Table 3 shows selected potential applications using the EDBPM.

3.1. Production of Acids and Bases

By now, one of the largest potential applications of EDBPM is the production of acids and bases from the corresponding salts. Conventionally, caustic soda is produced as a co-product with chlorine by electrolysis, resulting in environmental problems caused by many chlorinated hydrocarbons products. EDBPM is considered to be an alternative process to produce caustic soda and hydrochloric acid from the corresponding salts in an environmentally friendly way [69,70]. In spite of extensive applications, EDBPM is not used on a large commercial scale for the production of high purity concentrated mineral acids and bases due to insufficient membrane permselectivity. In order to increase the process feasibility of EDBPM, an integrated system should be considered such as pretreatment of a salt solution including multivalent cations and organic pollutants using pressure-driven membrane processes.

Table 3. Potential applications with EDBPM

Applications	Process development	Potential advantages	Problems	Ref.
Production of mineral acids and bases from salts	Pilot plant operation	Lower energy consumption	Contamination of products Low current utilization	3,66, 68
Recovery of organic acids from fermentation processes	Commercial and pilot plant operation	Integrated process, Lower costs	Unsatisfactory membrane stability Membrane fouling	30,35, 41,68, 73
Recovery of amino acids from fermentation processes	Laboratory and pilot plant tests	Integrated process, Lower costs	Relatively complex process, High investment costs	30,65, 77,78, 79
Removal of SO_2 from flue gas	Extensive pilot plant test	Decreased salt production and disposal costs	High investment costs, Long-term membrane stability	64,68
Recovering and recycling of H_2SO_4 and NaOH from waste waters	Laboratory and pilot plant tests	Savings in chemicals and sludge disposal costs	Few long-term experience, Membrane stability, Membrane fouling, High investment costs	1,22, 72
Recycling of HF and HNO_3 from steel pickling solutions	Commercial plants	Cost savings and decreased salt disposal	Relatively complex process, High investment costs	66,70
Food industry	Laboratory and pilot plant tests	Cost savings in chemicals, less by-products, Less salt production and disposal	Process and investment costs	30,58, 70,82

The EDBPM can be used effectively to recycle acids and bases recovered from the corresponding salts from a waste stream [66]. One of the successful applications in an industrial scale is the recovery of HF and HNO_3 from a stream containing KF and KNO_3 generated from a pickling bath in a steel plant. Figure 3 illustrates the process for the recycle of HF/HNO_3 steel pickling solution in the integrated EDBPM system. The spent pickling acid is neutralized with KOH to precipitate heavy metals, which are removed by filtration. Then, the remaining neutral solution of KF and KNO_3 is treated in a three compartment EDBPM cell, where the salts are split to form KOH. This KOH is directed to the neutralization tank and the mixed acids of HF and HNO_3 are directed to the pickling operation. The diluate salt solution from the salt compartment of the EDBPM process is then reconcentrated by the conventional ED and subsequently is returned to the EDBPM stack, while the diluate is used to rinse the filter cake [66].

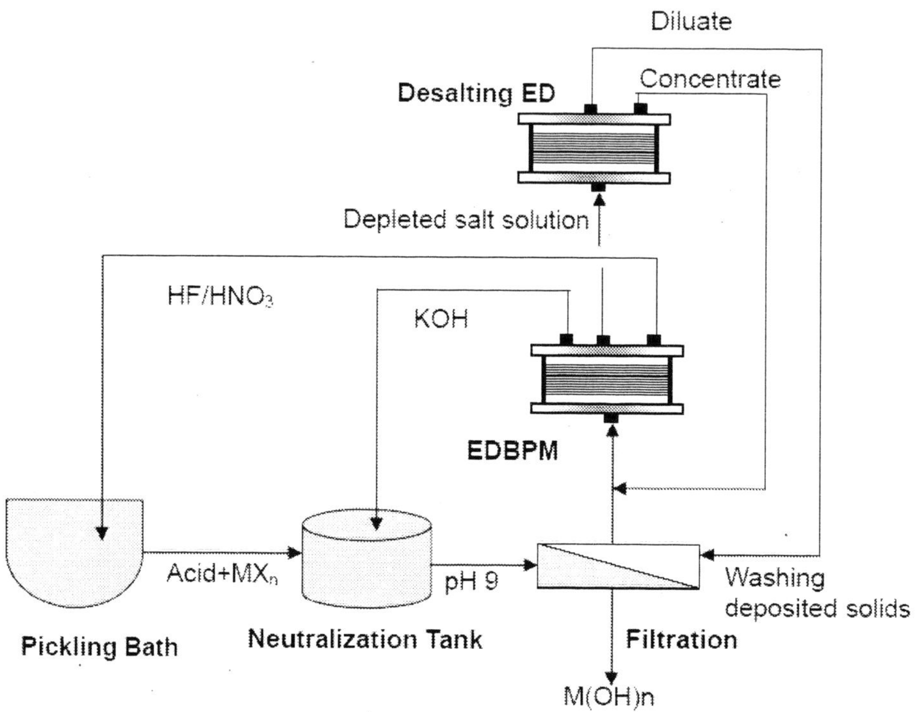

Figure 3. Integrated ED and EDBPM system for the recycle of HF/HNO_3 steel pickling solution.

3.2. Pollution Control Processes

Another interesting application of EDBPM is the treatment of alkaline or acid scrubber to remove components which are hazardous to the environmental such as SOx and NOx. In alkaline or acid scrubbers, large amounts of acids or bases are consumed, the salts being produced. The disposal of the produced salts is often difficult due to contamination of toxic materials such as heavy metal ions or organic compounds. The recovery of the base in the scrubber to remove SO_2 from flue gas of coal burning power plants by EDBPM has been extensively studied [64,68]. The integrated process of EDBPM consists of two basic steps for the removal of SO_2 from the flue gas. In the first step, NaOH is used as an adsorbent forming Na_2SO_4 with oxygen. Then, the base is regenerated form the salt solution in a three compartment of EDBPM, H_2SO_4 being produced as by-product. In the EDBPM stack, the Na_2SO_4 is fed and the Na^+ migrates through a cation exchange membrane forming NaOH with the OH^- generated on the bipolar membrane surface. Meanwhile, the SO_4^{2-} migrates through an anion exchange membrane and then forms H_2SO_4 with H^+ generated on the membrane surface.

A process for the adsorbing SO_2 in the treatment of the flue gas is known as flue gas desulfurication, so called the Soxal[R] process. The process is a regenerative wet scrubbing system producing a concentrated SO_2 stream shown in figure 4. The process is based on the use of a high pH solution as a scrubbing medium to remove the SO_2 from the flue gas. The basic scrubbing solution is generated by the water-splitting process using a bipolar membrane. In the first step, the flue gas passes through an absorber where SO_2 reacts with NaOH solution. Reactions in the SO_2 adsorption produce $NaHSO_3$, which is oxidized to Na_2SO_4 in the presence of oxygen in the flue gas. The spent solution of $NaHSO_3$ and Na_2SO_4 is treated in a two compartment EDBPM stack in the following step, regenerating SO_2, Na_2SO_4 and Na_2SO_3 in the Soxal[R] process [71].

Another application of EDBPM is an example for recovering and recycling an acid from the salt of a spent liquor bath in the rayon process. The process treating the effluent of the rayon spinning bath involves EDBPM to convert Na_2SO_4 to H_2SO_4 and NaOH, which is known as an economically promising process. The acid and base are recycled to the cellulose dissolution process. A substantial saving in chemical and water costs as well as decrease in waste salt disposal is achieved by recovering the spent acid and base from the spinning bath liquor [72]. A number of applications in the pollution control and resource recovery are summarized in table 4.

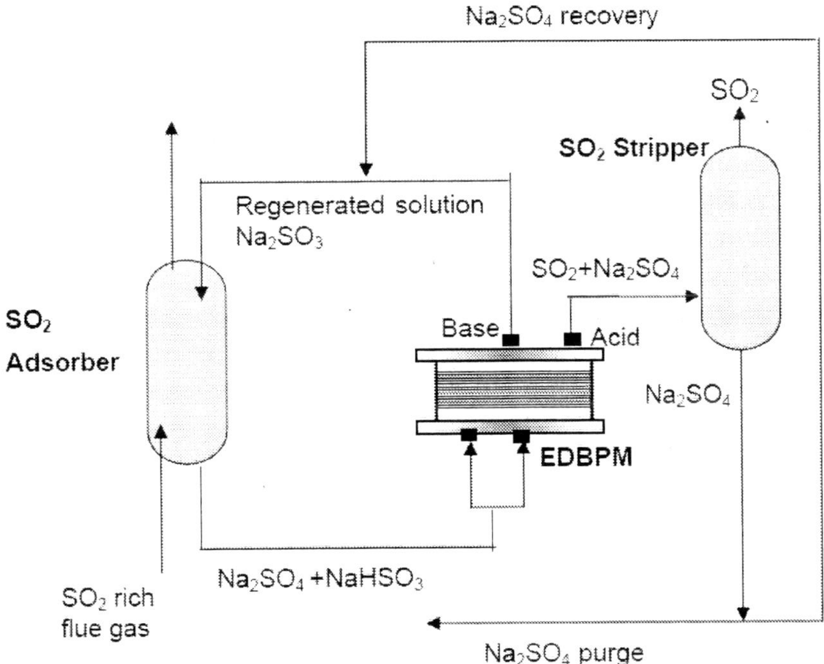

Figure 4. Soxal[R] flue gas desulfurizatio process using EDBPM.

Table 4. EDBPM applications for the pollution control and resource recovery

Application	KF conversion from alkylation process	Recovery of NH_3 and HNO_3 from a stream containing NH_4NO_3	Treatment of KNO_3 effluent generated from a uranium conversion
Scale	Industrial plant	Semi-industrial pilot	Semi-industrial pilot
Process description	The pretreated KF solution is fed in the EDBPM stack to recover HF and KOH.	After the stripping, the solution containing $NaNO_3$ is treated in EDBPM stack to recover HNO_3 and NaOH.	The EDBPM process produces HNO_3 and a mixture of KOH and KNO_3, recycled in the stripping step.
Process efficiency	Efficiency: Unknown	Desalination efficiency: 97 %	Regeneration efficiency: 92 %
Pretreatment	Removal of organic compounds	Stripping of NH_3 form NH_4OH using NaOH	Microfiltration and ion exchange process to remove divalent cations
Ref.	64	65, 129	1, 64

3.3. Food Industry and Biotechnology

3.3.1. Applications on the Recovery and Processing of Organic Acids

One of the promising EDBPM applications is the recovery of organic acids from fermentation process [66,73]. The performance of fermentation increases when the pH is slightly higher than the *pKa* of the organic acids to be produced. The pH value adjustment in the fermenter and the spent medium creates a substantial amount of salts mixed with the product, resulting in decrease in the purity of an organic acid and an additional waste disposal problem. In the fermentation of organic acid in the EDBPM system, the pH in the broth shifts to lower values due to the production of the acid on the bipolar membrane surface without addition of chemicals [74].

Of special interest is the production of lactic acid by the integrated system with the conventional ED and EDBPM [35,41]. The conventional process of the lactic acid production consists of neutralization, filtration, concentration, and ion exchange. Especially, the separation and purification of the lactic acid is achieved by ion exchange and requires different ion exchange steps resulting in a high volume of wastewater from regeneration salts. In the production process with integrated EDBPM, a minimum of ion exchange resin is needed in a final purification step. The concentration of the lactate salt is achieved by the conventional ED and the conversion of the lactate into lactic acid by the EDBPM [64]. Figure 5 illustrates the production process with integrated EDBPM and ED.

In the lactic acid recovery, the desalted broth usually contains a significant amount of hardness ions (mainly Ca^{2+} and Mg^{2+}) due to a poor rejection of ions in the conventional ED [35,75]. Since these hardness ions can foul bipolar membranes in EDBPM, an ion exchange step is required to remove hardness before EDBPM [35,41]. NF can be as a one of the pretreatment methods to remove hardness ions for the recovery of lactate from a fermentation broth, which obviates the need for an ion exchange step before EDBPM. Using NF, it was observed that the rejection of lactic acid was less than 5 %, with hardness removal efficiency being 40-45 %. In a subsequent EDBPM, sodium lactate conversion to lactic acid was reported to about 95 % [75,76].

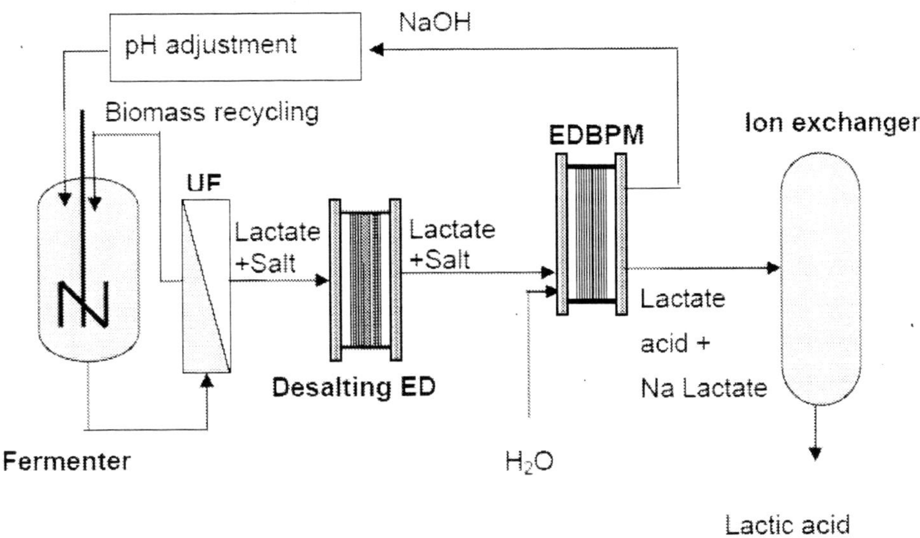

Figure 5. Continuous lactic acid fermeation process in the integrated EDBPM and ED system.

In the continuous fermentation with integrated EDBPM, the production of additional salts in the fermentation broth can be eliminated. The fermentation and the production of itaconic acid can be carried out more efficiently as one of the effective recovery processes in the integrated EDBPM system. In the production of itaconic acid by the continuous fermentation, the fermenter is continuously fed with substrate (See figure 6). The constituents of the feed are passed through an UF unit where biomass is separated from the product containing solution. The UF filtrate is fed to the EDBPM stack. The NH_4^+ is removed from the permeate, forming NH_4OH with OH^- generated on the bipolar membrane surface. Meanwhile, the itaconate migrates through an anion exchange membrane, forming the itaconic acid with H^+ generated on the bipolar membrane surface. Thus, itaconic acid is produced continuously by the fermentation with integrated EDBPM without addition of bases or acids and salts.

3.3.2. Applications on the Recovery of Amino Acids and Proteins

It is of importance to maintain the appropriate pH value for the separation of amino acids. Separation of amino acids and reconditioning of the salt in the process of the EDBPM is possible in a three compartment arrangement, resulting in the recovery of L type of amino acids [65,68,77,78]. Here, existing cations produces as forms of hydroxide while amino acids and some of organic acids are recovered as forms of free acids. Theses solutions can be recycled after a further treatment. EDBPM reduces salt waste and provides diluate products, which can be recycled.

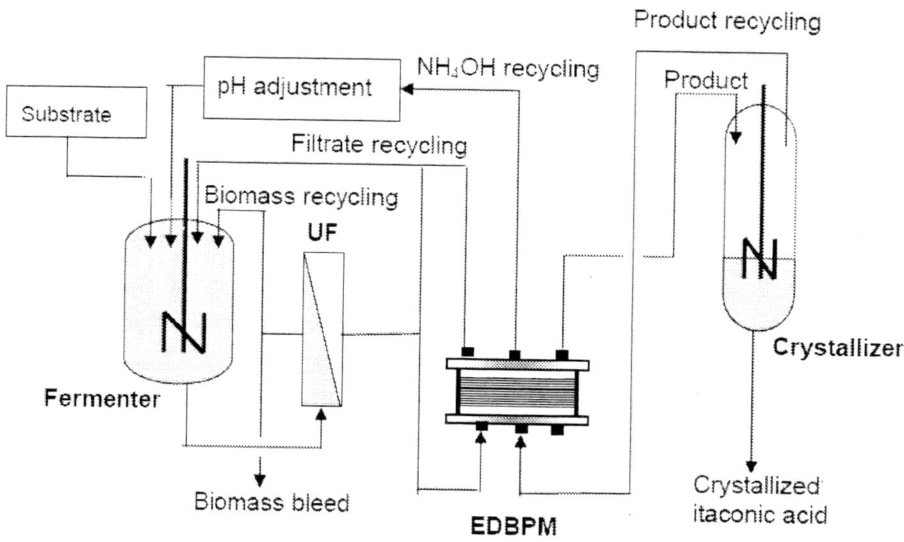

Figure 6. Flow diagram for the production of itaconic acid by the continuous fermentation with integrated EDBPM system.

One of the the integrated EDBPM processes is the recovery of an amino acid from a fermentation broth in an industrial plant. Enzymes are removed by UF and recycled to the fermenter after a diluate organic salt is prodced in a ferimenter. The clear organic salt is fed to a three compartment EDBPM stack, which is combined with the base stream and is recycled to the fermenter. The integration of the EDBPM decreased the losses of amino acid even at higher desalting degrees [77].

Grib *et al.* reported on the recovery of amino acids (alanine, glycine, phenylalanine, serine, and valine) which are a zwitterionic form at pH 5-6 after neutralization of protein hydrolysates in the EDBPM applications. In the EDBPM operation, the amino acids preferentially electromigrate through an anion exchange membrane towards the product compartment with a small amount of amino acid leakage [79]. In addition, EDBPM was applied to regenerate sodium *p*-toluenesulfonate in the production of *p*-hydroxyphenylglycine and the removal ratio of Na^+ was typically around 80% with an average current efficiency of 20-50% [78].

Soy protein concentrate (SPC) and its isolates (SPI) are increasingly used as ingredients in various types of prepared foods, meat analogues, dairy and bakery products. Now, SPC/SPI can be produced by the traditional acid precipitation and UF. A disadvantage of the conventional methods is that the SPC or SPI has poor protein solubility and functional properties due to extreme conditions such as extraction with alcohol or alkali, heat treatment, precipitation or centrifugation [80]. Meanwhile, the acidification by the H^+ ions produced on the bipolar membrane surface in the electric field causes the gradual acidification of the solution, resulting in protein precipitation which is recovered by the centrifugation [58]. The acidification using the bipolar membrane has important advantages because of the controlled acidification rate and *in-situ* generation of acid and base. In addition, the acidification of

SPC/SPI by a bipolar membrane resulted in a lower ash content than that of the conventional processes [58,81,82].

3.3.3. Applications on the Fruit Juices and Beverages

The deacidification by the EDBPM has some advantages over the above methods, giving a high quality juice in terms of physicochemical and sensorial analyses without added reagents. The EDBPM, equipped with an anion exchange membrane and two bipolar membrane constituting two compartments besides the electrode rinse compartments, can produce citric acid simultaneously during the deacification and provide good flavor due to a small leakage of organic anions (See figure 7).

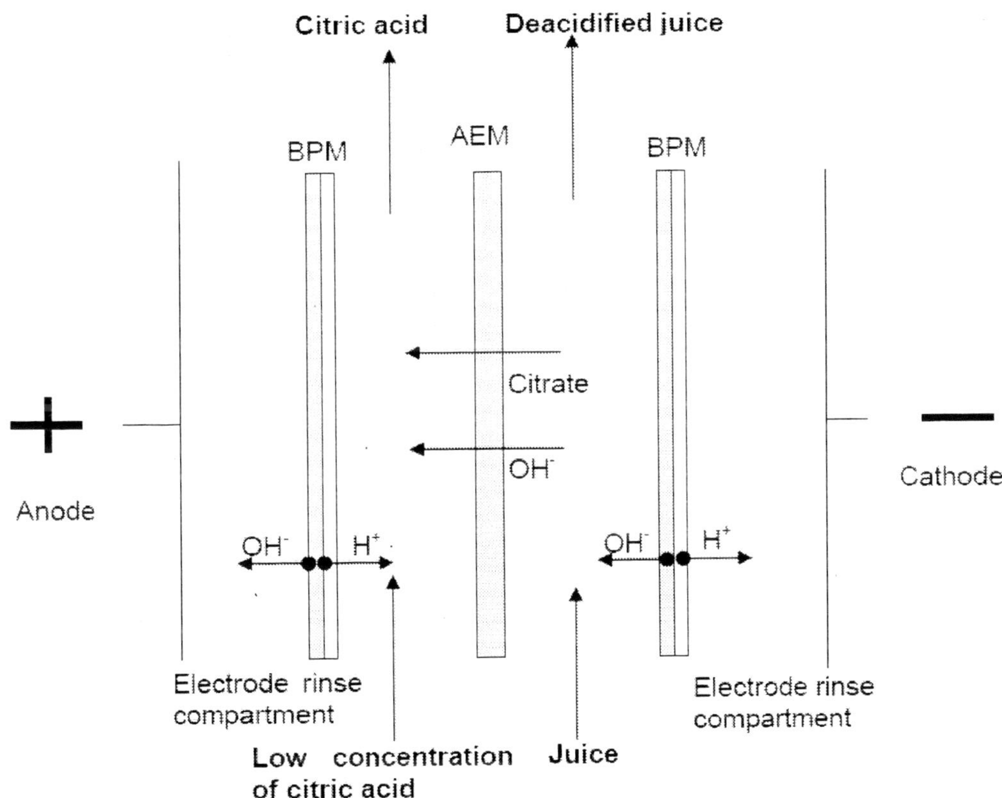

Figure 7. Illustration of deacidification from a fruit juice using EDBPM (AEM: anion exchange membrane, BPM: bipolar membrane).

In addition, a bipolar membrane was used to lower the pH of cloudy apple juice for the inhibition of enzyme browning [4,83]. By circulating apple juice on the cationic side of the bipolar membrane where the H^+ ions are generated, the pH was reduced from 3.5 to 2.0. This pH reduction completely inhibited polyphenol oxidase activity compared with the control. Following this temporary acidification, the pH was returned to its initial value by recirculating the juice on anionic side of the bipolar membrane. Although the pH readjustment partially reactivated polyphenol oxidase, the color of cloudy apple juice remained stable during storage. In acidic media, the free carboxyl groups of amino acids are protonated and the negative charges are neutralized. Inhibition of the polyphenol oxidase likely resulted from the change in tertiary structure of the protein caused by electrostatic repulsion between acids and positively charged amino groups [4]. EDBPM can be used to treat various liquid foods for which the enzymatic, chemical and microbiological stabilization is pH dependent in the food industries.

3.4. Other Applications of EDBPM

In wastewaters consisting of household wastewater, urine, industrial effluent and surface run-off water, the pollutants of the individual waste stream are diluted and mixed with a range of other substances. Treatment of urine in the mixed wastewater stream provides the possibility of the effective micro-pollutant removal at the source. The recovery and reuse of these nutrients can result in a more sustainable wastewater management. EDBPM is a possible technology for this separation process in order to remove and concentrate salts from a urine solution. The EDBPM consists of an additional mass transfer unit, a gas-filled hydrophobic membrane unit placed in between the circulating acid and basic concentrate streams, in order to render a product containing ammonium and phosphate at a low pH [84]. The use of ammonia-selective membranes for the transfer into the acid concentrate could provide a solution to generate an ammonium phosphate product at a high recovery in the low pH condition.

Interests on the reduction of greenhouse gas emissions from related sources have increased since the global warming problem has a significant impact on the earth's environment. Of greenhouse gases, CO_2 can be separately several approaches including absorption into liquid solvents, adsorption on solids, permeation through membranes, and chemical conversion [85-87]. One of the suitable processes is to remove CO_2 from high-volume waste gas streams by the absorption into liquid solvents. However, the process results in the thermal instability of chemical absorbents, corrosion of materials, and high-energy consumption. Kang *et al.* showed the feasibility of a new hybrid CO_2 separation process using a membrane contactor and EDBPM as an alternative method for the conventional CO_2 separation [88]. Bicarbonate (or carbonate) ions migrate into the acid compartment in the stack and are changed into gaseous CO_2 at low pH. The CO_2 reacting with absorbents are recovered in the base compartment simultaneously in the EDBPM system.

4. Integration of Reactors with Ion Exchange Membranes

4.1. Electrochemical Membrane Reactors

Toxic effluents generated by surface treatment and metal refining industry contain heavy metal ions (chromium, nickel, copper, and etc) as well as some other ionic and non-ionic components. Of heavy metals, compounds of Cr(VI) represents a significantly high toxicity mainly due to their confirmed carcinogenic action. In the conventional methods for the treatment of chromic acids, the direct elimination of Cr(VI) on appropriate anion exchange resins, liquid–liquid extraction, and liquid membrane techniques have been considered until now [89]. Among the possible alternatives to the treatment of industrial spent chromic acids, a conventional ED is a promising technique to recover valuable substances for further reuse. However, the separation efficiency of chromate anion and metal cations is very low in the conventional ED [90].

Electro-electrodialysis (EED) process, combining electrolysis and ED, has been considered as one of the promising technologies for the removal and recovery of metal ions [90,91]. The rinse water from the plating process containing chromic acid and metallic ions (copper, iron, zinc, aluminium, nickel and chromium) flows through the EED cell. Metallic ions migrate through a cation exchange membrane towards the cathode, forming dissolved metal sulphates. Meanwhile, chromic acid is formed by the chromate and dichromate anions migrating through an anion exchange membrane and by the H^+ generated at the anode. The treated rinse water is reused in the rinsing process and the pure chromic acid is returned to the plating bath. This process provides the purification of chromium plating solutions and the treatment of the rinse water in a single step without need of either water or chemicals [90,91].

The recovery of sodium hydroxide involves pretreatment processes where the sodium hydroxide-containing effluent is neutralized with an acidic gas, followed by the purification of the neutralized stream and cross-flow MF and NF. The sodium hydroxide is then recovered using a membrane electrolysis cell. Acidic gas is simultaneously evolved and is recycled to the neutralization stage, as shown in figure 8. The acidic gas can be either chlorine or carbon dioxide. Carbon dioxide system is preferred because the chlorine system is hazardous and can lead to the production of oxidized organic compounds. CO_2 neutralizes the sodium hydroxide solution, producing sodium bicarbonate solution to decrease pH. Cross-flow MF is used to remove suspended solids and the permeate is fed into the NF stage, which removes further organic material and some polyvalent ions. The NF permeate is introduced into the anode compartment of a two-compartment electrolysis cell containing a cation exchange membrane. In the electrolysis cell, the anodic and cathodic reactions result in oxygen gas production at the anode and hydrogen gas at the cathode, while the reaction in the anolyte leads to CO_2 gas evolution from the anolyte [92].

4.2. Ion Exchange Membrane Bioreactor (IEMBR)

Pollution of drinking water, ground water and surface water by nitrate presents a serious health hazard because of the acute asphyxiation, a syndrome known as methemoglobinemia in the circumstance of sufficient quantities of nitrate [93-95]. Several processes are available for the treatment of water sources with high nitrate concentration including physicochemical processes, such as ion exchange, RO and ED [96]. These processes are efficient in removing specific ions or salts; however, they produce large amounts of brines which are extremely concentrated not only in nitrate, but also in the various ions present in water. Meanwhile, the biological denitrification is based on the conversion of nitrate into gaseous nitrogen by an anoxic bacterial activity with an external carbon source. However, the process performances are varied depending on the water quality affecting the biological metabolism, external carbon source supply for the activity of microorganisms, and the auxiliary processes such as filtration and adsorption to remove some bacterial products [97].

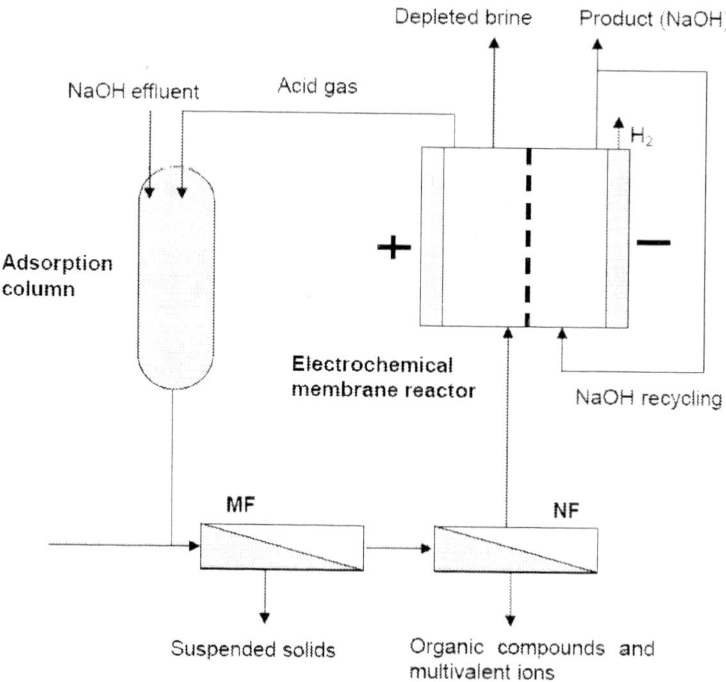

Figure 8. Recovery of sodium hydroxide from a sodium hydroxide-containing effluent in the combined system consisting of the electrochemical membrane reactor and pretreatment processes.

A novel technology combining an ED process and a biological denitrification, so called ion exchange membrane bioreactor (IEMBR), is considered for the removal of nitrate. Nitrate ions migrate through a mono-anion selective membrane to produce suitable drinking water or desirable industrial water because it facilitates the transport of nitrate and hinders diffusion of cationic and non-ionic substances. Meanwhile, the migration of nitrate ions to denitrifying biofilm enhances biological denitrification in the presence of an electron donor and nutrients. The IEMBR isolates the microbial culture in the biological compartment from the feed stream, avoiding contamination of the treated effluent with cells, excess carbon source, and metabolic by-products. The hydraulic residence time in the two compartments can be independently adjusted allowing optimization of the pollutant extraction degree. The mechanism of nitrate removal in the IEMBR is illustrated in figure 9. Chloride is added to the bioreactor feed to serve as the main counter-ion since it is not regulated by primary water quality standards. Ethanol is supplied as a carbon source and electron donor for denitrification because it is not readily fermentable, thus avoiding the growth of fermentative organisms and production of organic acids in the anoxic bioreactor.

The IEMBR provides several advantages, *i.e.*, (i) a complete elimination of nitrate, (ii) a complete separation of hydraulic retention time and suspended solids retention time, and (iii) the constant water production independent of the biological process or bacterial contaminants

[98-102]. Fonseca *et al.* used ethanol as the electron donor and subsequently opted for a membrane with the lowest permeability to ethanol. The subsequent apparent diffusion coefficient for ethanol through the membrane was found to be almost three orders of magnitude lower than that of ethanol in water [93]. Table 5 summarizes biological characteristics of IEMBR for nitrate removal from a drinking water

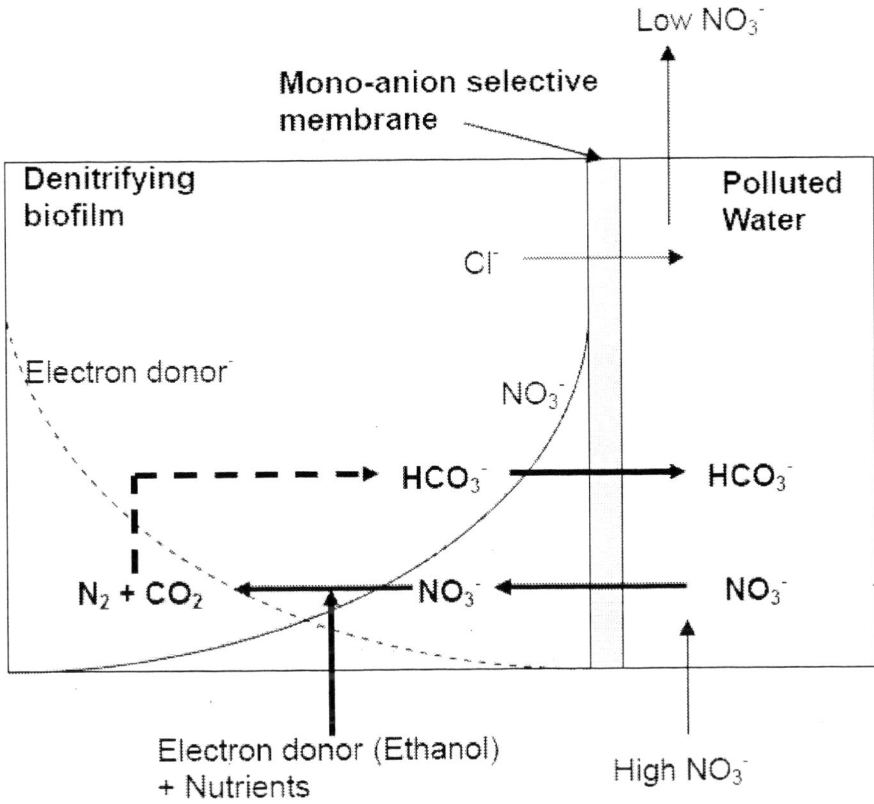

Figure 9. Nitrate removal mechanism in the ion exchange membrane bioreactor(IEMBR).

Table 5. Biological characteristics of IEMBR for nitrate removal from drinking water

Electron donor	Electron acceptor	Nitrate loading (as N)	Nitrate removal rate (as N)	Ref.
Ethanol	Nitrate	0.08 kg/m^3d	7.45 g/m^2d	101
		0.81 kg/m^3d	6.5-7.5 g/m^2d	93
		150 mg/L	33.6 g/m^2d	102
		60 mg/L	4.8 g/m^2d	100

5. Integration of Ion Exchanges with Ion Exchange Membranes

5.1. Production of High Purity Water

The demand of high purity water is rapidly increasing in the areas of boiler feed water, the production of semiconductors, chemical and biochemical laboratories, power generation plants, etc [103,104]. Especially the electronic industry, analytical laboratories and power generation plants require large quantities of water, which is free of particles and organic matters and has a very low conductivity [105]. Conventional systems for high purity water production contain several process steps in series such as MF and RO. Ion exchange with mixed beds can be applied as a final demineralization step, where the ion exchange resins have to be regenerated or replaced in certain time intervals.

In many cases, pure water is produced with EDI (electrodeionzation), combined ED with ion exchange process after raw water from well or surface water is purified in a series of processes consisting of water softening, MF, RO, and ion exchange, (See figure 10) [105]. The process line may be varied depending on the quality of the raw water and include precipitation or carbon adsorption, UF, or UV sterilization. The most important advantage of EDI is that no chemical regeneration of the ion exchange resin is required. Thus, it avoids contamination of the product water and reduces the need of chemicals. As a result, it provides a continuous process with a reduction of system components and regeneration chemicals, leading to a substantial reduction in the overall water production costs [106-110]. As one of the applications on a large scale, EDI is used to produce boiler feed water for the power generation plants, where the requirements for the water purity are not strict as in the electronic industry. The demand for high quality water in the steam generation plants has rapidly increased now.

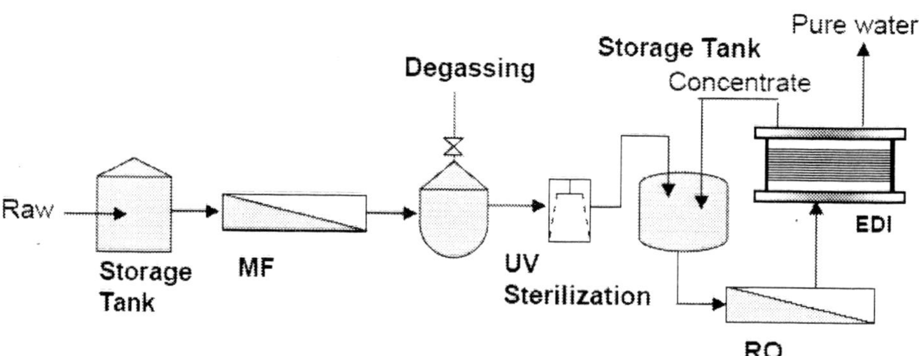

Figure 10. Integrated EDI system for the production of pure water.

5.2. Recovery of Heavy Metals from Industrial Effluents

The recovery of heavy metals from industrial process solutions has received great attention in recent years. This is mainly due to more stringent regulation for the environmental protection. ED is considered to be an alternative process of ion exchange in the treatment of heavy metals in industrial effluents. However, it cannot be carried out at high metal concentrations. The combined process of electromigration through ion exchange membranes and adsorption of ion exchanges can be used to recovery of heavy metals from industrial effluents as well as to produce high purity water. The secondary treatment processes such as ion exchange are operated to decrease below mg/L levels because the conventional techniques for the removal of metal ions show insufficient removal efficiency.

In the removal of heavy metals from dilute solutions in a continuous process, the compartment is located between two ion exchange membranes that essentially divide the cell into three separate compartments [111]. Water electrolysis occurs in the two external electrode compartments and H^+ formed at the anode is transferred to exchange pre-adsorbed heavy metals on the ion exchange resin surface. Use of the permselective cation exchange membranes increases the removal efficiency of heavy metals in the process. Of important applications on the heavy metal removal is divalent metals such as copper [112,113], Ni^{2+} ions [114-116], and other ions (Pb^{2+}, Zn^{2+} and Cd^{2+}) [117]. In the results, the electromigration rate of the ions in the resin bed is reported to strongly depend on the resin flexibility and the affinity between resins and ions as well as and the intensity of the electrical field [110,116,117].

6. Other Integration Processes Based on Ion Exchange Membranes

Of ion exchange membrane separation processes, ED and EDBPM are by far the most important industrial membrane separation processes. Other ion exchange membrane separation processes such as diffusion dialysis and Donnan dialysis are used in many areas. Diffusion dialysis, consisting of cell arrangement of the feed and the dialysate using an anion exchange membrane, is one of the oldest ion exchange membrane processes [118,119]. The individual ions in the mixture of salt and acid and a dialysate (mostly pure water) migrate from the feed solution to dialysate in a counter-current due to the concentration gradient. It is well established as a relevant commercial process with a number of applications mainly in the treatment of industrial effluents from the metal processing industry. One of the typical applications in wastewater treatment by diffusion dialysis is to recover an acidic from the ion exchange regeneration process [120,121].

One of the major commercial applications of diffusion dialysis in Japan, Europe and USA is the recovery of acids from pickling solutions used to remove metal oxides from steel parts before surface finishing. The recovery of HF and HNO_3 from spent pickling solutions has been especially attractive since these acids often cause severe pollution problems [122]. The spent pickling solution contains dissolved ferric, nickel, and chromium ions with a substantial amount of suspended metal particles from surface scale. The suspended solids are removed in a filtration and fed to diffusion dialysis stack through a catridge filter. Water heated to ca. 40 °C is pumped in counter-current flow as acid stripping solution through the stack. The recovered acid is recycled to the pickling bath without further treatment. The waste acid and 95 % of ferric ion in the stripped feed solution go directly to the waste disposal system.

In the Donnan dialysis, metal cations migrate through a cation exchange membrane due to difference of H^+ concentration. Accordingly, the process can be carried out with anions through anion exchange membranes [123,124]. So far, Donnan dialysis is not used on a large industrial scale. Most results concerning the performance and process cost estimation have been generated with laboratory or pilot plant scale equipments in relatively short term experiences. One of the major applications of the Donnan dialysis is water softening and the removal of certain heavy metal ions from wastewater [125,126].

7. Other Integrated Membrane Processes

Integrated ion exchange membrane separation processes are more environmentally friendly in the comprehensive integration with physicochemical processes such as pressure-driven membrane separation processes, electrochemical reaction, and adsorption. The integrated systems of ion exchange membrane process can be found in many areas such as production of water, the food and chemical industries, the demineralization of whey or deacidification of fruit juice. Also the process is applied to recover toxic or valuable components from effluents of galvanic and metal surface finishing processes [127,128].

Among integrated membrane processes, the integration of RO with other membrane or non-membrane processes have also used for desalination [136-139]. RO can combine with multistage flash distillation (MSF), which gives advantages including lower capital costs and energy requirements compared to that of a MSF plant with same capacity. Also, smaller feed water is required since the RO concentrate is used to the feed of MSF and electric power generation from MSF can be utilized in the RO plant. Sometimes, it is not desirable to use RO carry out in treatment of the wastewater streams because of high viscosity, concentration, and suspended solids. Example of the RO hybrid process are the combination of RO with PV for the removal of organics from water, RO with coupled or facilitated transport to remove heavy metals from water, RO with ED to treat streams with high salinity, and RO with crystallization for the recovery of valuable products from solution. When RO process is combined with other process in series or in a recycle streams, the advantage of process are the application on a broader range of feed concentrations than is expected with either RO or the other unit operation, the easy process optimization because of the adjustment of flow rate for each operation, increase of RO membrane and reduction of production cost [8,9,17] Nowadays, the number of water treatment plants equipped with boron and arsenic removal facilities is growing due to their adverse effects both on human/animal health and on agriculture. Usually it is difficult to remove them from water due to their size and charge. Therefore, the most part of the desalination/purification plants should use chemicals agents and RO systems with several pass-stages such as membrane distillation and ion exchange in order to obtain substantial arsenic and boron reduction in the treated water [140-142].

There are a great number of examples of UF combined with various conventional water treatment processes such as coagulation/flocculation, adsorption and biological processes [143,144]. The integrated UF processes with ozone, aeration, carbon adsorption, biological denitrification have been applied in the removal of dissolved organic pollutants, improving water quality and increasing process performance due to decrease of fouling potential. Also, some of integrated membrane processes, such as air separation via membrane permeation with pressure swing adsorption and cryogenics, CO_2 separation via membrane permeation with amine treating and cryogenics, and PV with distillation or condensation and phase separation, are of increasing interest [145,146]

8. Conclusion

Interests in the membrane separation process have increased due to various industrial applications such as biotechnology and membrane based energy devices as well as membrane separation and purification processes. The main membrane separation processes in industrial scale are reverse osmosis (RO), nanofiltration (NF), ultrafiltration (UF), microfiltration (MF), pervaporation (PV), and electrodialysis (ED).

Table 6. Process characterization and economics of EDBPM in the recovery of organic acids

Application	Scale	Process characterization	Energy consumption	Ref.
Production of vitamin C (Ascrobic acid) from sodium ascrobate	Lab scale or semi-industrial pilot	Two component EDBPM, Current density: 1000 A/m^2 Current efficiency: 75 % Acid concentration: 1 M	1.4-2.3 kWh/kg acid	64, 130, 131
Production of citric acid	Pilot scale	Two component EDBPM, Current density: 1000 A/m^2 Current efficiency: 75 % Acid concentration: 30 g/L	2-5 kWh/kg acid	120, 132
Production of salicylic acid	Laboratory	Three component EDBPM, Current density: 750 A/m^2 Current efficiency: 80-90 % Acid concentration:4.5 g/L	15-20 kWh/kg acid	133
Production of acetic acid from sodium acetate	Pilot scale	Five component EDBPM, Current density: 750 A/m^2 Current efficiency: above 97 % Acid concentration: 1 -1.5 M	1.3-2.2 kWh/kg acid	134, 135

In order to utilize the effectiveness of membrane processes, the process integration is of importance in the combination of membrane process with non-membrane processes or other membrane processes, which can achieve an efficiency level not achievable by either process used alone. The process integration can be found in membrane reactors using chemical and biochemical reactions, pressure-driven membrane processes (MF, UF, NF and RO), and chemical unit operations. A substantial number of commercial applications have been practiced in food industries, pollution control of industrial effluents and production of industrial waters treatment.

Often integrated membrane processes are employed as environmentally friendly and energy-efficient processes in a large various fields such as water desalination in brackish waters and seawaters, production of industrial waters, food industry, table salt production, production of acids and bases, elimination of toxic components from industrial effluents, ultra pure water production, etc. In addition, a substantial number of industrial applications in the integration processes of RO, UF, and PV as well as ED are found especially in biotechnology and chemical engineering as one of the promising potentials. The applications of integrated membranes have been expanded with the competitiveness over traditional, high energy intensive, environmentally undesirable and costly processes.

Acknowledgment

This work was supported by Energy Technology R&D program (2006-E-ID-11-P-14) under the Korea Ministry of Commerce, Industry and Energy (MOCIE).

References

[1] Strathmann, H. Ion-exchange membrane separation processes; Elsevier: Amsterdam, 2004.

[2] Winston Ho, W.S.; Sirkar, K.K. Membrane Handbook; Van Nostrand Reinhold: New York, 1992.

[3] Xu, T. Ion exchange membranes: State of their development and perspective. *J. Membr. Sci.* 263 (2005) 1-29.

[4] Girard, B.; Fukumoto, L. R.; Koseoglu, S.S. Membrane Processing of Fruit Juices and Beverages: A Review. *Crit. Rev. Biotechnol.* 20 (2000) 109-175.

[5] Vera, E.; Ruales, J.; Dornier, M.; Sandeaux, J.; Persin, F.; Pourcelly, G.; Vaillant, F.; Reynes, M. Comparison of different methods for deacidification of clarified passion fruit juice. *J. Food Eng.* 59(2003) 361-367.

[6] Vera, E.; Sandeaux, J.; Persin, F.; Pourcelly, G.; Dornier, M.; Ruales, J. Deacidification of clarified tropical fruit juices by electrodialysis. Part I. Influence of operating conditions on the process performances. J. *Food Eng.* 78 (2007) 1427-1438.

[7] Vera E.; Ruales, J.; Dornier, M.; Sandeaux, J.; Sandeaux, R.; Pourcelly, G. Deacidification of the clarified passion fruit juice (P. edulis f. flavicarpa). *Desalination,* 149 (2002) 357-361.

[8] Koprivnjak, J.F.; Perdue E.M.; Pfromm, P.H. Coupling reverse osmosis with electrodialysis to isolate natural organic matter from fresh waters. *Water Research,* 40 (2006) 3385-3392.

[9] Wilf, M. The guidebook to membrane desalination technology, Balaban Desalination Publications: Awerbuch, 2007.

[10] van der Bruggen, B.; Vandecasteele, C. Distillation vs. membrane filtration: overview of process evolutions in seawater desalination. *Desalination*, 143 (2002) 207-218.

[11] Wangnick, K. Desalination plants inventory report, The International Desalting Association, 2000.

[12] Turek, M. Cost effective electrodialytic seawater desalination. *Desalination*. 153 (2002) 371-376.

[13] Schmoldt, H.; Strathmann, H.; Kaschemekat, J. Desalination of sea water by an electrodialysis-reverse osmosis hybrid system. *Desalination*. 38 (1981) 567-582.

[14] Husseiny, A.A.; Hamester, H. L. Engineering design of a 6000m^3/day seawater hybrid RO-ED helio-desalting plant. *Desalination*. 39 (1981) 171-172.

[15] Kamel, I.; Emmermann, D.K.; Tusel, G.F. Influence of voltage and flow rate on electrodeionization (EDI) process efficiency. *Desalination*. 33 (1980) 229-239.

[16] Seigworth, A.; Ludlum, R.; Reahl, E. Case study: Integrating membrane processes with evaporation to achieve economical zero liquid discharge at the Doswell Combined Cycle Facility. *Desalination*. 102 (1995) 81-86.

[17] Scott, K. Handbook of Industrial membranes, 2nd ed.; Elsevier: Amsterdam, 1998.

[18] Grebenyuk, V. D.; Chebotareva, R. D.; Linkov, N. A.; Linkov, V. M. Electromembrane extraction of Zn from Na-containing solutions using hybrid electrodialysis-ion exchange method, *Desalination*. 115 (1998) 255-263.

[19] Grimm, J.; Bessarabov, D.; Sanderson, R. Review of electro-assisted methods for water purification. *Desalination*. 115 (1998) 285-294.

[20] van der Hoek, J.P.; Rijnbende, D.O.; Lokin, C. J. A.; Bonné, P. A. C. ; Loonen. M. T.; Hofman, J. A. M. H. Electrodialysis as an alternative for reverse osmosis in an integrated membrane system. *Desalination*. 117 (1998) 159–172.

[21] Turek, M.; Dydo, P.; Was J. High efficiency electrodialysis reversal of concentrated calcium sulfate and calcium carbonate solutions. *Desalination*. 205 (2007) 62-66.

[22] Mallevialle, J.; Odendaal, P.E.; Wiesner, M.R. Water treatment membrane processes, McGraw-Hill: New York, 1996.

[23] Yeon, K.H.; Song, J.H.; Shim, J.; Moon, S.H.; Jeong, Y.U.; Joo, H.Y. Integrating electrochemical processes with electrodialysis reversal and electro-oxidation to minimize COD and T-N at wastewater treatment facilities of power plants. *Desalination*. 22 (2007) 400-410.

[24] Gómez Gotor, A. ; Perez Baez, S.O. ; Espinoza, C.A. ; Bahir, S.I. Membrane processes for the recovery and reuse of wastewater in agriculture. *Desalination*. 137 (2001) 187–192.

[25] Ito, S.; Nakamura, I.; Kawahara T. Electrodialytic recovery process of metal finishing waste water. *Desalination*. 32 (1980) 383-389.

[26] Paquay, E.; Clarinval, A. M.; Delvaux, A.; Degrez, M.; Hurwitz, H. D. Applications of electrodialysis for acid pickling wastewater treatment. *Chem. Eng. J.* 79 (2000) 197-201.

[27] Schmidt, B.; Wolters, R.; Kaplin, J.; Schneiker, T.; de los Angeles Lobo-Recio, M.; López, F., López-Delgado, A., José Alguacil, F.J. Rinse water regeneration in stainless steel pickling. *Desalination*. 211 (2007) 64-71.

[28] Korngold, E.; Kock, K.; Strathmann, H. Electrodialysis in advaned waste water treatment. *Desalination.* 24 (1978) 129-139.

[29] Rapp, H.J.; Pfromm, P.H. Electrodialysis for chloride removal from the chemical recovery of a Kraft pulp mill. *J. Membr. Sci.* 49 (1998) 1-13.

[30] Huang, C.; Xu, T.; Zhang, Y.; Xue, Y.; Chen, G. Application of electrodialysis to the production of organic acids: State-of-the-art and recent developments. *J. Membr. Sci.* 2007, 288, 1-12.

[31] Anastas, P.T.; Breen, J.J. Design for the environment and green chemistry: the heart and soul of industrial ecology. *J. Cleaner Prod.* 5 (1997) 97–102.

[32] Takahashi, H.; Ohba. K.; Kikuchi, K.I. Sorption of di- and tricarboxylic acids by an anion-exchange membrane. *J. Membr. Sci.* 222(2003) 103–111.

[33] Huang, C.; Xu, T. Electrodialysis with bipolar membranes for sustainable development. *Environ. Sci. Technol.* 40 (2006) 5233–5243.

[34] Lee, H.J.; Oh, S.J.; Moon, S.H. Removal of hardness in fermention broth by electrodialysis. *J. Chem. Technol. Biotechnol.* 77 (2002) 1005-1012.

[35] Lee, E.G.; Moon, S.H.; Chang, Y.K.; Yoo, I.; Chang, H.N. Lactic acid recovery using two-stage electrodialysis and its modeling. *J. Membr. Sci.* 145 (1998) 53-66.

[36] Danner, H.; Madzingaidzo, L.; Holzer, M.; Mayrhuber, L.; Braun, R. Extraction and purification of lactic acid from silages. *Bioresour. Technol.* 75 (2000) 181–187.

[37] Madzingaidzo, L.; Danner, H.; Braun, R. Process development and optimization of lactic acid purification using electrodialysis. *J. Biotechnol.* 96 (2002) 223–239.

[38] Hábová, V.; Melzoch, K.; Rychtera, M.; Sekavová, B. Electrodialysis as a useful technique for lactic acid separation from a model solution and a fermentation broth. *Desalination.* 163 (2004) 361–372.

[39] Bailly, M.; Roux-de Balmann, H.; Aimar, P.; Lutin, F.; Cheryan, M. Production processes of fermented organic acids targeted around membrane operations: design of the concentration step by conventional electrodialysis. *J. Membr. Sci.* 191 (2001) 129–142.

[40] Datta, R.; Tsai, S.P.; Bonsignore, P.; Moon, S.H.; Frank, J.R. Technology and economic potential of poly(lactic acid) and lactic acid derivatives. *FEMS Microbiol. Rev.* 16 (1995) 221–231.

[41] Kim, H.Y.; Moon, S.H. Lactic acid recovery from fermentation broth using one-stage electrodialysis. *J. Chem. Technol. Biotechnol. 76* (2001) 169-178.

[42] Thang, V. H.; Koschuh, W.; Kulbe, K.D.; Novalin, S. Detailed investigation of an electrodialytic process during the separation of lactic acid from a complex mixture. *J. Membr. Sci.* 249 (2005) 173-182.

[43] Audinos, R. Ion-exchange membrane processes for clean industrial chemistry. *Chem. Eng. Technol.* 1997, 20, 247–258.

[44] A°kerberg, C.; Zacchi, G. An economic evaluation of the fermentative production of lactic acid from wheat flour. *Bioresour. Technol.* 75 (2000) 119–126.

[45] Martinez, D.; Sandeaux, R.; Sandeaux, J.; Gavach, C. Electrotransport of alanine through ion-exchange membranes. *J. Membr. Sci.* 69 (1992) 273-281.

[46] Elisseeva, T. V.; Shaposhnik, V. A.; Luschik, I. G. Demineralization and separation of amino acids by electrodialysis with ion-exchange membranes. *Desalination.* 149 (2002) 405-409.

[47] Atungulu, G.; Koide, S.; Sasaki, S.; Cao, W. Ion-exchange membrane mediated electrodialysis of scallop broth: Ion, free amino acid and heavy metal profiles. *J. Food Eng.* 78 (2007) 1285-1290.

[48] Sandeaux, J.; Sandeaux, R.; Gavach, C.; Grib, H.; Sadat, T.; Belhocine, D.; Mameri, N. Extraction of amino acids from protein hydrolysates by electrodialysis. *J. Chem. Technol. Biotechnol.* 71 (1998) 267-273.

[49] Resbeut, S.; Pourcelly, G.; Sandeaux, R.; Gavach, C. Electromembrane processes for waste stream treatment: electrodialysis applied to the demineralization of phenylalanine solutions. *Desalination.* 120 (1998) 235-245.

[50] Montiel, V.; García-García, V.; González-García, J.; Carmona, F.; Aldaz, A. Recovery by means of electrodialysis of an aromatic amino acid from a solution with a high concentration of sulphates and phosphates. *J. Membr. Sci.* 140 (1998) 243-250.

[51] Poquis, J. A.; García-García, V.; González-García, J.; Montiel, V.; Aldaz, A. Partial electro-neutralisation of D-α-p-hydroxyphenylglycine in sulphuric acid medium. *J. Membr. Sci.* 2000, 170, 225-233.

[52] Choi, Y.J.; Park, J.M.; Yeon K.H.; Moon, S.H. Electrochemical characterization of poly(vinyl alcohol)/formyl methyl pyridinium (PVA-FP) anion-exchange membranes. *J. Membr. Sci.* 250 (2005) 295-304.

[53] Sandeaux, J.; Fares, A.; Sandeaux, R.; Gavach, C. Transport properties of electrodialysis membranes in the presence of arginine I. Equilibrium properties of a cation exchange membrane in an aqueous solution of arginine chlorhydrate and sodium chloride. *J. Membr. Sci. 89* (1994) 73-81.

[54] Ahlgreen, R.M. Electromembrane processing of cheese whey in: Industrial processing with membranes; Lacey, E.R.; Loeb. S.; Ed.; Wiley Interscience: New York, 1972; pp. 57-69.

[55] Andres, Z.; Riera, F.A.; Alvarez, R. Skimmed milk demineralization by electrodialsyis: Conventional versus selective membranes. *J. Food Eng.* 26 (1995) 57-66.

[56] Bleha, M.; Tishenko, G.; Sumberoa, V.; Kudela, V. Characterization of the critical state of membranes in electrodialysis desalination of milk whey. *Desalination.* 86 (1992) 173-186.

[57] Greiter, M.; Novalin, S.; Wendland, M.; Kulbe, K.D.; Fisher, J. Desalination of whey by electrodialysis and ion-exchange resins: Analysis of both processes with regard to sustainability by calculating their cumulative energy demand. *J. Membr. Sci.* 210 (2002) 91-102.

[58] Alibhai, Z.; Mondor, M.; Moresoli, C.; Ippersiel, D.; Lamarche, F. Production of soy protein concentrates/isolates: traditional and membrane technologies. *Desalination.* 191 (2006) 351-358.

[59] Elmidaoui, A.; Lutin, F.; Chay, L.; Taky, M.; Tahaikt, M.; Alaoui Hafidi, M. R. Removal of melassigenic ions for beet sugar syrups by electrodialysis using a new anion-exchange membrane. *Desalination.* 148 (2002)143-148.

[60] Elmidaoui, A.; Chay, L.; Tahaikt, M.; Menkouchi Sahli, M.A.; Taky, M.; Tiyal, F.; Khalidi, A.; Alaoui Hafidi, M.R. Demineralisation for beet sugar solutions using an electrodialysis pilot plant to reduce melassigenic ions. *Desalination*, 2006, 189, 209-214.

[61] Elmidaoui, A.; Chay, L.; Tahaikt, M.; Menkouchi Sahli, M.A.; Taky, M.; Tiyal, F.; Khalidi, A.; Alaoui Hafidi, M.R. Demineralisation of beet sugar syrup, juice and molasses using an electrodialysis pilot plant to reduce melassigenic ions. *Desalination*. 165 (2004) 435

[62] Jiao, B.; Cassano, A.; Drioli, E. Recent advances on membrane processes for the concentration of fruit juices: a review. *J. Food Eng.* 63 (2004) 303-324.

[63] Voss, H. Deacidification of citric acid solutions by electrodialysis. *J. Membr. Sci.* 27 (1987) 165-171.

[64] Kemperman, A.J.B. Handbook on bipolar membrane technology, Twente Univ. Press: Enschede, 2000.

[65] Xu, T.; Yang W. Effect of cell configurations on the performance of citric acid production by a bipolar membrane electrodialysis. *J. Membr. Sci.* 203 (2002) 145-153.

[66] Mani, K.N. Electrodialysis water-splitting technology. *J. Membr. Sci.* 58 (1991) 117-138.

[67] Liu, K.J.; Chlanda, F.P.; Nagasubramanian, K.J. Use of bipolar membranes for generation of acid and base: an engineering and economic analysis. *J. Membr. Sci.* 2 (1977) 109-124.

[68] Huang, C.; Xu, T.; Yang, X. Regenerating Fuel-Gas Desulfurizing Agents by Using Bipolar Membrane Electrodialysis (BMED): Effect of Molecular Structure of Alkanolamines on the Regeneration Performance. *Environ. Sci. Technol.* 41 (2007) 984-989.

[69] Nagarale, R.K.; Gohil, G.S.; Shahi, V.K. Recent development on ion-exchange membranes and electro-membrane processes. *Adv. Colloid Interf. Sci.* 119 (2006) 97-130.

[70] Raissouni, I.; Marraha, M.; Azmani, A. Effect of some parameters on the improvement of the bipolar membrane electrodialysis process. *Desalination*. 208 (2007) 62-72.

[71] Liu, K.J.; Nagasubramanian, K.; Chlanda, F. P. Membrane electrodialysis process for recovery of sulfur dioxide from power plant stack gases. *J. Membr. Sci.* 3 (1978) 71-83.

[72] Mani, K. N.; Chlanda, F. P.; Byszewski, C. H. Aquatech membrane technology for recovery of acid/base values for salt streams. *Desalination*. 68 (1988) 149-166.

[73] Pinacci, P.; Radaelli, M. Recovery of citric acid from fermentation broths by electrodialysis with bipolar membranes. *Desalination*. 148 (2002) 177-179.

[74] Moresi, M.; Sappino, F. Electrodialytic recovery of some fermentation products from model solutions: techno-economic feasibility study. *J. Membr. Sci.* 164 (2000) 129-140.

[75] Lee, E.G.; Kang, S.H.; Kim, H.H.; Chang, Y.K. , Recovery of lactic acid from fermentation broth by the two-stage process of nanofiltration and water-splitting electrodialysis. *Biotechnol. Bioprocess Eng.* 11 (2006) 313-318.

[76] Kang, S. H.; Chang, Y. K.; Chang, H. N. Recovery of ammonium lactate and removal of hardness from fermentation broth by nanofiltration. *Biotechnol. Prog.* 20 (2004) 764-770.

[77] Cauwenberg, V.; Peels, J.; Resbeut, S.; Pourcelly, G.. Application of electrodialysis within fine chemistry. *Sep. Purif. Technol.* 22-23 (2001) 115-121.

[78] Yu, L.; Su, J.; Wang, J. Bipolar membrane-based process for the recycle of p-toluenesulfonic acid in D-(-)-p-hydroxyphenylglycine production. *Desalination.* 177 (2005) 209-215.

[79] Grib, H.; Bonnal, L.; Sandeaux, J.; Sandeaux, R.; Gavach, C.; Mameri, N. Extraction of amphoteric amino acids by an electromembrane process. pH and electrical state control by electrodialysis with bipolar membranes. *J. Chem. Technol. Biotechnol.* 73 (1998) 64-70.

[80] Fisher, R.R.; Glatz, C.E.; Murphy, P.A. Effects of mixing during acid addition on fractionally precipitated protein. *Biotech. Bioeng.* 28 (1986) 1056–1063.

[81] Bazinet, L.; Lamarche, F.; Labrecque, R.; Toupin, R.; Boulet, M.; Ippersiel, D. Electro-acidification of soybean proteins for the production of isolate. *Food Technol.* 51 (1997) 52–60.

[82] Mondor, M.; Ippersiel, D.; Lamarche, F.; Boye, J.I. Effect of electro-acidification treatment and ionic environment on soy protein extract particle size distribution and ultrafiltration permeate flux. *J. Membr. Sci.* 231 (2004) 169–179.

[83] Trone, J. S.; Lamarche, F.; Makhlouf, J. Effeet of pH variation by electrodialysis on the inhibition tion of enzymatic browning in cloudy apple juice. *J. Agric. Food Chem.* 46 (1998) 829-833.

[84] Pronk W.; Biebow M.; Boller P. Treatment of source-separated urine by a combination of bipolar electrodialysis and a gas transfer membrane. *Water Sci. Technol.* 53 (2006) 139-146.

[85] Kreulen, H.; Smolders, C.A.; Versteeg, G.F.; van Swaaij, W.P.M. Microporous hollow fibre membrane modules as gas–liquid contactors. Part 2. Mass transfer with chemical reaction. *J. Membr. Sci.* 78 (1993) 217–238.

[86] Bos, A.; Pünt, I.G.M.; Wessling, M.; Strathmann, H. CO_2-induced plasticization phenomena in glassy polymers. *J. Membr. Sci.* 155 (1999) 67–78.

[87] Nam, S.S.; Kim, H.; Kishan, G.; Choi, M.J.; Lee, K.W. Catalytic conversion of carbon dioxide into hydrocarbons over iron supported on alkali ion-exchanged Y-zeolite catalysts. *Appl. Catal.* A 179 (1999) 155–163.

[88] Kang, M.S.; Moon, S.H.; Park, Y.I.; Lee. K.H. Development of carbon dioxide separation process using continuous hollow-fiber membrane contactor and water-splitting electrodialysis. *Sep. Sci. Technol.* 37 (2002) 1789-1806.

[89] Hansen, H.K.; Ribeiro, A.B.; Mateus, E.P.; Ottosen, L.M. Diagnostic analysis of electrodialysis in mine tailing materials. *Electrochim. Acta.* 52 (2007) 3406-3411.

[90] Frenzel, I.; Holdik, H.; Stamatialis, D.F.; Pourcelly, G.; Wessling, M. Chromic acid recovery by electro-electrodialysis: I. Evaluation of anion-exchange membrane. *J. Membr. Sci.* 261 (2005) 49 57.

[91] Khan, J.; Tripathi, B. P.; Saxena, A.; Shahi, V. K. Electrochemical membrane reactor: In situ separation and recovery of chromic acid and metal ions. *Electrochimica Acta*, 52 (2007) 6719-6727.

[92] Simpson, A. E.; Buckley, C. A. The treatment of industrial effluents containing sodium hydroxide to enable the reuse of chemicals and water. *Desalination*. 67 (1987) 409-429.

[93] Fonseca, A.D.; Crespo, J.G.; Almeida, J.S.; Reis, M.A. Drinking water denitrification using a novel ion-exchange membrane bioreactor. *Environ. Sci. Technol.* 34 (2000) 1557–1562.

[94] Della Rocca, C.; Belgiorno. V.; Meriç, S. Overview of *in-situ* applicable nitrate removal processes. *Desalination*. 204 (2007) 46-62.

[95] McAdam, E.J.; Judd, S.J. A review of membrane bioreactor potential for nitrate removal from drinking water. *Desalination*. 196 (2006) 135-148.

[96] Wisniewski, C.; Persin, F.; Cherif, T.; Sandeaux, R.; Grasmick, A.; Gavach, C.; Lutin, F. Use of a membrane bioreactor for denitrification of brine from an electrodialysis process. *Desalination*. 149 (2002) 331-336.

[97] Wisniewski, C.; Persin, F.; Cherif, T.; Sandeaux, R.; Grasmick, A.; Gavach, C. Denitrification of drinking water by the association of an electrodialysis process and a membrane bioreactor: feasibility and application. *Desalination*. 139 (2001) 199-205.

[98] Kapoor, A.; Viraraghavan, T. Nitrate removal from drinking water – review. *J. Environ. Eng.* 123 (1997) 371-380.

[99] Delanghe, B.; Nakamura, F.; Myoga, H.; Magara, Y. Biological denitrification with ethanol in a membrane bioreactor. *Environ. Technol.* 15 (1994) 61-70.

[100] Velizarov, S.; Reis, M.A.; Crespo, J.G. Removal of trace mono-valent inorganic pollutants in an ion exchange membrane bioreactor: analysis of transport rate in a denitrification process. *J. Membr. Sci.* 217 (2003) 269–284.

[101] Velizarov, S.; Rodrigues, C.M.; Reis, M.A.; Crespo, J.G. Mechanism of charged pollutants removal in an ion exchange membrane bioreactor: drinking water denitrification. *Biotechnol. Bioeng.* 71 (2000/2001) 245–254.

[102] Velizarov, S.; Crespo, J.G.; Reis, A.M. Ion exchange membrane bioreactor for selective removal of nitrate from drinking water: control of ion fluxes and process performance. *Biotechnol. Prog.* 2002, 18, 296–302.

[103] Song, J.H.; Yeon, K.H.; Moon, S.H. Transport characteristics of Co^{2+} through an ion exchange textile in a continuous electrodeionization (CEDI) system under electro-regeneration. *Sep. Sci. Technol.* 39 (2004) 3601-3618.

[104] Yeon, K.H.; Song, J.H.; Moon, S.H. A study on stack configuration of continuous electrodeionization for removal of heavy metal ions from the primary coolant of a nuclear power plant. *Water Res.* 38 (2004) 1911-1921.

[105] Dey A.; Thomas, G. Electronics grade water preparation; Tall Oaks Publishing: Littleton, 2003.

[106] Ganzi. G. C.; Jha, A.D.; Wood J.H. Theory and practice of continuous electrodeionization. *Ultra Pure Water*. 14 (1997) 64-69.

[107] Grabowski, A.; Zhang, G.; Strathmann, H.; Eigenberger, G. The production of high purity water by continuous electrodeionization with bipolar membranes: Influence of the anion-exchange membrane permselectivity. *J. Membr. Sci.* 281 (2006) 297-306.

[108] Lee, J.W.; Yeon, K.H.; Song, J.H.; Moon, S.H. Characterization of electroregeneration and determination of optimal current density in continuous electrodeionization. *Desalination.* 207 (2007) 276-285.

[109] Bouhidel, K.E.; Lakehal, A. Influence of voltage and flow rate on electrodeionization (EDI) process efficiency. *Desalination.* 193 (2006) 411-421.

[110] Song, J.H.; Yeon, K.H.; Moon, S.H. Effect of current density on ionic transport and water dissociation phenomena in a continuous electrodeionization (CEDI). *J. Membr. Sci.* 291 (2007) 165-171.

[111] Basta, K.; Aliane, A.; Lounis, A.; Sandeaux, R.; Sandeaux, J.; Gavach, C. Electroextraction of Pb^{2+} ions from diluted solutions by a process combining ion-exchange textiles and membranes. *Desalination.* 120 (1998) 175-184.

[112] Wesselingh, J.A.; Vonk, P.; Kraaijeveld, G. Exploring the Maxwell-Stefan description of ion exchange. *Chem. Eng. J.* 57 (1995) 75-89.

[113] Mohmoud, A.; Muhr, L.; Vasiluk, S.; Aleknikoff, A.; Lapocque, F. Investigation of transport phenomena in a hybrid ion exchange-electrodialysis system for the removal of copper ions. *J. Appl. Electrochem.* 33 (2003) 875-884.

[114] Spoor, P.B.; ter Veen, W.R.; Janssen, L.J.J. Electrodeionization 1: Migration of Nickel ions absorbed in a rigid, Macroporous Cation-exchange Resin. *J. Appl. Electrochem.* 31 (2001) 523-530.

[115] Linkov, N.A.; Smit, J.J.; Linkov, V.M.; Grebenyuk, V.D. Electroadsorption of Ni^{2+} ions in an electrodialysis chamber containing granulated ion-exchange resin. *J. Appl. Electrochem.* 28 (1998) 1189-1193.

[116] Spoor, P. B.; Grabovska, L.; Koene, L.; Janssen, L. J. J.; ter Veen, W. R. Pilot scale deionisation of a galvanic nickel solution using a hybrid ion-exchange/electrodialysis system. *Chem. Eng. J.* 89 (2002) 193-202.

[117] Smara, A.; Delimi, R:; Chainet, E.; Sandeaux, J. Removal of heavy metals from diluted mixtures by a hybrid ion-exchange/electrodialysis process. *Sep. Purif. Technol.* 57 (2007) 103-110.

[118] Oh, S.J.; Moon, S.H.; Davis T. Effects of metal ions on diffusion dialysis of inorganic acids. *J. Membr. Sci.* 169 (2000) 95-105.

[119] Kang, M.S.; Yoo, K.S.; Oh, S.J.; Moon, S.H. A lumped parameter model to predict hydrochloric acid recovery in diffusion dialysis. *J. Membr. Sci.* 188 (2001) 61-70.

[120] Xu, T.; Yang W. Tuning the diffusion dialysis performance by surface cross-linking of PPO anion exchange membranes—simultaneous recovery of sulfuric acid and nickel from electrolysis spent liquor of relatively low acid concentration. *J. Hazard. Mater.* 109 (2004) 157-164.

[121] Xu, T.; Yang W. Industrial recovery of mixed acid ($HF + HNO_3$) from the titanium spent leaching solutions by diffusion dialysis with a new series of anion exchange membranes. *J. Membr. Sci.* 220 (2003) 89-95.

[122] Kobuchi, Y.; Motomura, H.; Noma, Y.; Hanada, F. Application of ion exchange membranes to the recovery of acids by diffusion dialysis. *J. Membr. Sci.* 27 (1986) 173-179.

[123] Ruiz, T.; Persin, F.; Hichour, M.; Sandeaux, J. Modelisation of fluoride removal in Donnan dialysis. *J. Membr. Sci.* 212 (2003) 113-121.

[124] Akretche, D.E.; Kerdjoudj, H. Donnan dialysis of copper, gold and silver cyanides with various anion exchange membranes. *Talanta.* 51 (2000) 281-289.

[125] Tor, A.; Qengeloglu, Y.; Ersoz, M.; Arstan, G.; Transport of chromium through cation-exchange membranesby Donnan dialysis in the presence of some metals of different valences. *Desalination.* 170 (2004) 151-159.

[126] Hichour, M.; Persin, F.; Sandeaux, J.; Gavach, C. Fluoride removal from waters by Donnan dialysis. *Sep. Purif. Technol.* 18 (2000) 1-11.

[127] Judd, S.; Jefferson, B. Membrane for industrial wastewater recovery and re-use; Elsevier: Oxford, 2003.

[128] Tanaka, Y. Ion exchange membranes: Fundamentals and applications; Elsevier: Amsterdam, 2007.

[129] Graillon, S.; Persin, F.; Pourcelly, G.; Gavach, C. Development of electrodialysis with bipolar membrane for the treatment of concentrated nitrate effluents. *Desalination.* 107 (1996) 159-169.

[130] Yu, L.; Lin, A.; Zhang, L.; Chen, C.; Jiang, W. Application of electrodialysis to the production of Vitamin C. *Chem. Eng. J.* 78 (2000) 153–157.

[131] Novalic, S.; Kongbangkerd, T.; Kulbe, K.D. Recovery of organic acids with high molecular weight using a combined electrodialytic process. *J. Membr. Sci.* 166 (2000) 99–104.

[132] Novalic, S.; Kulbe, K.D. Separation and concentration of citric acid by means of electrodialytic bipolar membrane technology. *Food Technol. Biotechnol.* 36 (1998) 193–195.

[133] Alvarez, F.; Alvarez, R.; Coca, J.; Sandeaux, J.; Sandeaux, R.; Gavach, C. Salicylic acid production by electrodialysis with bipolar membranes. *J. Membr. Sci.* 123 (1997) 61–69.

[134] Trivedi, G.S.; Shah, B.G.; Adhikary, S.K.; Indusekhar, V.K.; Rangarajan, R. Studies on bipolar membranes. Part II—conversion of sodium acetate to acetic and sodium hydroxide. React. *Funct. Polym.* 32 (1997) 209–215.

[135] Yu, L.; Guo, Q.; Hao, J.H.; Jiang, W. Recovery of acetic acid from dilute wastewater by means of bipolar membrane electrodialysis. *Desalination.* 129 (2000) 283–288.

[136] Rahardianto, A.; Gao, J.; Gabelich, C.J.; Williams, M.D.; Cohen, Y. High recovery membrane desalting of low-salinity brackish water: Integration of accelerated precipitation softening with membrane RO, *J. Membr. Sci.* 289 (2007) 123-137.

[137] Rautenbach, R.; Linn, T.; Eilers, L. Treatment of severely contaminated waste water by a combination of RO, high-pressure RO and NF — potential and limits of the process. *J. Membr. Sci.* 174(2000) 231-241.

[138] M'nif, A.; Bouguecha, S.; Hamrouni, B.;Dhahbi, M. Coupling of membrane processes for brackish water desalination. *Desalination.* 203 (2007) 331-336.

[139] El-Sayed, E.; Abdel-Jawad, M.; Ebrahim, S.; Al-Saffar, A. Performance evaluation of two RO membrane configurations in a MSF/RO hybrid system. *Desalination.* 128 (2000) 231-245.

[140] Macedonio, F.; and Drioli, E. Pressure-driven membrane operations and membrane distillation technology integration for water purification. *Desalination.* 223 (2008) 396-409

[141] Jacob, C. Seawater desalination: Boron removal by ion exchange technology. *Desalination.* 205 (2007) 47-52.

[142] Turek, M.; Bandura, B.; Dydo, P. Electrodialytic boron removal from SWRO permeate. *Desalination.* 223 (2008) 17-22.

[143] Galaverna, G.; Silvestro, G.D.; Cassano, A.; Sforza, S.; Dossena, A.; Drioli E.; Marchelli, R. A new integrated membrane process for the production of concentrated blood orange juice: Effect on bioactive compounds and antioxidant activity. *Food Chemistry.* 106 (2008) 1021-1030.

[144] Wang, Z.; Fan,Z.; Xie L.; Wang, S. Study of integrated membrane systems for the treatment of wastewater from cooling towers. *Desalination.* 191 (2006) 117-124.

[145] Noworyta, A.; Trusek-Hołownia, A.; Mielczarski, S.; Kubasiewicz-Ponitka, M. An integrated pervaporation–biodegradation process of phenolic wastewater treatment. *Desalination.* 198 (2006) 191-197.

[146] Lipnizki, F.; Field, R.W.; Ten, P. Pervaporation-based hybrid process: a review of process design, applications and economics. *J. Membr. Sci.* 153(1999) 183-210.

In: Advances in Membrane Science and Technology
Editor: Tongwen Xu
ISBN: 978-1-60692-883-7

Chapter VIII

Membrane Controlled Release

Liang-Yin Chu *

School of Chemical Engineering, Sichuan University;
Chengdu, Sichuan, 610065, China

Abstract

As a kind of functional barriers, membranes have been successfully developed for controlled release. Recently, as a new type of membranes, environmental stimuli-responsive membranes have been proposed. These functional membranes can control the diffusion permeability of solutes responding to environmental stimuli, such as temperature, pH, glucose concentration, special chemical substances and so on. They have been considered to be very useful for controlled release especially drug delivery, because they can be used to achieve stimuli-responsive rate-/time-programmed and/or site-specific drug delivery. The environmental stimuli-responsive membrane systems can just release chemicals or drugs as designed at a specific site where an environmental condition, such as pH or temperature, is different from that at other sites. For an example, they might release drug in a body when the membrane carrier systems passing through tumor cells that have a considerably lower pH than normal tissues. Because the release rate from membrane systems is generally controlled by the diffusion rate of the permeants across the thin membranes, a quicker change in the release rate responding to the stimuli may be expected, compared to cross-linked gels or microspheres. Therefore, much attention has been paid to the environmental stimuli-responsive membrane controlled release. In this Chapter, several kinds of environmental stimuli-responsive membrane controlled release systems and processes, such as thermo-responsive, pH-responsive, glucose-responsive, and molecular-recognizable ones, were introduced and discussed. Finally, the prospect for controlled release based on functional membrane technology will be outlined.

Keywords: Membrane; Controlled release; Drug delivery; Stimuli-response; Permeability; Diffusion.

* Correspondence concerning this chapter should be addressed to Prof Liang-Yin CHU; at chuly@scu.edu.cn

1. Introduction

The target of a controlled drug delivery system is for an improved drug treatment (outcome) through rate- and time-programmed and site-specific drug delivery. Environmental stimuli-responsive controlled-release systems have been developed specifically for this purpose. These environmental stimuli-responsive release systems can release specified chemicals or drugs at a particular site where an environmental condition, such as temperature, pH or other information, is different from that at other sites.

In developing environmental stimuli-responsive controlled-release systems, especially pulsate release systems with an on-off switching response, one of the important parameters is to reduce the response time of the release rate to stimuli. As the release rate from membrane systems is generally controlled by the rate of diffusion of the permeants across the thin membrane, the release rate in response to stimuli may be expected to be fast. Therefore, stimuli-responsive membranes have been developed for controlled-release. Usually, the stimuli-responsive membranes for controlled-release are composed of porous membrane substrate and functional polymeric gates, as shown in figure 1. The functional polymeric gates are usually grafted "smart" polymers, whose volume could change abruptly in response to gentle changes in environmental temperature, pH, ionic strength, glucose concentration, electric field, light, oxidoreduction, or substance species. Because of the abrupt volume change of the polymeric gates in the membrane pores, the permeation properties can be controlled or adjusted by the gates according to the external chemical and/or physical environment. Recently, by introducing the functional polymeric gates into the porous membrane of hollow microcapsules, we have developed several kinds of stimuli-responsive microcapsule membranes for controlled-release, as shown in figure 2.

In this Chapter, thermo-responsive, pH-responsive, glucose-responsive, and molecular-recognizable membranes for controlled-release will be introduced, and the prospect of the controlled release based on stimuli-responsive membrane technology will be discussed.

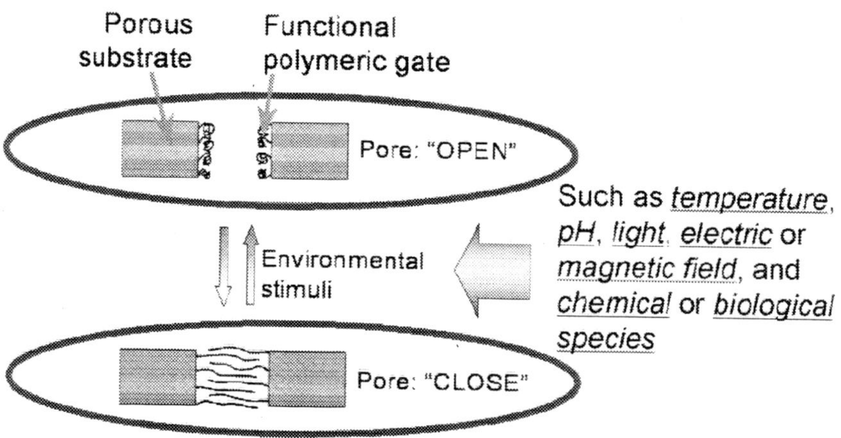

Figure 1. A schematic representation of the stimuli-responsive gating membrane for controlled-release.

2. Thermo-Responsive Membrane Controlled Release

Because there are many cases in which environmental temperature fluctuations occur naturally, and in which the environmental temperature stimuli can be easily designed and artificially controlled, in recent years much attention has been focused on thermo-responsive membrane controlled-release systems.

Among thermo-responsive polymers, poly(*N*-isopropylacrylamide) (PNIPAM) is one of the most studied materials and exhibits a lower critical solution temperature (LCST) due to the presence of both hydrophilic amide groups and hydrophobic isopropyl groups in its side-chains. The PNIPAM hydrogel exhibits a reversible volume change in response to the change in surrounding temperature. When the temperature is below the LCST, the hydrogel is swollen and hydrophilic; while above the LCST, the hydrogel shrinks and turns into a collapsed and hydrophobic state. Due to the dramatic thermo-responsive properties, PNIPAM hydrogels have attracted great interests for a wide variety of applications. In developing thermo-responsive gating membranes for controlled-release, PNIPAM is also mostly used as the thermo-responsive gate material.

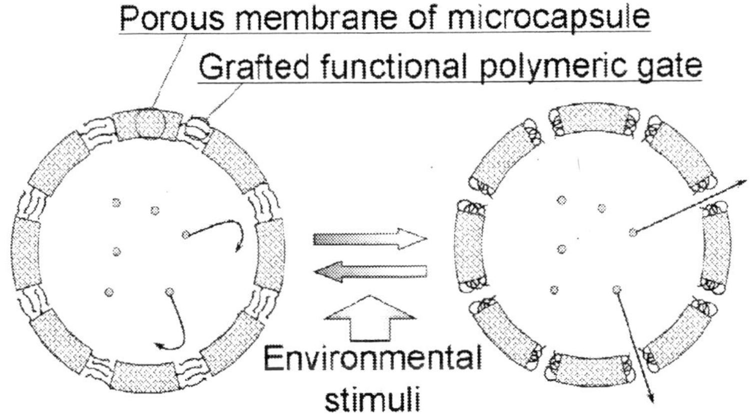

Figure 2. A schematic illustration of stimuli-responsive release principle of hollow microcapsules with a porous membrane and stimuli-responsive polymeric gates.

By applying the plasma-graft pore-filling polymerization method to graft a linear thermo-responsive PNIPAM polymer inside the porous membrane of a core-shell microcapsule, we successfully developed a thermo-responsive core-shell microcapsule with a porous membrane and thermo-responsive polymeric gates [1]. This is the first time that the method of plasma-graft pore-filling polymerization has been applied to graft polymers inside the porous walls of core-shell particles. As shown in figure 2, the grafted functional polymeric gates are PNIPAM chains. The core-shell polyamide microcapsules with porous membranes were prepared from ethylenediamine and terephthaloyl dichloride by interfacial polymerization, and plasma-graft pore-filling polymerization was employed to graft linear PNIPAM chains into the pores in the microcapsule membrane. When the graft yield was proper, the membrane exhibited satisfactory thermo-responsive controlled-release property. Below the LCST, the linear grafted PNIPAM on the inner pore surface was in the swollen state, and the pores in the membrane were closed by the PNIPAM gate; as a result the permeability coefficient was low. In contrast, the grafted PNIPAM on the inner pore surface was in the shrunken state at temperatures above the LCST, and therefore the pores in the membrane were open, and a higher permeability coefficient was the result. The thermo-responsive release of the PNIPAM-grafted microcapsules was found to be satisfactorily reversible and reproducible, suggesting that the grafted PNIPAM gates retained their thermal swelling/shrinking and hydrophilic/hydrophobic properties intact, even though they underwent repeated temperature changes across the LCST. That is to say, a repeatable thermo-responsive release can be effectively achieved.

For drug delivery systems (DDS), monodispersed small-size microcapsules are preferable to polydispersed large-size microcapsules. A uniform microcapsule particle size is important, because the distribution of the microcapsules within the body, and the interaction with biological cells, is greatly affected by the particle size. In addition, if monodispersed microcapsules are available, the drug release kinetics can be manipulated, thereby making it easier to formulate more sophisticated systems. For DDS, there is a size limit for particles to traverse certain organs, and a small particle size minimizes any potential irritant reaction at the injection site. Therefore, the fabrication of small size monodispersed microcapsules (or hollow reservoirs) that enable the encapsulation of various materials is of both scientific and technological interest. Recently, we have successfully prepared monodispersed thermo-responsive core-shell microcapsules with a mean diameter of about 4 μm with a porous membrane, and with linear-grafted PNIPAM chains in the membrane pores acting as thermo-responsive gates [2]. The preparation was carried out by using a Shirasu porous glass (SPG) membrane emulsification technique to prepare small-sized monodispersed oil-in-water emulsions, and using interfacial polymerization to prepare the core-shell microcapsules with porous membranes. Plasma-graft pore-filling polymerization was used to graft linear PNIPAM chains into the pores of the microcapsule membranes. The as-prepared PNIPAM-grafted monodispersed microcapsule membranes with mean diameter of about 4 μm, showed satisfactorily reversible and reproducible thermo-responsive release characteristics. The release of both NaCl and VB_{12} from the PNIPAM-grafted microcapsules was slow at 25 °C and fast at 40 °C, which is due to the close/open state of the grafted "gates". The "on/off" ratio of the release rate of VB_{12} from the PNIPAM-grafted microcapsules was much larger than that of NaCl [2].

In order to understand the morphological control of the porous membranes with linear-grafted PNIPAM gates, the effect of membrane morphology on the thermo-responsive transport behavior, and the prediction of the thermo-responsive transport behavior with morphological parameters, thermo-responsive flat membranes with a porous membrane substrate and grafted PNIPAM gates were prepared using the plasma-graft pore-filling polymerization method [3,4]. The grafted polymer formation profile of the flat porous substrate was measured using the microscopic Fourier transform infrared (FT-IR) mapping method. The pore sizes of the PNIPAM-grafted flat PE membranes at temperatures above the LCST of PNIPAM were measured by the mercury intrusion method. The pore condition of the PNIPAM-grafted PE membranes at temperatures below the LCST of PNIPAM was determined by the bubble point method. The solute diffusion experiments of the flat membranes were carried out using a standard side-by-side diffusion cell. The diffusion cell was located in a constant-temperature incubator to keep the diffusional temperature constant. The results showed that PNIPAM was grafted homogeneously onto the porous membrane substrates, not only in the direction of the membrane thickness, but also in the direction of the membrane surface. Regardless of the solute molecular size, temperature had an opposite effect on the diffusion coefficients of the solute across the PNIPAM-grafted membranes with low graft yields as opposed to those with high graft yields. When the pore-filling ratio was below ~30%, the diffusional coefficients of the solutes across the PNIPAM-grafted PE membranes were higher at temperatures above the LCST than those below the LCST. In contrast, when the pore-filling ratio was higher than ~30%, the diffusional coefficients were lower at temperatures above the LCST than those below the LCST [3].

To investigate the thermo-responsive gating characteristics of thermo-responsive membranes more comprehensively, a series of thermo-responsive gating membranes, with a wide range of grafting yields, were prepared by grafting PNIPAM onto porous polyvinylidene fluoride (PVDF) membrane substrates with the plasma-induced pore-filling polymerization method, and the effect of grafting yield on the gating characteristics of thermo-responsive gating membranes was investigated systematically [5]. The results showed that the grafting yield heavily affected both water flux responsiveness coefficient and thermo-responsivity of membrane pore size. When the grafting yield was smaller than 2.81%, both flux responsiveness coefficient and thermo-responsivity of pore size increased with increasing the grafting yield; however, when the grafting yield was higher than 6.38%, both flux responsiveness coefficient and thermo-responsivity of membrane pore size were always equal to 1.0, i.e., no gating characteristics existed any more. Diffusional permeation experiments showed that two distinct types of temperature responses were observed, depending on the grafting yield. Diffusional coefficient of solute across membranes with low grafting yields increased with temperature, while that across membranes with high grafting yields decreased with temperature. In order to get a desired or satisfactory thermo-responsive gating performance, the membranes should be designed and prepared with a proper grafting yield [5].

Furthermore, to investigate the pore microstructure inside the membrane more directly, PNIPAM was grafted on the surfaces and in the pores of polycarbonate track-etched (PCTE) porous membranes by plasma-graft pore-filling polymerization method, and the microstructure of the PNIPAM-*g*-PCTE membrane was investigated systematically by employing XPS, SEM, FT-IR, AFM, contact angle instrument and water flux experiments [6]. The results showed that, the grafted PNIPAM polymers were formed inside the pores throughout the entire membrane thickness, and there was not a dense PNIPAM layer formed on the membrane surface even at a pore-filling ratio as high as 76.1%. With the pore-filling ratio increasing, the pore diameters of PNIPAM-grafted membranes became smaller. When the pore-filling ratio was smaller than 44.2%, the pores of PNIPAM-*g*-PCTE membranes showed thermo-responsive gating characteristics because of the conformational change of grafted PNIPAM in the pores. On the other hand, when the pore-filling ratio was larger than 44.2%, the pores of membranes immersed in water were choked by the volume expansion of the grafted PNIPAM polymers, and the membranes did not show thermo-responsive gating characteristics any longer. The critical pore-filling ratio for choking the membrane pores was in the range from 30% to 40% [6].

Actually, the physical and chemical properties of porous membrane substrates also affect the stimuli-responsive gating characteristics of grafted thermo-responsive membranes to a certain extent. Both the hydrophilic Nylon-6 membranes and hydrophobic PVDF membranes with a wide range of grafting yield of PNIPAM were prepared by plasma-graft pore-filling polymerization method, and the effect of physical and chemical properties of substrates on the thermo-responsive gating characteristics of PNIPAM-grafted membranes was experimentally investigated [7]. For both PVDF and Nylon-6 membranes, grafted PNIPAM polymers were found not only on the membrane outer surface but also on the inner surfaces of the pores throughout the entire membrane thickness. The thermo-responsive gating characteristics of the PNIPAM-grafted membranes were heavily affected by the physical and chemical properties of the porous membrane substrates. The PNIPAM-*g*-Nylon6 membranes exhibited much larger thermo-responsive gating coefficient than the PNIPAM-*g*-PVDF membranes. Besides, to achieve the largest thermo-responsive gating coefficient, the corresponding optimum grafting yield of PNIPAM for the PNIPAM-*g*-Nylon6 membranes was also larger than that of the PNIPAM-*g*-PVDF membranes [7].

Up to now, most of the functional gates for thermo-responsive membranes were constructed from the PNIPAM. Therefore, the response temperature of these membranes is always around 32 °C, which is determined by the LCST of PNIPAM. In most applications, however, higher or lower response temperatures of the thermo-responsive membranes are preferred. Recently, systematical investigations have been carried out on the preparation of thermo-responsive gating membranes with controllable response temperature [8]. Plasma-induced grafting polymerization method was introduced to graft thermo-responsive polymers onto porous membrane substrates, and the response temperature of thermo-responsive gating membranes was adjusted or manipulated by adding hydrophilic or hydrophobic monomers into the NIPAM monomer solution in the fabrication of the thermo-responsive gates. It has been reported that the addition of hydrophilic monomer in the NIPAM-based copolymer could result in a shift of the LCST to a higher temperature, and the addition of hydrophobic monomer could result in a lower LCST of the copolymer. The microstructures and chemical compositions of the prepared membranes were characterized by SEM and XPS, and the effect of the dosage of hydrophilic monomer and that of hydrophobic monomer on the response temperature of the prepared gating membranes was systematically studied by carrying out water flux experiments. The thermo-dependent hydrophilic/hydrophobic transition of membrane surface around response temperature of grafted membranes was investigated by measuring the contact angle. The water flux experimental results showed that the response temperatures of grafted gating membranes were linearly increased with increasing the molar ratio of hydrophilic monomer acrylamide (AAM) in the NIPAM co-monomer solution, but linearly decreased with increasing the molar ratio of hydrophobic monomer butyl methacrylate (BMA) in the NIPAM co-monomer solution. The response temperature of PNA-*g*-PVDF membrane was raised to 40 °C when 7 mol.% of AAM was added into the NIPAM co-monomer solution, and that of PNB-*g*-N_6 membrane was reduced to 17.5 °C when 10 mol.% of BMA was introduced into the NIPAM co-monomer solution [8].

In general, currently-existed thermo-responsive controlled-release systems can be mainly classified as reservoir-type systems and matrix-type ones. To achieve such reservoir-type systems, functional membranes with thermo-responsive gates have been considered as efficient tools; on the other hand, to achieve the matrix-type systems, thermo-responsive hydrogels have been selected as the main material candidates. Unfortunately, due to the inherent physical and structural characteristics, both of the above-mentioned systems are always accompanied with some inherent disadvantages, such as the drug-security problem of the reservoir-type systems and the mechanical-strength problem of the hydrogel matrix-type systems. These inherent weaknesses restrict the practical applications of the current thermo-responsive controlled-release systems to a certain extent. Recently, we have developed a novel composite thermo-responsive membrane system for improved controlled-release [9]. The proposed composite system is mainly composed of a thermo-responsive gating membrane with grafted PNIPAM gates and a crosslinked PNIPAM hydrogel inside the reservoir. The release of the loaded drug is therefore dependent on the cooperative control by the grafted PNIPAM gates and the crosslinked PNIPAM hydrogel. When the environmental temperature is lower than the LCST of PNIPAM, the membrane pores are closed due to the swollen PNIPAM chains grafted onto the membrane and the drugs are entrapped in the swollen PNIPAM hydrogel inside the reservoir, resulting in a low release rate of drugs from the reservoir; on the other hand, when the environmental temperature is increased to be higher than the LCST, the drugs are squeezed out from the crosslinked hydrogel due to the shrinking effect and the membrane pores open owing to the coil-globule transition of the grafted PNIPAM chains, and then a high release rate is resulted. Because of the existence of inner crosslinked hydrogel, the drug-security problem of the reservoir-type systems could be avoided; meanwhile, due to the protection function of the outer reservoir, the mechanical-strength problem of the hydrogel matrix-type systems could also be solved. More importantly, the thermo-responsive controlled-release of solutes from the proposed composite membrane system in this study significantly exhibited better performance than those currently-existed single-functional systems, because of the cooperative action of the gating membrane and the crosslinked hydrogel.

Up to 2005, almost all of the thermo-responsive gating membranes have been featured with positively thermo-responsive characteristics, i.e., the membrane permeability increases with increasing the environmental temperature, because all of the thermo-responsive functional gates were constructed from PNIPAM, and therefore the membrane pores changed from "closed" situation into "open" situation when the environmental temperature increased from a lower one that below the LCST of PNIPAM to a higher one that above the LCST, due to the swelling/shrinking conformational change of PNIPAM. In certain applications, however, an inverse mode of the thermo-responsive gating behavior for the membranes is preferred. Recently, we have developed a novel family of thermo-responsive gating membranes, which is featured with negatively thermo-responsive gating characteristics, i.e., the membrane pores' "opening" is induced by a decrease rather than an increase in temperature [10].

The functional gates of the developed negatively thermo-responsive gating membrane are constructed from thermo-responsive interpenetrating polymer networks (IPNs) that composed of poly(acrylamide) (PAAM) and poly(acrylic acid) (PAAC) [10]. The thermo-responsive membranes with PAAM/PAAC-based IPN gates were prepared by a method of sequential IPN synthesis, in which the crosslinked PAAM that grafted in the pores of porous nylon 6 (N6) membranes and PAAC gels were synthesized as initial gel matrix and as secondary gels, respectively. It is known that PAAM and PAAC form polycomplexes in solution through hydrogen bonding. By the cooperative "zipping" interactions between the molecules that result from hydrogen bonding, it has been found that the PAAM/PAAC-based IPN hydrogels are featured with a thermo-responsive volume phase transition characteristic that is on the reverse of that of PNIPAM, i.e., the hydrogel swelling is induced by an increase rather than a decrease in temperature. When the environmental temperature is lower than the upper critical solution temperature (UCST) of the PAAM/PAAC-based IPN gel, PAAC forms intermolecular hydrogen bonds with PAAM, and the IPN hydrogels keep shrinking state by the interaction between two polymer chains or so-called chain-chain zipper effect; on the other hand, when the environmental temperature is higher than the UCST of the IPN gel, PAAC dissociates intermolecular hydrogen bonds with PAAM, and the IPN hydrogels keep swelling state by the relaxation of the two polymer chains. Therefore, the developed membrane gates shrink at temperatures below the UCST due to the complex formation by hydrogen bonding, and swell at temperatures above the UCST due to PAAM/PAAC complex dissociation by the breakage of hydrogen bonds. As a result, the membrane pores change from "open" situation into "closed" situation when the temperature increased from a lower one that below the UCST to a higher one that above the UCST. The prepared membranes with PAAM/PAAC-based IPN gates showed a satisfactorily reversible and reproducible thermo-responsive permeation characteristic [10]. Such gating membranes provide a new mode of the phase transition behavior for thermo-responsive gating membranes.

3. pH-Responsive Membrane Controlled Release

It has been well-known that the interstitial fluids of a number of tumors or inflamed sites in humans and animals have an ambient pH that is considerably lower than that of normal tissues, and the pH values are different from place to place in the gastrointestinal tract. Consequently, pH-responsive controlled-release delivery systems enable drugs to be targeted specific areas of the body such as tumors, sites of inflammation and inflection, or the colonic region. Therefore, the development of pH-responsive controlled-release delivery systems is of both scientific and technological interest.

pH-responsive membranes for controlled-release are usually prepared by grafting pH-responsive polymers, which exhibit abrupt volume change around the corresponding pKa point. Two pH-responsive polymers with opposite tendency of the pH-responsive volume change, poly(methacrylic acid) (PMAA) and poly(*N,N*-dimethylaminoethyl methacrylate) (PDM), are shown in figure 3. Such kinds of polymers can be used as polymeric gates in the pH-responsive gating membranes.

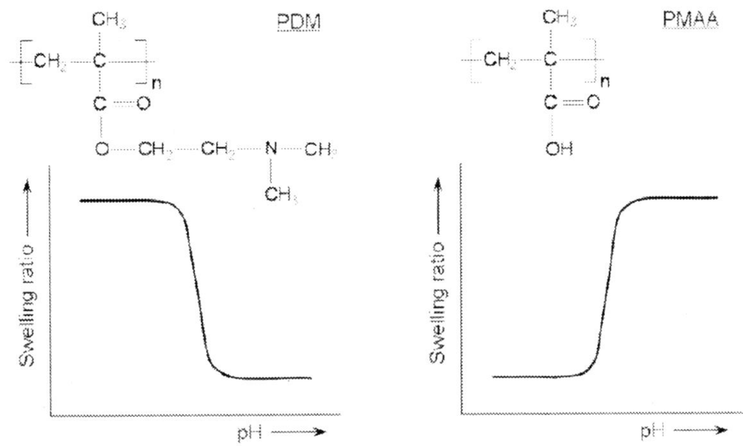

Figure 3. Opposite tendency of the pH-responsive volume change of PDM and PMAA.

For the environmental stimuli-responsive controlled-release systems, it is very important to manipulate them to respond as quickly as possible upon receiving environmental signals, because the fast response is the key for their successful applications. Up to now, most environmental stimuli-responsive controlled-release systems were fabricated by stimuli-responsive polymeric hydrogels. To increase the response dynamics of stimuli-responsive polymeric hydrogels, several strategies have been explored, such as improving the internal architecture or structure of the hydrogels, developing microgels or hydrogel particles with micro- or nano-dimensions, and introducing linear-grafted hydrogel chain-configurations with freely mobile ends. However, the previous investigations were just concentrated mainly on the improvement of the response time of phase transitions of hydrogels themselves, but nearly not on the improvement of the stimuli-responsive release rate. It is equally important that the delivery systems release drugs as quickly as possible upon meeting environmental stimuli. For example, although the response time or the responsive conformational change of linear-grafted hydrogel gates of gating membrane systems to environmental stimuli is much faster than that of crosslinked hydrogel membrane systems because of the chain-configurations with freely mobile ends, the maximum release rate of drugs from the gating membrane systems is still limited by the concentration-driven diffusion. The limitation of the release rate restrains the development of fast-response rate-programmed drug delivery systems.

To overcome the above-mentioned limitation, we developed a novel composite system for pH-responsive controlled-release, which is coupled with a positively pH-responsive linear-grafted gating membrane and a negatively pH-responsive crosslinked hydrogel inside the reservoir [11]. The proposed system is featured with a fast response because of the linear-grafted gates, and more importantly with a large responsive release rate that goes effectively beyond the limit of concentration-driven diffusion due to the pumping effects of the negatively responsive hydrogel. The developed pH-responsive gating membrane system with pumping effects is composed of a gating membrane with linear-grafted poly(methacrylic acid) (PMAA) gates and a crosslinked poly(*N,N*-dimethylaminoethyl methacrylate) (PDM) hydrogel inside the reservoir. The pH-responsive gating membranes were prepared by grafting PMAA linear chains onto porous PVDF membrane substrates using plasma-graft pore-filling polymerization, and the crosslinked PDM hydrogels were synthesized by the well-known free radical polymerization [11]. The phase transition characteristics of PMAA and PDM have been verified to be just opposite, as shown in figure 3. PMAA hydrogels are featured with a positively pH-responsive volume phase transition characteristic, *i.e.*, the hydrogel swelling is induced by an increase in the environmental pH; on the contrary, PDM hydrogels show a negatively pH-responsive volume phase transition characteristic, *i.e.*, the hydrogel swelling is induced by an decrease in the environmental pH. The linear-grafted positively pH-responsive PMAA polymers in the membrane pores act as pH-sensitive "gates", and the crosslinked negatively pH-responsive PDM hydrogel in the reservoir acts as a pH-sensitive "pumping element". The effective dissociation constant (pK) of PMAA (pK_{PMAA}) has been reported to be 4.65 ~ 5.35. Although the pK value of the DM monomer is around 8.00, the pK value of crosslinked PDM networks (pK_{PDM}) has been found to be much lower (about 4.5 ~ 5.5). The swelling/deswelling behaviors of both PMAA polymer and crosslinked PDM networks have been verified to be consistent with their pK values. When the environmental pH is higher than the value of max(pK_{PMAA}, pK_{PDM}), the inner PDM hydrogel shrinks, at the same time the grafted PMAA swells and then closes the membrane pores, as a result the release rate is slow; On the other hand, when the ambient pH is decreased to be lower than the value of min(pK_{PMAA}, pK_{PDM}), the grafted PMAA shrinks and consequently the membrane pores open, at the same time the PDM hydrogel swells and then accelerates the release rate. Therefore, the maximum release rate of drugs from the proposed systems responding to environmental pH changes can be effectively improved, because the limitation of release rate that restricted by the concentration-driven diffusion can be broken through due to the pumping effects of the negatively pH-responsive hydrogel inside the reservoir. The proposed system provides a new mode for pH-responsive "smart" or "intelligent" controlled-release systems.

The proposed system could also be easily applied to micro-scale systems by fabricating negatively responsive microgels or hydrogel nano-particles inside the hollow microcapsules with porous membranes and functional gates, as illustrated in figure 4. In such a microcapsule system, the inner PDM hydrogel would respond to the pH variation of the environment (externally to the outer PMAA-grafted microcapsule membrane), so quickly to be effective in the delivery of the active substance: when the system goes from an environment with higher pH toward the environment with lower pH, the grafted PMAA shrinks, the membrane pores open, the external solution diffuses inside the microcapsule, the pH inside the system lowers and then the PDM hydrogel responds with swelling and pumping the drug solutes externally to the system, through the pores of the microcapsule membrane. Similar thermo-responsive microcapsule systems can also be developed by introducing two kinds of thermo-responsive polymers with opposite phase transition behaviors for the gate and pumping element respectively, i.e., one is featured with swelling upon heating, but the other one is featured with swelling upon cooling.

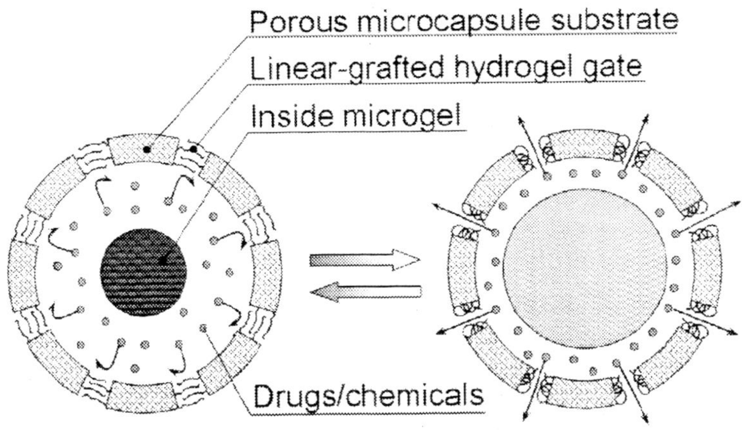

Figure 4. A schematic representation of a composite stimuli-responsive controlled-release system with functional gating and pumping effects.

4. Glucose-Responsive Membrane Controlled Release

The development of a glucose-sensitive insulin-releasing system for diabetes therapy is a long-standing challenging for biomedical engineers [12]. Although diabetes mellitus is a major cause of death in industrialized countries, periodical parenteral injections of insulin are currently the standard treatment for insulin-dependent diabetic patients. However, poor control of blood glucose level and poor patient compliance are associated with this method. Therefore, there is a need for self-regulated delivery systems having the capability of adapting the insulin release rate in response to changes in glucose concentration in order to keep the blood glucose levels within the normal range.

Up to now, several kinds of glucose-responsive insulin delivery systems have been devised. However, none of these systems could fully mimic the physiology of insulin secretion as yet. Therefore, better ways of glucose-responsive self-regulated administration of insulin delivery are still being sought. For glucose-responsive self-regulated insulin release systems, stability and responsivity of the system are very important and essential, because only stable system can ensure the safety during therapy and only fast response can ensure the self-regulated insulin-release exactly during changes in glucose concentration. To meet both stability and responsivity, glucose-responsive gating membranes with porous substrates and linear-grafted functional polymeric gates are competent. The porous membrane substrates can provide mechanical strength and dimensional stability. As the linear grafted polymeric chains have freely mobile ends, which are different from the typical crosslinked network structure of the hydrogels that gives rise to relatively immobile chain ends, the responsiveness of the prepared membranes to the environmental stimuli could therefore be faster than that of their corresponding homogeneous analogs, owing to the more rapid conformational changes of the functional polymers. By grafting pH-responsive polymeric chains onto porous membrane substrates and immobilizing glucose oxidase (GOD) onto the grafted polymers, glucose-responsive gating membranes have been prepared for insulin release, and the response time of the membrane to glucose was reported as fast as 16 s or even shorter.

To comprehensively understand the gating characteristics of this kind of glucose-responsive membranes and the self-regulated control of the membrane pore size and the permeability response, we prepared glucose-responsive gating membranes with grafted poly(acrylic acid) (PAAC) gates and covalently bound GOD, by grafting PAAC onto porous PVDF membrane substrates with a plasma-graft pore-filling polymerization method, and immobilizing GOD onto the grafted membranes with a carbodiimide method [13]. The principle of glucose-responsive control of the permeation through the gating membrane is schematically illustrated in figure 5. The linear grafted PAAC chains in the membrane pores acted as the pH-responsive gates or actuators. The immobilized GOD acted as the glucose sensor and catalyzer; it was sensitive to glucose and catalyzed the glucose conversion to gluconic acid. At neutral pH in the absence of glucose, the carboxyl groups of the grafted PAAC chains are dissociated and negatively charged, therefore the membrane gates "closed" because the repulsion between negative charges make the PAAC chains extended; when glucose concentration increased, GOD catalyzes the oxidation of glucose into gluconic acid, thereby lowering the local pH in the microenvironment, protonating the carboxylate groups of the grafted PAAC chains, therefore the gates "open" because of the reduced electrostatic repulsion between the grafted PAAC chains in the pores.

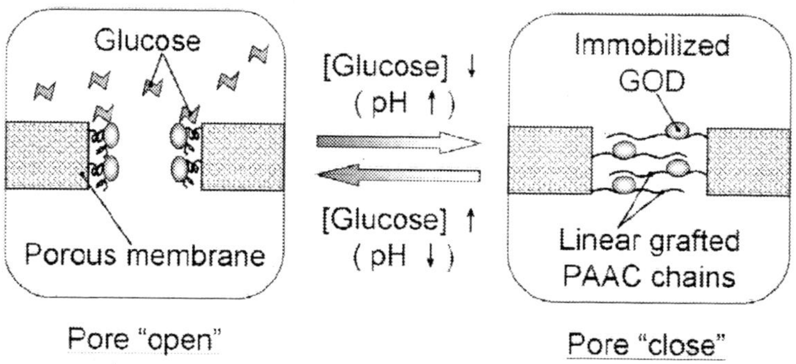

Figure 5. A schematic illustration of the principle of glucose-responsive control of the permeation through the gating membrane.

The experimental results showed that, the glucose-responsivity of the solute diffusional permeability through the prepared membranes was heavily dependent on the PAAC grafting yield, because the pH-responsive change of pore size governed the glucose-responsive diffusional permeability. It is very important to design a proper grafting yield for obtaining an ideal gating response. For the proposed gating membrane with a PAAC grafting yield of 1.55%, the insulin permeation coefficient after the glucose addition (0.2 mol/L) was about 9.37 times that in the absence of glucose, presenting an exciting result on glucose-sensitive self-regulated insulin permeation.

With the same glucose-responsive functional gates, a glucose-sensitive microcapsule was also successfully prepared [14]. Polyamide microcapsules with a porous membrane were prepared by interfacial polymerization, PAAC chains were grafted into the pores of the microcapsule membrane by plasma-graft pore-filling polymerization, and GOD enzymes were immobilized onto the PAAC-grafted microcapsules by the carbodiimide method [14]. The release rates of model drug solutes from the fabricated microcapsule membranes were significantly sensitive to the existence of glucose in the environmental solution. In solution, the release rate of either sodium chloride or VB_{12} molecules from the microcapsule membranes was low but increased dramatically in the presence of 0.2 mol/L glucose. The prepared PAAC-grafted and GOD-immobilized microcapsule membranes showed a reversible glucose-sensitive release characteristic. The developed microcapsule membranes provide a new mode for injection-type self-regulated drug delivery systems having the capability of adapting the release rate of drugs such as insulin in response to changes in glucose concentration, which is highly attractive for diabetes therapy.

5. Molecular-Recognizable Membrane Controlled Release

Some of "smart" polymers have been developed for ion-responsive sensors to detect special metal cations such as K^+ and Na^+ as they are important for chemical signal transduction in biological systems. Such polymers have been designed for sensing these metal ions mainly via phase transition. They have been synthesized by attaching functional groups such as crown ethers or cyclodextrins to the polymer chains, which could interact subtly with specific metal ions or molecules and demonstrate their potential applications in numerous fields, such as artificial biomembranes. The ion-responsive phase transition behavior of such a polymer is usually associated with the ion-induced shift of the LCST of thermo-responsive polymer. As shown in figure 6, the LCST value of the poly(*N*-isopropylacrylamide-*co*-benzo-18-crown-6-acrylamide) (poly(NIPAM-*co*-BCAm)) shifts to higher temperature when the host molecules in the polymer capture the specific metal ions, due to the enhancement of hydrophilicity of the copolymer when specific ions are captured.

By introducing poly(NIPAM-*co*-BCAm) as the functional gate materials, we developed a molecular recognition microcapsule membrane for environmental stimuli-responsive controlled-release [15]. The molecular recognition microcapsule is composed of a core-shell porous membrane and linear-grafted poly(NIPAM-*co*-BCAm) chains in the pores, acting as molecular recognition gates. Benzo-18-crown-6-acrylamide (BCAm), a crown ether receptor, allows ionic molecular recognition. PNIPAM has a lower critical solution temperature (LCST), and when the receptor BCAm captures a specific metal ion, the LCST can be shifted, as shown in figure 6. Therefore, the phase transition of the grafted hydrogel gates can occur isothermally as the result of a specific metal ion signal at a temperature between the two LCSTs. The linear-grafted PNIPAM polymer with pendant crown ether was fixed on the pore surfaces of the core-shell porous microcapsule membrane. When there exist specific molecules in the environmental solution, the pores of the microcapsule membrane closed by the swelling of the gel gates, resulting in a low release rate of solute from the microcapsule; on the other hand, when the recognizable specific molecules do not exist in the environmental solution, the pores open due to the shrinking of the gel gates, and then a high release rate is resulted. That is, the encapsulated solutes can be controlled to release only at those situations that the recognizable molecules do not exist in the surrounding solution. The experimental results showed that, the release of solute from the prepared microcapsules was significantly sensitive to the existence of Ba^{2+} ions in the environmental solution, but was not sensitive to Na^{+} [15]. When $BaCl_2$ molecules did not exist in the environmental solution, the release of vitamin B_{12} (VB_{12}) from the microcapsule was fast; in contrast, when $BaCl_2$ molecules existed in the surrounding solution, the release rate became very low. The prepared microcapsule membranes showed a satisfactorily reversible and reproducible molecular recognition stimuli-responsive release characteristic.

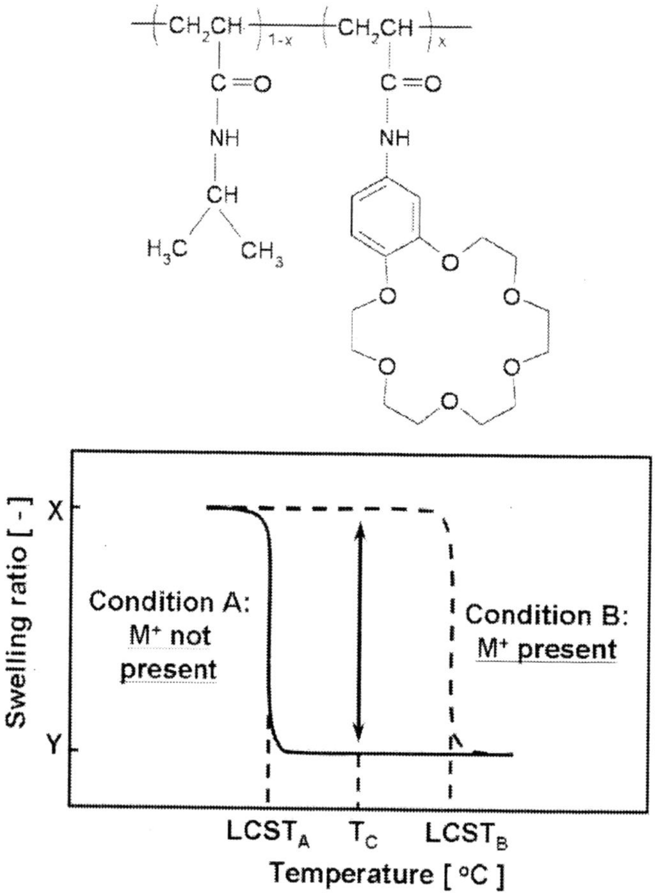

Figure 6. Schematic representation of the LCST shift of poly(NIPAM-*co*-BCAm) triggered by molecular recognition.

6. Summary and Perspective

Environmental stimuli-responsive membranes have been developed as a new type of membranes for controlled-release. Such functional membranes can control the diffusion permeability of solutes responding to environmental stimuli, such as temperature, pH, glucose concentration, and special chemical substances and so on. Up to now, a lot of achievements have been reported on the environmental stimuli-responsive membrane controlled release. The environmental stimuli-responsive membrane systems, which can just release chemicals or drugs as designed at a specific site or time where or when an environmental condition is changed, have been considered as satisfactory candidates for site-targeting or self-regulated "smart" controlled-release systems.

Of course, there will still be a long way for those stimuli-responsive membrane systems to go before they can be applied widely in biomedical fields. Among a lot of problems that needed to be solved, the following issues might be the most important ones: 1) Developing biodegradable and biocompatible both membrane substrates and stimuli-responsive polymer materials; 2) increasing the responsiveness of the membrane comtrolled-release systems in response to environmental stimuli; 3) increasing the drug-loading efficiency; 4) increasing the controllability of the drug release from membrane systems; and 5) developing new stimuli-responsive "smart" materials and new membrane controlled-release systems. We believe that such stimuli-responsive controlled-release membranes will find wide applications in biomedical fields in the future.

References

[1] Chu LY, Park SH, Yamaguchi T, Nakao S. Preparation of thermo-responsive core-shell microcapsules with a porous membrane and poly(*N*-isopropylacrylamide) gates. *Journal of Membrane Science*, 192(2001) 27-39.

[2]. Chu LY, Park SH, Yamaguchi T, Nakao S. Preparation of small-sized monodispersed thermo-responsive core-shell microcapsules. *Langmuir*, 18(2002) 1856-1864.

[3] Chu LY, Niitsuma T, Yamaguchi T, Nakao S. Thermo-responsive transport through porous membranes with grafted PNIPAM gates. *AIChE Journal*, 49(2003) 896-909.

[4] Chu LY, Zhu JH, Chen WM, Niitsuma T, Yamaguchi T, Nakao S. Effect of graft yield on the thermo-responsive permeability through porous membranes with plasma-grafted poly(*N*-isopropylacrylamide) gates. *Chinese Journal of Chemical Engineering*, 11(2003) 269-275.

[5] Li Y, Chu LY, Zhu JH, Wang HD, Xia SL, Chen WM. Thermo-responsive gating characteristics of poly(*N*-isopropylacrylamide)-grafted porous polyvinylidene fluoride membranes. *Industrial and Engineering Chemistry Research*, 43(2005) 2643-2649.

[6] Xie R, Chu LY, Chen WM, Xiao W, Wang HD, Qu JB. Characterization of microstructure of poly(N-isopropylacrylamide)-grafted polycarbonate track-etched membranes prepared by plasma-graft pore-filling polymerization. *Journal of Membrane Science*, 258(2005) 157-166.

[7] Yang M, Chu LY, Li Y, Zhao XJ, Song H, Chen WM. Thermo-responsive gating characteristics of poly(*N*-isopropylacrylamide)-grafted membranes. *Chemical Engineering and Technology*, 29(2006) 631-636.

[8] Xie R, Li Y, Chu LY. Preparation of thermo-responsive gating membranes with controllable response temperature. *Journal of Membrane Science*, 289(2007) 76-85.

[9] Hu L, Chu LY, Yang M, Yu J, Wang HD. A composite thermo-responsive membrane system for improved controlled-release, *Chemical Engineering and Technology*, 30(2007) 523-529.

[10] Chu LY, Li Y, Zhu JH, Chen WM. Negatively thermoresponsive membranes with functional gates driven by zipper-type hydrogen-bonding interactions. *Angewandte Chemie International Edition*, 44(2005) 2124-2127.

[11] Qu JB, Chu LY, Yang M, Xie R, Hu L, Chen WM. A pH-responsive gating membrane system with pumping effects for improved controlled-release, *Advanced Functional Materials*, 16(2006) 1865-1872.

[12] Chu LY. Controlled release systems for insulin delivery. *Expert Opinion on Therapeutic Patents*, 15(2005) 1147-1155.

[13] Chu LY, Li Y, Zhu JH, Wang HD, Liang YJ. Control of pore size and permeability of a glucose-responsive gating membrane for insulin delivery. *Journal of Controlled Release*, 97(2004) 43-53.

[14] Chu LY, Liang YJ, Chen WM, Ju XJ, Wang HD. Preparation of glucose-sensitive microcapsules with a porous membrane and functional gates. *Colloids and Surfaces B: Biointerfaces*, 37(2004) 9-14.

[15] Chu LY, Yamaguchi T, Nakao S. A molecular recognition microcapsule for environmental stimuli-responsive controlled-release. *Advanced Materials*, 14(2002) 386-389.

Index

A

AAC, 286, 290
ABC, 34
absorbents, 143, 260
absorption, 19, 24, 172, 260
Abu Dhabi, 227
academic, 27, 153
acceptor, 181, 263
accounting, 223
acetate, 6, 7, 61, 68, 184, 185, 187, 188, 197, 199, 241, 267, 276
acetic acid, 72, 129, 196, 267, 276
acetylation, 32
acid, 22, 24, 25, 28, 29, 30, 32, 39, 42, 43, 48, 49, 50, 51, 52, 53, 55, 56, 58, 59, 64, 66, 67, 68, 69, 70, 72, 128, 129, 131, 135, 136, 139, 144, 146, 149, 150, 152, 154, 156, 157, 160, 161, 172, 176, 177, 178, 179, 196, 198, 214, 240, 248, 250, 251, 252, 253, 254, 255, 256, 257, 258, 259, 260, 261, 265, 267, 269, 270, 271, 272, 273, 274, 275, 276, 286, 287, 290
acidic, 29, 30, 47, 48, 126, 155, 259, 261, 265
acidification, 58, 66, 67, 72, 250, 258, 259, 273
acidity, 51, 252
acrylate, 29, 38, 39, 129, 130, 131, 132, 155, 199
acrylic acid, 68, 70, 177, 286, 290
acrylonitrile, 29, 187, 188, 195
ACS, 19
activated carbon, 218, 241, 247
activated sludge flocs, 213, 214, 239
activation, 42
activation energy, 42
actuators, 290
acute, 8, 261
additives, 49, 176
adhesion, 175, 208, 214
adhesives, 151
adjustment, 176, 256, 266
administration, viii, 289
adsorption, 9, 10, 15, 16, 24, 44, 47, 48, 70, 144, 163, 164, 183, 188, 202, 209, 214, 218, 219, 239, 241, 243, 244, 249, 250, 251, 255, 260, 262, 264, 265, 266
aerobic, 202, 203, 204, 205, 214, 224, 227, 230, 233, 234, 235
aerobic bacteria, 202
aerobic granulation, 205
AFM, 283
agent, 31, 32, 50, 59, 136, 154, 155, 182
agents, 4, 5, 16, 26, 60, 66, 67, 73, 123, 147, 153, 154, 177, 218, 266
aggregates, 208, 215
aggregation, 212, 214, 215
aging, 125, 126, 251
agriculture, 47, 266, 269
aid, 169, 173, 176, 181
air, 10, 11, 12, 58, 167, 175, 176, 188, 192, 194, 202, 208, 218, 267
alanine, 250, 258, 271
alcohol, 12, 21, 27, 55, 57, 61, 62, 73, 137, 143, 164, 169, 172, 177, 178, 179, 180, 183, 185, 190, 191, 192, 194, 195, 196, 197, 198, 199, 258, 271
alcohols, 25, 60, 169, 172, 177, 179, 186, 188, 190, 196
algae, 208
Alginate, 212
alkali, 24, 258, 273
alkaline, 9, 37, 43, 46, 51, 54, 69, 72, 235, 254
alkylation, 256
allylamine, 178

alternative, 11, 32, 42, 154, 250, 252, 260, 265, 269
alternatives, 189, 261
aluminium, 39, 261
aluminum, 48, 49, 50, 52, 54, 55, 58
Aluminum, 49, 54
aluminum surface, 49
ambiguity, 122
amide, 135, 156, 181, 184, 281
amine, 32, 33, 35, 44, 131, 133, 134, 139, 145, 147, 154, 267
amines, 33, 38, 41, 70, 133, 159
amino, 42, 43, 44, 52, 53, 58, 130, 137, 154, 177, 212, 249, 250, 253, 257, 258, 259, 271, 273
amino acid, 52, 53, 58, 131, 154, 212, 249, 250, 253, 257, 258, 259, 271, 273
amino acids, 52, 53, 131, 154, 212, 249, 250, 253, 257, 258, 259, 271, 273
amino groups, 42, 43, 44, 137, 260
ammonia, 10, 44, 260
ammonium, 32, 33, 41, 43, 50, 57, 128, 154, 260, 273
ammonium chloride, 32, 128, 154
amorphous, 143
amphoteric, 22, 24, 25, 47, 72, 250, 273
Amsterdam, 19, 64, 70, 191, 195, 268, 269, 276
anaerobic, 205, 234, 240
animal health, 266
animal waste, 225
animals, 286
Anion, 72
ANN, 66
annealing, 196
anode, 246, 261, 265
anoxic, 224, 225, 234, 235, 261, 262
antioxidant, 277
application, vii, 1, 2, 4, 6, 8, 9, 10, 12, 13, 15, 16, 17, 18, 21, 23, 25, 30, 47, 48, 51, 52, 57, 58, 59, 62, 63, 64, 65, 67, 69, 70, 72, 120, 123, 128, 130, 141, 144, 145, 146, 147, 148, 152, 158, 159, 161, 163, 164, 173, 181, 184, 188, 189, 198, 202, 203, 204, 205, 218, 229, 238, 239, 241, 247, 248, 249, 250, 254, 255, 266, 274
applied research, 15
aqueous solution, 12, 25, 36, 50, 54, 143, 156, 163, 177, 182, 194, 197, 198, 271
aqueous solutions, 12, 36, 177, 182, 197, 198
Archaea, 208
arginine, 250, 271
aromatic polyimide, 179, 196
aromatics, 186
arsenic, 39, 266
artificial organs, 2, 4, 5, 18
ascorbic, 58
ascorbic acid, 58
ash, 259
Asian, 9
Asian countries, 9
aspect ratio, 172
assessment, 1, 2
atmosphere, 30, 129, 144
atoms, 127
attachment, 207, 215
Australia, 237
availability, 15
awareness, 10, 64
azeotropic point, 164

B

bacteria, 18, 202, 206, 208, 209, 212, 213, 214, 218, 239, 249, 250
bacterial, 204, 214, 261, 262
bacterial cells, 204
band gap, 46
barrier, 5, 243
barriers, 3, 279
basic research, 14, 25
baths, 48, 52
batteries, 17, 48, 146, 244
battery, 48, 248
beet sugar, 251, 272
behavior, 46, 67, 133, 145, 146, 149, 151, 156, 159, 160, 191, 240, 282, 285, 286, 291
Belgium, 229, 230
benchmark, 184
benefits, 40, 51, 203
benign, 143
benzene, 39, 184, 186, 187, 188, 192, 194, 197, 198, 199
benzoyl peroxide, 28, 31
beverages, 249
bias, 40, 45
bicarbonate, 261
binding, 39, 130, 208, 212
bioactive compounds, 277
biochemistry, 120
biocompatibility, 147
biocompatible, 293

biodegradable, 157, 178, 210, 293
biodegradation, 240, 277
biofilms, 239, 240
biofuel, 188, 189
biofuels, 163
biogas, 10
biological macromolecules, 47
biological media, 205
biological processes, 266
biological systems, 291
biomass, 192, 203, 206, 207, 208, 209, 210, 218, 239, 257
biomass growth, 208, 209
biomedical applications, 12
biomimetic, 152
biomolecules, 152
biopolymer, 212
bioreactor, 67, 201, 202, 203, 204, 206, 215, 218, 220, 226, 227, 229, 230, 234, 235, 236, 237, 238, 239, 240, 241, 262, 263, 274
Bioreactor, v, 201, 202, 224, 261
bioreactors, vii, 17, 237, 238, 239, 240, 241
biosensors, 160
biotechnology, 8, 17, 52, 57, 199, 239, 243, 244, 245, 249, 267, 268
bipolar, 17, 21, 22, 24, 25, 27, 28, 39, 40, 41, 42, 43, 44, 45, 46, 47, 48, 58, 61, 62, 63, 64, 65, 66, 69, 70, 71, 72, 73, 251, 252, 254, 255, 256, 257, 258, 259, 270, 272, 273, 275, 276
bipolar membrane, 17, 21, 22, 24, 25, 27, 28, 39, 40, 41, 42, 43, 44, 45, 46, 47, 48, 58, 61, 62, 63, 64, 65, 66, 69, 70, 71, 72, 73, 251, 252, 254, 255, 256, 257, 258, 259, 270, 272, 275, 276
bipolar membrane (BPM), 39
birth, 123
bladder, 5
blends, 30, 37, 38, 196
block polymers, 29
blood, 9, 277, 289
blood glucose, 289
blood stream, 9
bloodstream, 8
BMA, 284
boiling, 143, 163, 164
Boltzmann constant, 223
bonding, 44, 66, 121, 122, 123, 126, 134, 135, 140, 159, 171, 175, 179, 211, 213, 286, 294
bonds, 121, 122, 125, 186, 286
Boron, 277
Boston, 190
boundary conditions, 221
bovine, 66
breakdown, 204
bromination, 33, 36, 37, 38
bromine, 34, 41
bubble, 283
bubbles, 202, 208
building blocks, 69, 149
burning, 254
butadiene, 39, 187
butyl ether, 186, 198
butyl methacrylate, 284
by-products, 1, 2, 9, 17, 61, 253, 262

C

cables, 49
cadmium, 248, 249
calcium, 65, 211, 212, 218, 240, 250, 251, 269
calcium carbonate, 269
Canada, 169, 226, 237
candidates, 145, 285, 293
capacity, 12, 28, 39, 46, 58, 64, 69, 127, 133, 143, 224, 225, 226, 227, 229, 231, 233, 234, 235, 236, 246, 266
capillary, 3, 4, 9, 206
capital cost, 246, 266
carbohydrate, 206, 210, 214, 215, 218
carbohydrates, 209, 211, 212, 215
carbon, 9, 15, 55, 59, 139, 143, 145, 149, 151, 157, 173, 186, 187, 188, 192, 199, 202, 218, 241, 247, 261, 262, 264, 266, 273
carbon dioxide, 59, 192, 202, 261, 273
carbon molecule, 143
carbon nanotubes, 139, 157, 186
carbon-fiber, 145
carboxyl, 177, 212, 259, 290
carboxyl groups, 212, 259, 290
carboxylic, 29, 42, 43, 139, 149
carcinogenic, 32, 188, 260
carrier, 3, 46, 279
CAS, 202, 203, 205
casein, 66
cast, 30, 31, 129, 130, 176, 193
casting, 27, 30, 41, 48, 129, 131, 134, 139, 174, 176, 205
catalysis, 45, 147
catalyst, 31, 36, 126, 131, 134, 135, 136
catalytic activity, 42

catalytic effect, 43, 45
catalytic properties, 144
cathode, 261
cation, 30, 39, 40, 41, 42, 43, 45, 47, 56, 58, 63, 65, 67, 68, 131, 210, 211, 212, 218, 239, 250, 251, 252, 254, 261, 265, 266, 271, 276
cell, vii, 5, 12, 51, 58, 64, 68, 69, 72, 128, 145, 146, 149, 152, 160, 161, 204, 206, 209, 210, 211, 213, 214, 217, 220, 252, 253, 261, 265, 272, 283
cell surface, 209, 213
cellular phone, 145
cellular phones, 145
cellulose, 6, 7, 120, 184, 185, 187, 195, 197, 199, 241, 249, 255
Cellulose, 152
ceramic, 3, 18, 120, 125, 137, 140, 157, 158, 159, 184, 185, 197, 198
ceramics, 122
cerium, 153
chain mobility, 181
charge density, 30, 47
charged particle, 212
cheese, 52, 67, 249, 250, 251, 271
chemical bonds, 121
chemical engineering, 149, 243, 268
chemical industry, 17, 18, 48, 57, 163
chemical interaction, 121, 123
chemical properties, 284
chemical reactions, 15
chemical stability, 16, 18, 25, 26, 41, 46, 251
chemical structures, 121, 123, 148
chemical vapor deposition, 139, 140, 158, 181
chemicals, 1, 2, 18, 24, 25, 42, 55, 57, 66, 188, 243, 244, 248, 253, 256, 261, 264, 266, 274, 279, 280, 293
chimera, 19
China, viii, 21, 30, 39, 49, 50, 58, 64, 73, 119, 148, 189, 232, 237, 279
chitin, 178
chitosan, 143, 150, 156, 172, 178, 186, 187, 192, 194, 195, 198
Chitosan, 66, 143
chloride, 9, 31, 39, 54, 55, 57, 58, 128, 210, 249, 270, 271, 291
Chloride, 54, 262
chlorinated hydrocarbons, 252
chlorine, 9, 188, 244, 246, 252, 261
chlorobenzene, 188
chloroform, 188
chromatography, 147
chromium, 72, 248, 260, 261, 265, 276
chronic renal failure, 8
circulation, 47, 59
Claisen, 62
classical, 124, 125, 212
classification, 22, 119, 120, 121, 122, 124
clay, 124
cleaning, 66, 137, 204, 206, 218, 220, 241
clusters, 135, 136, 172
coagulation, 234, 251, 266
coagulation process, 234
coal, 58, 254
coatings, 123, 146, 154
cohesion, 208
coil, 285
colloidal particles, 125, 136
colloidal theory, 212
colloids, 207, 208
commercialization, 120
communities, 206, 213
community, 207
compatibility, 3, 132
compensation, 176
competence, vii, 52
competition, 18
competitiveness, 243, 244, 268
complexity, 136, 144, 148, 176, 181, 204
compliance, 289
components, 3, 5, 8, 9, 12, 15, 25, 26, 27, 31, 46, 56, 121, 123, 124, 125, 128, 139, 140, 141, 147, 148, 164, 166, 167, 168, 169, 173, 186, 189, 202, 206, 207, 208, 209, 214, 218, 243, 244, 245, 248, 254, 260, 264, 266, 268
composites, 124, 151
composition, 10, 134, 164, 167, 173, 174, 175, 176, 178, 179, 209, 212, 213, 239, 251
compounds, 46, 62, 121, 123, 136, 147, 150, 182, 188, 202, 203, 214, 215, 218, 260, 277
concentrates, vii, 43, 66, 145, 246, 252, 271
concentration, 1, 2, 4, 5, 6, 10, 11, 16, 17, 21, 23, 24, 36, 43, 44, 45, 47, 48, 50, 52, 53, 55, 58, 60, 72, 164, 165, 166, 167, 168, 169, 170, 177, 179, 180, 183, 184, 185, 187, 188, 195, 199, 206, 207, 209, 211, 214, 215, 218, 221, 235, 236, 239, 241, 244, 246, 247, 250, 252, 256, 261, 265, 266, 267, 270, 271, 272, 275, 276, 279, 280, 287, 289, 290, 291, 292
condensation, 24, 58, 61, 62, 123, 126, 130, 132, 135, 151, 153, 267

conditioning, 5, 172
conductance, 57
conductive, 27, 69, 154
conductivity, 6, 25, 32, 37, 41, 58, 124, 127, 145, 146, 264
configuration, 58, 60, 64, 173, 189, 204, 206, 213, 218, 274
conflict, 213
conservation, 252
construction, 14
consumption, 2, 11, 54, 60, 63, 203, 204, 247, 251, 252, 253, 260, 267
contaminant, 188
contaminants, 202, 262
contamination, 54, 254, 262, 264
control, 4, 6, 17, 30, 50, 57, 68, 126, 157, 176, 177, 188, 201, 208, 215, 218, 221, 224, 239, 255, 256, 259, 267, 274, 279, 282, 285, 289, 290, 292
convective, 5, 176
conversion, 2, 4, 5, 16, 17, 24, 25, 29, 36, 45, 58, 188, 244, 256, 257, 260, 261, 273, 276, 290
conversion rate, 58, 188
cooling, 246, 247, 277, 288
copolymer, 29, 68, 133, 156, 185, 284, 291
copolymerisation, 68
copolymerization, 26, 32, 38, 39, 68, 133, 137
copolymers, 69, 132, 133, 153, 184
copper, 145, 147, 154, 248, 249, 260, 261, 265, 275, 276
core-shell, 281, 282, 291, 293
corona, 140
corona discharge, 140
correlation, 172, 191, 215
corrosion, 246, 260
Corynebacterium, 213
cost saving, 243
cost-effective, 61
costs, 4, 7, 8, 10, 11, 12, 13, 16, 17, 18, 57, 58, 206, 245, 246, 247, 248, 253, 255, 264, 266
coupling, 67, 123, 153, 154, 155, 172, 191
covalent, 121, 122, 123, 136, 140, 143, 155
covalent bond, 122, 123, 136, 140, 143
covalent bonding, 140
CRC, 191
critical state, 271
crosslinking, 30, 31, 32, 33, 36, 39, 42, 69, 72, 159, 175, 176, 177, 181, 182, 184, 189
crude oil, 188, 247
cryogenic, 10
crystalline, 135, 149, 192, 251
crystalline solids, 149
crystallization, 58, 246, 250, 266
culture, 239, 262
culture conditions, 239
curing, 124, 129, 130, 142, 154, 178
curing process, 124
CVD, 139, 140, 158
cycles, 247
cyclodextrin, 198
cyclodextrins, 198, 291
cyclohexane, 186, 187, 192, 194, 198, 199

D

dairy, 47, 65, 72, 258
dairy industry, 65, 72
danger, 217
Darcy, 216, 219
database, 217
death, 289
decay, 209, 210
decomposition, 53, 135, 202, 250
defects, 176, 177, 178, 181, 182
definition, 22, 168, 172, 208
degradation, 140
degradation process, 140
dehalogenation, 57
dehydration, 125, 156, 163, 164, 169, 170, 172, 175, 177, 178, 179, 181, 182, 183, 184, 186, 188, 189, 190, 191, 192, 193, 194, 195, 196, 197
delivery, 5, 279, 280, 286, 287, 288, 289, 294
demineralized, 247, 251
denaturation, 250
dendrimers, 43, 71
denitrification, 225, 240, 261, 262, 267, 274
denitrifying, 262
density, 4, 6, 16, 29, 41, 42, 43, 44, 45, 60, 120, 125, 145, 173, 267, 275
deposition, 24, 136, 139, 140, 158, 174, 181, 195, 215
deposits, 251
depressed, 143
derivatives, 153, 270
desalination, vii, 1, 2, 6, 7, 8, 9, 12, 14, 15, 16, 18, 21, 24, 52, 64, 70, 120, 238, 244, 245, 246, 247, 251, 266, 268, 269, 271, 277
detachment, 220
detection, 145

dextrose, 58
diabetes, 289, 291
diabetes mellitus, 289
diabetic patients, 289
dialysis, 8, 14, 15, 21, 27, 48, 49, 50, 51, 63, 69, 72, 147, 244, 265, 266, 275, 276
diamines, 178, 179, 181, 182
Diamond, 158
dianhydrides, 137, 179
dichloroethane, 32
dielectric constant, 130
diffusion, 5, 27, 48, 49, 50, 51, 54, 63, 69, 72, 120, 123, 142, 143, 145, 147, 150, 163, 164, 165, 166, 167, 171, 172, 173, 174, 175, 188, 190, 191, 192, 195, 218, 221, 222, 244, 262, 263, 265, 275, 276, 279, 280, 283, 287, 292
diffusion permeability, 279, 292
diffusion process, 145
diffusion time, 123, 150
diffusional permeability, 290
diffusivity, 15, 184
dimethacrylate, 68
discharges, 21
discrimination, 144
dispersion, 121, 134, 171
displacement, 212
dissociation, 21, 25, 40, 41, 42, 43, 45, 57, 58, 60, 61, 71, 72, 275, 286, 288
dissolved oxygen, 213, 225
distillation, 5, 6, 9, 11, 12, 24, 163, 164, 188, 190, 245, 250, 266, 267, 277
distribution, 66, 207, 214, 240, 273, 282
diversity, 119, 147
DNA, 209
doctors, 16
donor, 181, 262, 263
dopamine, 12
doped, 146, 151, 154, 155
dosage, 284
double bonds, 186
downsizing, 246
dream, 7, 9
drinking, 17, 67, 261, 262, 263, 274
drinking water, 17, 67, 261, 262, 263, 274
drug delivery, 5, 18, 279, 280, 282, 287, 291
drug delivery systems, 5, 18, 282, 287, 291
drug release, 282, 293
drug treatment, 280
drugs, 1, 2, 12, 279, 280, 285, 286, 287, 288, 291, 293
drying, 9, 58, 125, 126, 129, 131, 134, 139, 176, 177
durability, 144, 147, 174, 184
dyes, 139

E

E. coli, 225
earth, vii, 5, 48, 50, 51, 158, 260
ecology, 270
economics, 17, 26, 32, 50, 52, 194, 249, 267, 277
ecosystems, 40
efficiency level, 244, 267
effluent, 40, 203, 208, 209, 218, 245, 246, 248, 249, 250, 252, 255, 256, 260, 261, 262
effluents, 1, 2, 4, 9, 17, 21, 48, 57, 236, 238, 240, 245, 248, 260, 265, 266, 267, 268, 274, 276
egg, 212
elaboration, 159
electric charge, 252
electric circuit, 47
electric field, 70, 252, 258, 280
electric power, 266
electrical properties, 46
electrical resistance, 24, 25, 26, 28, 37, 43, 55, 63
electricity, 58
Electroanalysis, 160
electrochemical impedance, 145
electrochemical reaction, 213, 244, 266
electrochemistry, 160
electrodeposition, 180, 181
electrodes, 24, 58, 64, 144, 152, 160
electrodialysis with bipolar membrane, 18, 21, 40, 58, 59, 64, 65, 66, 243, 244, 272, 273, 276
electrolysis, 9, 24, 48, 51, 72, 244, 252, 261, 265, 275
electrolyte, 6, 13, 23, 40, 42, 47, 55, 57, 60, 64, 65, 68, 69, 70, 145, 146, 149, 151, 152, 161
electrolytes, 5, 47, 65, 145, 151, 161, 244
electromigration, 250, 265
electron, 181, 199, 210, 262, 263
electron beam, 199
electrophoresis, 56
elongation, 142, 177
emission, 145, 188, 189
empirical methods, 220
emulsification, 214, 282
emulsions, 282
encapsulated, 291
encapsulation, 282

encouragement, viii
energy, vii, 1, 2, 4, 5, 7, 10, 11, 13, 16, 17, 18, 21, 24, 51, 55, 58, 60, 63, 120, 143, 145, 146, 163, 164, 172, 173, 184, 189, 203, 204, 243, 244, 245, 247, 252, 253, 260, 266, 267, 268, 271
energy consumption, 2, 11, 60, 63, 204, 247, 252, 253, 260
energy efficiency, vii, 2, 163, 184, 243
energy supply, vii
entanglement, 176
enthusiasm, 7
environment, 3, 46, 51, 61, 66, 143, 172, 191, 209, 248, 260, 270, 273, 288
environmental awareness, 10, 64
environmental impact, 9
environmental protection, vii, 24, 63, 120, 265
environmental stimuli, 279, 280, 287, 289, 291, 292, 293, 294
enzymatic, 260, 273
enzyme immobilization, 140, 141, 158
enzymes, 139, 202, 290
EPA, 228
epoxy, 29, 38, 39, 129, 130, 132, 133, 155
epoxy groups, 38, 39, 130, 133
equilibrium, 5, 19, 23, 65, 167, 169, 172, 192
ester, 57, 61, 62, 63, 67, 131, 188
esters, 188
etching, 49, 52, 72
ethane, 145, 147
ethanol, 5, 55, 56, 131, 143, 156, 164, 171, 172, 178, 181, 184, 188, 189, 190, 191, 194, 195, 197, 198, 263, 274
Ethanol, 129, 180, 183, 185, 199, 262, 263
ethanolamine, 39
ethers, 188, 291
ethyl acetate, 184, 188
ethylene, 132, 135, 156, 158, 178, 194
ethylene glycol, 156, 178, 194
ethylene oxide, 135, 156
ethylenediamine, 44, 281
Europe, 9, 202, 265
europium, 151
evaporation, 57, 65, 176, 193, 246, 269
evolution, 172, 261
exchange rate, 177
exclusion, 24, 244
expert, iv
extraction, 24, 48, 51, 210, 211, 213, 215, 250, 258, 261, 262, 269
extrusion, 41, 175

F

fabric, 45, 46, 47, 188
fabricate, 134, 139, 178, 182
fabrication, 124, 174, 175, 176, 177, 181, 182, 183, 189, 197, 282, 284
failure, 9
family, 285
fat, 52, 249
feedstock, 40, 246
fermentation, 25, 47, 48, 51, 52, 56, 58, 64, 66, 163, 188, 189, 192, 199, 249, 250, 253, 256, 257, 258, 270, 272, 273
fermentation broth, 25, 52, 64, 66, 163, 188, 189, 250, 256, 257, 258, 270, 272, 273
fermentation technology, 58
ferric ion, 266
fiber, 4, 6, 9, 45, 64, 142, 147, 150, 152, 158, 176, 177, 178, 179, 181, 182, 192, 193, 196, 204, 226, 227, 228, 229, 230, 231, 232, 233, 235, 236, 237, 238, 273
fiber membranes, 150, 192, 193, 196, 204, 229, 231, 235, 236
fibers, 3, 14, 45, 142, 175, 176, 178, 179, 181, 182, 192, 193, 194, 196, 204, 229
filler particles, 182
fillers, 142, 150, 184
film, 19, 30, 31, 39, 48, 70, 139, 142, 146, 152, 187
films, 3, 29, 68, 70, 144, 149, 156, 174, 179, 182
filters, 147
filtration, 4, 8, 142, 159, 201, 204, 205, 215, 216, 217, 220, 223, 224, 237, 238, 240, 241, 246, 247, 253, 256, 262, 265, 269
flavor, 252, 259
flexibility, vii, 31, 32, 46, 120, 121, 140, 265
floating, 228
flocculation, 55, 208, 212, 213, 214, 218, 239, 266
flow, 11, 18, 46, 47, 50, 59, 142, 150, 165, 174, 176, 204, 206, 216, 218, 222, 223, 229, 233, 234, 238, 261, 265, 266, 269, 275
flow rate, 176, 266, 269, 275
fluctuations, 203, 281
flue gas, 59, 253, 254, 255
fluid, 164, 175, 176, 177, 192, 241
fluorescence, 181
fluoride, 185, 276, 283, 293

fluorinated, 65
fluorine, 9
flushing, 18
food, vii, viii, 2, 14, 15, 16, 17, 21, 24, 47, 48, 52, 57, 65, 66, 241, 244, 246, 249, 250, 251, 252, 260, 266, 267, 268
food industry, 48, 52, 66, 249, 250, 251, 268
food processing industry, 18
Ford, 157
formaldehyde, 24, 178, 210
fouling, 3, 4, 16, 17, 46, 47, 72, 142, 144, 159, 201, 203, 204, 206, 207, 208, 215, 216, 217, 218, 219, 220, 238, 239, 240, 241, 245, 249, 250, 251, 253, 267
Fourier, 215, 283
fractionation, 67, 207, 220
fragility, 139
France, 58, 236, 237, 238
free energy, 13
free radical, 288
free volume, 30, 143, 156, 169, 172, 173, 179, 186
free-radical, 137, 138
fresh water, 13, 17, 66, 268
fruit juice, 48, 52, 244, 249, 251, 252, 259, 266, 268, 272
fruit juices, 52, 249, 252, 268, 272
FTIR, 215
FT-IR, 283
fuel, vii, 2, 10, 17, 24, 30, 37, 45, 68, 69, 72, 73, 119, 120, 128, 145, 146, 149, 151, 152, 153, 158, 160, 161, 244
fuel cell, vii, 2, 10, 17, 24, 30, 37, 45, 68, 69, 72, 119, 120, 128, 145, 146, 149, 151, 152, 153, 158, 160, 161, 244
fumaric, 177, 178
functionalization, 30, 120, 124
funding, 189
fungi, 208
fusion, 31

G

gas, 4, 10, 11, 12, 15, 17, 18, 19, 59, 73, 119, 120, 121, 123, 140, 141, 142, 148, 149, 150, 152, 154, 156, 157, 158, 159, 164, 167, 174, 176, 178, 181, 182, 192, 193, 194, 195, 196, 246, 252, 253, 254, 255, 260, 261, 273
gas diffusion, 120
gas phase, 167
gas separation, 4, 10, 12, 15, 17, 19, 119, 121, 123, 141, 142, 148, 149, 150, 154, 156, 157, 158, 159, 174, 176, 178, 181, 182, 192, 193, 194, 196
gases, 5, 10, 19, 120, 142, 167, 189, 244, 252
gastrointestinal, 286
gastrointestinal tract, 286
gel, 65, 119, 124, 125, 126, 127, 129, 130, 131, 132, 133, 134, 135, 136, 142, 143, 144, 145, 147, 150, 152, 160, 161, 208, 212, 214, 219, 286, 291
gel permeation chromatography, 214
gelatin, 30, 43, 71
gelation, 125, 126, 147
gels, 19, 65, 139, 212, 279, 286
generation, 44, 45, 71, 246, 250, 252, 258, 264, 266, 272
generators, 145
Georgia, 235
Germany, 1, 189, 226, 229, 235, 237
Gibbs, 13
glass, 15, 125, 142, 145, 147, 150, 151, 152, 282
glass transition, 15
glass transition temperature, 15
glasses, 151
glassy polymers, 273
global warming, 260
glucoamylase, 158
glucose, 52, 58, 152, 279, 280, 289, 290, 292, 294
glucose oxidase, 152, 289
glutamic acid, 250
glutamine, 52
glutaraldehyde, 177, 178
glycine, 52, 250, 258
glycogen, 204
glycol, 43, 155
GLYMO, 128
GPC, 214
grafting, 27, 29, 32, 39, 68, 70, 135, 136, 137, 141, 177, 184, 283, 284, 286, 287, 289, 290
grape juice, 251
graph, 11
graphite, 145, 160, 173, 186, 187, 192
Gravitation, 177
gravity, 202
greenhouse, 260
greenhouse gas, 260
greenhouse gases, 260
ground water, 261

groups, 22, 23, 24, 26, 27, 29, 31, 33, 35, 36, 38, 39, 41, 42, 43, 44, 45, 47, 48, 62, 65, 71, 121, 122, 123, 125, 127, 130, 131, 132, 133, 134, 136, 137, 139, 140, 142, 143, 147, 149, 173, 177, 179, 181, 186, 212, 259, 281, 290, 291
growth, 18, 20, 135, 139, 143, 202, 208, 209, 213, 226, 239, 262
growth rate, 202
guidelines, 168

H

handling, 121, 137
hardness, 124, 247, 256, 270, 273
health, 17, 32, 261, 266
health care, 17
health care system, 17
heart, 120, 270
heat, 12, 40, 60, 130, 164, 175, 178, 179, 181, 182, 210, 247, 258
heating, 59, 129, 134, 139, 140, 183, 197, 288
heavy metal, 9, 42, 45, 66, 70, 248, 249, 253, 254, 260, 264, 265, 266, 271, 274, 275
heavy metals, 66, 70, 248, 249, 253, 260, 264, 265, 266, 275
hemodialysis, 1, 2, 4, 6, 8, 9, 14, 15
heptane, 186, 198
heterogeneous, 13, 15, 18, 23, 123, 150
hexane, 186, 198
high pressure, 11
high temperature, vii, 125, 140, 145, 174, 179, 182, 183, 247
higher quality, 64, 202
histidine, 250
homogeneity, 132, 134
homogenous, 23, 31, 69, 176
hospitals, 15
host, 121, 122, 135, 291
household, 260
household waste, 260
housing, 6, 231
human, vii, 32, 120, 124, 251, 266
humans, 286
humic acid, 209, 240
humidity, 146, 179
hybrid, vii, 5, 21, 25, 46, 65, 68, 72, 119, 121, 122, 123, 124, 125, 126, 127, 128, 129, 130, 131, 132, 133, 134, 135, 139, 140, 141, 142, 143, 144, 145, 146, 147, 148, 149, 150, 151, 152, 153, 154, 155, 156, 157, 158, 160, 161, 164, 189, 194, 199, 241, 246, 260, 266, 269, 275, 277
hybrids, 23, 46, 130, 147, 148, 152, 153, 155, 160
hydro, 12, 44, 47, 172, 177, 178, 182, 183, 186, 188, 190, 194, 196, 213, 214, 239, 252, 273, 281, 284
hydrocarbon, 22, 23, 29, 36, 69, 120, 151, 214
hydrocarbons, 12, 186, 214, 273
hydrochloric acid, 49, 129, 253, 275
hydrodynamic, 16, 206
hydrodynamics, 238
hydrogels, 281, 285, 286, 287, 288, 289
hydrogen, 10, 40, 42, 44, 61, 120, 121, 123, 129, 134, 135, 137, 139, 143, 149, 152, 158, 171, 175, 179, 244, 261, 286, 294
hydrogen bonds, 286
hydrogen gas, 261
hydrogen peroxide, 129, 137, 152
hydrolysates, 258, 271
hydrolysis, 123, 125, 126, 131, 135, 147, 151, 179, 240
hydrolyzed, 136, 209
Hydrometallurgy, 50
hydrophilic, 44, 47, 172, 177, 178, 182, 183, 186, 188, 190, 194, 196, 213, 214, 239, 281, 284
Hydrophilic, 164, 177, 196
hydrophilic groups, 177
hydrophilic materials, 172, 177, 182, 188, 214
hydrophilicity, 44, 127, 133, 143, 172, 179, 181, 182, 213, 291
hydrophobic, 12, 39, 178, 182, 184, 198, 206, 213, 214, 260, 281, 284
Hydrophobic, 182, 239
hydrophobic interactions, 213
hydrophobic properties, 281
hydrophobicity, 130, 172, 182, 206, 207, 213, 214, 215, 218
hydrostatic pressure, 5, 6, 13, 47
hydrothermal, 183
Hydrothermal, 197
hydrothermal synthesis, 183
hydroxide, 52, 54, 58, 210, 257, 261, 262, 274, 276
hydroxyl, 37, 42, 44, 136, 137, 177
hydroxyl groups, 44, 136, 137
hysteresis, 191
Hyundai, 227

I

identification, 120
imide rings, 179
imidization, 179
impedance spectroscopy, 145
impregnation, 139
impurities, 48, 50, 58, 164
in situ, 46, 135, 145, 157, 159, 174, 178
in vitro, 65
incompressible, 216
Indiana, 237
indicators, 169
indium, 42
industrial, vii, 1, 2, 4, 6, 9, 11, 14, 17, 18, 20, 21, 24, 27, 40, 48, 49, 57, 63, 70, 73, 120, 174, 177, 188, 197, 225, 227, 229, 230, 236, 240, 241, 243, 244, 245, 246, 247, 248, 249, 253, 256, 258, 260, 261, 262, 264, 265, 266, 267, 268, 270, 274, 276
industrial application, viii, 11, 24, 174, 177, 243, 244, 245, 267, 268
industrial production, 1, 17, 197
industrial sectors, 14
industrial wastes, 70
industrialized countries, 289
industry, vii, 1, 2, 6, 7, 8, 9, 12, 13, 14, 15, 16, 17, 18, 48, 49, 51, 52, 57, 65, 66, 72, 163, 186, 188, 240, 245, 248, 249, 250, 251, 253, 260, 264, 265, 268
inefficiency, 12
inert, 29, 32, 39, 182, 209, 210, 239
infinite, 11, 215
inflammation, 286
infrared, 215, 283
inhibition, 203, 210, 259, 273
inhibitory, 188
initiation, 123
injection, 282, 291
injections, 289
inorganic, 23, 25, 46, 51, 52, 53, 55, 64, 67, 119, 120, 121, 122, 123, 124, 125, 126, 129, 131, 134, 135, 136, 137, 138, 139, 140, 141, 142, 143, 144, 145, 146, 147, 148, 149, 150, 151, 152, 153, 154, 155, 156, 157, 158, 159, 160, 161, 182, 183, 184, 185, 186, 189, 192, 194, 274, 275
inorganic filler, 142, 148, 156, 194
inorganic fillers, 142, 148, 156, 194
inorganic salts, 53, 55
insight, 169, 186, 189
instability, 175, 178, 179, 183, 188, 260
institutions, 14, 27
insulin, 12, 289, 290, 291, 294
integration, vii, 17, 24, 25, 48, 139, 166, 175, 243, 244, 249, 258, 266, 267, 268, 277
integrity, 174, 175
intensity, 207, 210, 265
interaction, 39, 44, 135, 166, 167, 171, 172, 173, 182, 186, 212, 282, 286
interactions, 15, 44, 121, 123, 135, 163, 166, 170, 171, 172, 173, 175, 179, 189, 191, 206, 212, 213, 215, 286, 294
intercalation, 46
interdisciplinary, 64
interface, 44, 46, 71, 72, 139, 143, 167, 175
interfacial layer, 42, 43, 44
intermolecular, 286
interphase, 124
interpretation, 5, 19, 44, 168, 213
interstitial, 166, 173, 178, 182, 286
interval, 168
intrinsic, 1, 36, 37, 38, 41, 144, 147, 168, 169, 174, 189, 219
inversion, 31, 176
investment, 13, 57, 253
ion transport, 24, 43, 56
Ion-exchange, 17, 19, 46, 64, 68, 72, 73, 268, 270, 271
ionic, vii, 21, 22, 23, 25, 26, 27, 29, 66, 121, 123, 124, 135, 146, 153, 212, 246, 250, 260, 262, 273, 275, 280, 291
ionic solutions, 26
ionics, 150
ionizable groups, 71
ions, 5, 13, 22, 26, 40, 42, 45, 47, 50, 53, 55, 57, 61, 131, 147, 154, 211, 212, 218, 244, 246, 248, 249, 250, 251, 252, 254, 256, 258, 259, 260, 261, 262, 265, 266, 272, 274, 275, 291, 292
iron, 152, 261, 273
irradiation, 68, 70, 199
isoelectric point, 53, 250
isoleucine, 250
isomers, 186, 189, 190, 198
isothermal, 166
isotope, 120
Italy, 226
ITT Industries, 228
I-V curves, 46, 69, 70

J

Japan, 9, 30, 189, 202, 231, 232, 246, 265
Jefferson, 237, 238, 276
Juices, 251, 259, 268
Jung, 160

K

kidney, 8, 10, 19
kidneys, vii, 2, 9, 15
kinetics, 59, 126, 147, 240, 282
KOH, 253, 256
Korea, 67, 201, 224, 227, 228, 229, 230, 234, 235, 240, 241, 243, 268

L

L2, 217
labor, 203
laboratory studies, 12
lactic acid, 52, 58, 250, 256, 257, 270, 273
lactose, 250
Lafayette, 237
lamina, 222
laminar, 222
laminated, 22
land, 149
Langmuir, 72, 157, 293
large-scale, 6, 7, 12, 13
law, 216, 241
leaching, 50, 51, 276
leakage, 51, 52, 54, 55, 258, 259
lenses, 151
leucine, 250
life expectancy, 9
ligands, 12, 145, 154
limitation, 1, 30, 44, 63, 212, 213, 287
limitations, 10, 145
linear, 25, 172, 196, 223, 281, 282, 287, 289, 290, 291
linkage, 131, 137, 139, 140
lipids, 67
liquid film, 3
liquid phase, 244
liquid water, 12
liquids, 12, 120, 164, 244
liquor, 48, 49, 50, 51, 69, 72, 206, 215, 251, 255, 275
lithium, 67
localization, 209
London, 19, 64, 153, 160, 237
losses, 258
Low cost, 4
low molecular weight, 8, 12, 130, 139
low temperatures, 124
low-temperature, 161, 182
LTC, 65, 72
LTD, 227
luminescence, 46, 124, 151
lysine, 52, 250
lysis, 209, 210

M

macromolecules, 43, 44, 47, 147, 240
magnesium, 211, 212
magnetic, iv, 46, 124, 215, 216
magnetic properties, 46
maintenance, 4, 51, 202, 203, 205, 218
malic, 252
management, 260
man-made, 120
manufacturer, 226
manufacturing, 15
MAO, 152
mapping, 283
market, 1, 2, 4, 9, 11, 12, 13, 14, 15, 17, 18, 224
market segment, 13
marketing, 15
markets, 9, 14, 15, 18
mass transfer, 145, 165, 168, 172, 221, 260
material resources, 10
materials science, 148
matrix, 12, 13, 22, 23, 26, 29, 30, 31, 42, 47, 121, 122, 123, 134, 135, 139, 141, 143, 145, 146, 147, 148, 150, 158, 159, 173, 174, 178, 182, 185, 186, 192, 193, 194, 196, 206, 207, 211, 212, 213, 239, 285, 286
Maya, viii
MDI, 175, 191, 193
measurement, 214
measures, 53
meat, 258
mechanical properties, 24, 142, 157
MED, 245
media, 5, 18, 25, 63, 67, 142, 205, 216, 259
medicine, 6, 17
melt, 139, 151

membrane permeability, 19, 167, 215, 241, 285
membrane separation processes, 9, 16, 19, 52, 64, 144, 164, 202, 244, 265, 266, 267, 268
mercury, 248, 283
metabolic, 262
metabolism, 209, 261
metal ions, 50, 147, 248, 261, 265, 274, 275, 291
metal oxide, 265
metal oxides, 265
metal processing industries, 48
metal recovery, 48
metals, 48, 50, 52, 248, 260, 265, 276
methacrylic acid, 29, 68, 135, 286, 287
methanol, 25, 30, 37, 40, 45, 57, 61, 62, 67, 68, 69, 129, 143, 145, 150, 152, 158, 160, 186, 188, 198
methemoglobinemia, 261
methyl group, 32, 36, 136
methyl groups, 32, 136
methyl methacrylate, 151
methyl viologen, 145, 160
MFC, 128, 145
MFI, 216, 217
microbes, 183, 212
microbial, 203, 204, 206, 208, 209, 211, 215, 239, 240, 241, 262
Microbial, 201, 208, 234, 239
microelectrode, 145, 160
microemulsions, 157
microenvironment, 290
microgels, 287, 288
microorganism, 206, 207, 213, 241
microorganisms, 202, 203, 204, 208, 209, 262
microscopy, 148
microspheres, 279
microstructure, 283, 293
microstructures, 284
microwave, 197
microwave heating, 197
migration, 5, 56, 262
milk, 52, 66, 249, 251, 271
minerals, 238, 250
mining, 47, 48
Mitsubishi, 232
mixing, 13, 17, 139, 214, 273
mobility, 25, 46, 52, 67, 166, 181
modeling, 16, 173, 189, 270
models, 16, 159, 165, 191, 201, 218, 223
modules, 4, 6, 14, 17, 174, 202, 204, 227, 229, 232, 273
modulus, 142, 173
moieties, 27, 29, 44, 182
moisture, 126
molar ratio, 284
molar volume, 166
molasses, 52, 249, 272
mold, 33
mole, 168, 170
molecular structure, 73, 179
molecular weight, 8, 12, 44, 72, 130, 139, 155, 157, 202, 207, 214, 217, 240, 241, 276
molecular weight distribution, 207, 214, 240
molecules, 5, 40, 42, 44, 47, 52, 61, 135, 136, 137, 142, 147, 155, 165, 166, 167, 171, 172, 173, 179, 183, 208, 286, 291
molybdenum, 52
monoamine, 152
monoamine oxidase, 152
monolayer, 136, 159
monomer, 26, 27, 29, 30, 31, 39, 68, 69, 130, 137, 173, 284, 288
monomers, 28, 29, 31, 32, 38, 39, 68, 70, 123, 132, 137, 139, 140, 284
Montenegro, 152
morphological, 68, 148, 282
morphology, 30, 139, 148, 156, 157, 163, 172, 174, 176, 177, 196, 282
mosaic, 13, 21, 22, 24, 25, 47, 72
motion, 166, 169, 173, 204
MPS, 129, 130, 133, 134
MTBE, 186, 188, 198
MWD, 207, 214, 218

N

Na2SO4, 61, 254, 255
NaCl, 52, 54, 144, 244, 246, 249, 282
Nafion, 24, 135, 136, 145, 146, 157, 158, 160
nanocomposites, 151, 152, 155, 157, 160, 161
nanometers, 135, 136, 144
nanoparticles, 128, 144, 149, 150, 158, 159, 178, 182
Nanostructures, 159
nanotube, 173, 186, 187, 192, 199
nanotubes, 139, 157, 186
National University of Singapore, 163
natural, 5, 10, 18, 66, 196, 208, 268
natural environment, 208
natural gas, 10, 18, 196
natural resources, 18

Netherlands, 8, 233, 237
network, 30, 68, 122, 125, 126, 127, 129, 132, 133, 136, 140, 142, 143, 145, 151, 153, 176, 289
neutralization, 57, 64, 248, 249, 253, 256, 258, 261
next generation, 47
nickel, 50, 51, 72, 240, 248, 260, 261, 265, 275
Nielsen, 208, 239
nightmares, 7
nitrate, 42, 261, 262, 263, 274, 276
nitric acid, 48, 129
nitrification, 225, 240
nitrifying bacteria, 202
nitrogen, 10, 11, 233, 237, 261
NMR, 148, 215
non-crystalline, 149
nonelectrolytes, 47
nonlinear, 151
non-Newtonian, 223, 241
non-Newtonian fluid, 223, 241
normal, 26, 204, 217, 231, 235, 251, 279, 286, 289
North America, 202
NTU, 229, 234, 236
nuclear, 215, 274
nuclear magnetic resonance, 215, 216
nuclear power, 274
nuclear power plant, 274
nucleation, 135
nuclei, 36
nucleic acid, 208
nutrient, 213
nutrients, 240, 250, 260, 262
NVD, 70
nylon, 150, 286

O

octane, 42
ODS, 137
oil, 142, 157, 159, 188, 189, 247, 282
oligomer, 157
oligomeric, 132, 151, 156, 157
oligomers, 123, 131, 132, 147
Oman, 231
online, 1
optical, 124, 127, 130, 147, 151, 196
optical properties, 127, 151, 196
optimization, 17, 64, 171, 262, 266, 270
orange juice, 277
organic chemicals, 25
organic compounds, 16, 17, 165, 188, 197, 214, 254, 256, 261
organic matter, 66, 240, 264, 268
organic polymers, 25, 46, 120, 121, 124, 125, 139, 143, 147, 148, 149, 194
organic solvent, 10, 17, 25, 147, 163, 172, 173, 177, 179, 182, 184, 188, 194
organic solvents, 10, 17, 25, 147, 172, 173, 177, 179, 182, 184, 188, 194
organism, 5, 16
organization, 15
orientation, 177
osmosis, 2, 4, 5, 6, 7, 8, 9, 12, 13, 14, 16, 17, 19, 24, 47, 50, 56, 66, 128, 148, 160, 218, 243, 244, 245, 267, 268, 269
osmotic, 5, 12, 13, 193
osmotic pressure, 5, 12, 13
oxidation, 58, 59, 61, 123, 140, 152, 161, 202, 208, 252, 269, 290
oxidative, 140
oxide, 31, 33, 36, 41, 42, 49, 69, 70, 122, 123, 126, 131, 132, 135, 136, 139, 146, 152, 155, 156, 157, 158, 159, 160
oxides, 67, 125, 146
oxygen, 10, 11, 12, 203, 204, 208, 209, 213, 214, 218, 225, 244, 254, 255, 261
oxygen consumption, 203
ozonation, 247
ozone, 266

P

PAA, 172, 177
palladium, 120, 161
PAN, 181, 188, 196
paper, 18, 120, 214, 240, 249
parameter, 171, 172, 215, 275
parenteral, 289
Paris, 19, 198
particles, 18, 47, 125, 134, 135, 136, 139, 143, 146, 151, 173, 178, 182, 189, 204, 206, 212, 220, 222, 223, 224, 248, 249, 264, 265, 281, 282, 287, 288
patents, 28, 72
pathways, 47
patients, 9, 289
patterning, 153
PBI, 175, 176, 178, 179, 181, 193, 196

PCP, 140
PDMS, 123, 140, 142, 159, 184, 185, 197, 198
PEEK, 182
peptide, 67
peptides, 67
performance, 1, 2, 3, 7, 12, 16, 42, 43, 69, 71, 72, 73, 120, 121, 149, 150, 154, 163, 167, 168, 169, 170, 171, 172, 173, 174, 175, 176, 177, 178, 179, 180, 181, 182, 183, 184, 185, 186, 187, 188, 189, 190, 191, 193, 195, 196, 197, 221, 225, 229, 231, 236, 238, 240, 241, 249, 250, 256, 266, 267, 272, 274, 275, 283, 285
permeability, 3, 19, 26, 27, 37, 38, 47, 65, 72, 120, 121, 139, 142, 143, 144, 149, 156, 157, 159, 166, 167, 168, 169, 173, 177, 178, 183, 184, 189, 192, 194, 215, 216, 241, 263, 279, 281, 285, 290, 292, 293, 294
permeant, 163, 165, 167, 168, 170, 171, 172, 189
permeation, 13, 123, 142, 143, 145, 150, 156, 157, 159, 168, 173, 174, 183, 184, 186, 190, 192, 193, 195, 197, 198, 199, 214, 260, 267, 280, 283, 286, 290
peroxidation, 70
personal, 145
pesticide, 9
petrochemical, 12, 18, 185, 186, 188
petroleum, 188, 189
pH, 16, 26, 47, 53, 54, 132, 133, 147, 212, 215, 228, 247, 250, 255, 256, 257, 258, 259, 260, 261, 273, 279, 280, 286, 287, 288, 289, 290, 292, 294
pH values, 286
pharmaceutical, viii, 2, 18, 21, 52, 188
pharmaceutical industry, 2
pharmaceuticals, 47, 152
phase inversion, 31, 176, 177, 193, 195
phase transitions, 287
phenol, 24, 197
phenolic, 42, 277
phenylalanine, 154, 250, 258, 271
pheromone, 9
phosphate, 25, 213, 250, 260
Phosphate, 213
phosphates, 271
phosphorous, 233
photobleaching, 151
photoconductivity, 46
photovoltaic, 152
physical environment, 280
physical properties, 147
physicochemical, 46, 167, 169, 171, 172, 206, 211, 252, 259, 261, 266
physicochemical properties, 46, 169, 171, 172
physics, 148
physiological, 206, 238
physiology, 289
pig, 5
planar, 145
plants, 7, 10, 15, 16, 57, 120, 188, 227, 246, 253, 264, 266, 269
plasma, 68, 139, 140, 158, 159, 281, 282, 283, 284, 287, 290, 293
plasticization, 169, 172, 196, 273
plasticizer, 31
plastics, 154
platinum, 161
play, vii, 1, 2, 141, 145, 165, 204, 211, 248
PMA, 131, 155
PMDA, 180
PMMA, 135, 156
PNA, 284
polar groups, 186
polarity, 169, 171, 172, 191
polarization, 4, 6, 13, 16, 17, 165, 184, 206, 221, 222, 241
pollutant, 61, 260, 262
pollutants, 253, 260, 267, 274
pollution, 40, 57, 58, 60, 165, 188, 204, 248, 255, 256, 265, 267
poly(dimethylsiloxane), 123, 185, 197
poly(methyl methacrylate), 135
poly(vinyl chloride), 31
polyamide, 120, 144, 173, 184, 185, 193, 195, 198, 281
polyamides, 6
polybutadiene, 184, 197
polycarbonate, 154, 283, 293
polycondensation, 125, 126, 131, 132, 179
polycondensation process, 125
polydimethylsiloxane, 159
Polydimethylsiloxane, 198
Polyelectrolyte, 157, 161
polyester, 44, 71, 187, 199
polyesters, 44
polyether, 120, 155, 184
polyetheretherketone, 42, 131
polyethylene, 39, 43, 68, 131, 132, 146, 155, 156, 232

polyimide, 120, 138, 139, 152, 156, 158, 172, 174, 175, 179, 180, 181, 182, 187, 191, 192, 193, 194, 195, 196, 199
polyimides, 149, 179, 195, 196
polymer, 3, 6, 9, 13, 17, 22, 23, 27, 28, 29, 30, 31, 36, 44, 45, 46, 47, 64, 65, 68, 69, 71, 120, 121, 122, 123, 126, 129, 132, 133, 134, 135, 136, 137, 139, 140, 142, 143, 145, 146, 149, 150, 151, 152, 153, 157, 158, 159, 161, 166, 169, 171, 172, 173, 174, 176, 177, 178, 179, 181, 182, 186, 189, 191, 194, 214, 281, 282, 286, 288, 291, 293
polymer blends, 27
polymer chains, 30, 123, 132, 134, 137, 139, 142, 166, 169, 173, 174, 179, 181, 182, 286, 291
polymer electrolytes, 161
polymer film, 23, 27, 29, 47, 159
polymer films, 23, 29
polymer materials, 120, 140, 293
polymer matrix, 13, 22, 23, 30, 47, 134, 139, 143, 146, 150, 173, 174, 178, 182
polymer membranes, 17, 23, 135, 145, 150, 157
polymer networks, 286
polymer solutions, 139
polymer structure, 143, 171
polymer swelling, 191
polymeric chains, 289
polymeric materials, 120, 153, 175, 178, 184, 218
polymeric membranes, 121, 139, 142, 149, 166, 174, 183, 193, 197
polymerization, 28, 29, 30, 31, 39, 46, 68, 70, 129, 132, 137, 138, 139, 140, 142, 147, 158, 159, 160, 161, 174, 175, 180, 181, 195, 281, 282, 283, 284, 288, 290, 293
polymerization process, 142
polymerization time, 30
polymers, 6, 15, 22, 23, 25, 27, 29, 30, 32, 36, 39, 46, 69, 71, 120, 121, 122, 123, 124, 125, 126, 131, 132, 134, 135, 139, 141, 142, 143, 145, 147, 148, 149, 151, 160, 171, 178, 184, 194, 198, 208, 213, 239, 240, 273, 280, 281, 283, 284, 286, 288, 289, 291
polyolefin, 231
polypropylene, 146
polysaccharide, 52, 212, 240, 249
polysaccharides, 208, 212, 240
polysiloxanes, 146
polystyrene, 22, 28, 30, 32, 39, 68
polyurethane, 71, 184, 199
polyurethanes, 187
polyvinyl alcohol, 43, 71, 150, 164, 172, 194
polyvinylchloride, 186, 187
polyvinylpyrrolidone, 142, 159
poor, 1, 2, 7, 63, 172, 206, 252, 256, 258, 289
poor performance, 1, 2
population, 7
pore, 6, 29, 30, 46, 68, 72, 125, 144, 157, 165, 181, 204, 206, 215, 218, 219, 220, 223, 227, 228, 230, 231, 232, 233, 281, 282, 283, 284, 288, 290, 291, 293, 294
pores, 25, 125, 139, 150, 174, 183, 202, 206, 218, 220, 223, 224, 229, 280, 281, 282, 283, 284, 285, 286, 288, 290, 291
porosity, 46, 130, 147, 206, 208, 220, 223
porous, 3, 6, 29, 39, 65, 68, 70, 72, 125, 136, 137, 139, 140, 142, 150, 154, 157, 159, 166, 174, 175, 216, 218, 219, 280, 281, 282, 283, 284, 286, 287, 288, 289, 290, 291, 293, 294
porous media, 216
potassium, 251
powder, 31
power, 19, 49, 58, 145, 241, 245, 254, 264, 266, 269, 272, 274
power plant, 245, 254, 269, 272
power plants, 245, 254, 269
power-law, 241
PPO, 33, 36, 37, 41, 49, 69, 70, 72, 131, 146, 155, 159, 275
praxis, 13
precipitation, 16, 24, 126, 214, 246, 248, 250, 252, 258, 264, 276
prediction, 172, 282
predictive model, 220
preference, 186
pressure, 5, 7, 8, 11, 12, 13, 19, 24, 31, 40, 47, 64, 142, 144, 164, 166, 168, 169, 170, 204, 206, 215, 216, 217, 219, 220, 222, 223, 229, 234, 241, 244, 250, 253, 266, 267, 277
prevention, 40, 165
prices, 189
private, 14
producers, 14
production, vii, 1, 2, 3, 4, 5, 6, 7, 8, 9, 10, 12, 13, 14, 15, 16, 17, 18, 19, 24, 40, 54, 55, 56, 58, 59, 60, 61, 62, 63, 64, 66, 67, 71, 72, 119, 121, 147, 149, 164, 188, 197, 199, 205, 208, 209, 227, 239, 243, 244, 245, 246, 247, 248, 249, 250, 252, 256, 257, 258, 261, 262, 264, 266, 267, 268, 270, 272, 273, 275, 276, 277

production costs, 7, 11, 16, 17, 248, 264
productivity, 55, 165, 243, 250
profit, 15
program, 7, 268
prokaryotic, 208
promote, 140, 212, 218
promoter, 17
propane, 39, 161, 182
property, iv, 30, 41, 47, 59, 121, 139, 148, 151, 156, 174, 181, 182, 183, 208, 213, 281
proportionality, 166
propylene, 161
protection, vii, 24, 63, 120, 265, 285
protective coating, 5, 158
proteic, 213
protein, 39, 52, 66, 210, 214, 215, 218, 240, 250, 258, 260, 271, 273
protein aggregation, 215
protein binding, 39
protein hydrolysates, 258, 271
protein-protein interactions, 215
proteins, 139, 206, 208, 211, 212, 215, 240, 249, 250, 273
Proteins, 158, 257
protic, 55
protocol, 210, 213
protons, 42
PSI, 168
PTFE, 29, 68, 182
PTMO, 123, 126, 127, 146
public, 14
publishers, 153
pulp mill, 270
pumping, 9, 51, 287, 288, 289, 294
pure water, 5, 12, 64, 144, 221, 264, 265, 268
purification, 1, 2, 15, 48, 51, 58, 149, 188, 240, 243, 244, 247, 250, 256, 261, 266, 267, 269, 270, 277
PVA, 43, 71, 143, 156, 164, 169, 172, 173, 177, 178, 186, 187, 188, 190, 194, 198, 271
PVC, 30, 31, 187
pyrolysis, 139, 140, 158
pyruvate, 52

Q

quality control, 72
quartz, 125
quaternary ammonium, 41

R

R&D, 268
race, 186
radiation, 68, 70
Radiation, 68
radical polymerization, 288
radius, 220, 223
range, 10, 13, 26, 44, 46, 129, 144, 146, 164, 171, 212, 228, 236, 246, 247, 260, 266, 277, 283, 284, 289
rare earth, 48, 50, 51
raw material, 10, 12, 64
reactants, 36, 54, 125
reaction rate, 203
reaction temperature, 36
reactive groups, 131
reactivity, 126
reagent, 39
reagents, 39, 259
reclamation, 66, 202, 241
recognition, 144, 291, 292, 294
reconditioning, 257
recovery, 10, 11, 12, 48, 49, 50, 51, 52, 66, 69, 72, 188, 189, 192, 245, 247, 248, 249, 250, 253, 254, 255, 256, 257, 258, 260, 261, 264, 265, 266, 267, 269, 270, 272, 273, 274, 275, 276
recovery processes, 257
recycling, vii, 2, 50, 57, 60, 165, 247, 248, 253, 255
reduction, 7, 159, 163, 176, 193, 201, 218, 246, 247, 252, 259, 260, 264, 266
refining, 260
refractive index, 130
regenerate, 258
regeneration, 40, 48, 51, 59, 60, 73, 188, 203, 247, 252, 256, 264, 265, 270, 274
regulation, 264
rehabilitation, 58
rejection, 6, 49, 50, 144, 256
relationship, 63, 168, 172, 174, 201, 209, 213, 215
relationships, 206, 213, 218
relaxation, 181, 286
relevance, vii, 1, 2, 7, 13, 18, 39, 48, 63
reliability, 6, 57, 245, 246
renal, 8
renal failure, 8
renewable resource, 178

research, viii, 2, 4, 7, 14, 15, 16, 17, 21, 25, 27, 61, 64, 119, 121, 122, 123, 124, 132, 143, 147, 164, 177, 179, 203, 204, 205, 207, 208, 215, 217, 218, 220
research and development, 4, 7, 14, 16, 17
researchers, vii, 64, 123, 148, 165, 205, 206, 207, 208, 209, 215
reservoir, 285, 287
reservoirs, 54, 282
resin, 22, 41, 58, 142, 210, 214, 239, 247, 256, 264, 265, 275
resins, 22, 26, 66, 247, 250, 252, 261, 264, 265, 271
resistance, 17, 24, 25, 26, 28, 36, 37, 42, 43, 55, 63, 72, 130, 174, 175, 178, 179, 183, 184, 192, 193, 196, 204, 206, 207, 216, 217, 218, 219, 220, 222
resources, 10, 18, 21, 40, 57, 60, 178, 249
response time, 280, 287, 290
responsiveness, 283, 289, 293
retention, 7, 8, 46, 203, 206, 213, 237, 262
rigidity, 127, 130, 147
risk, 211
room temperature, 28, 32, 130, 136, 146
roughness, 181
rubber, 5, 39, 185, 187
rural, 238
ruthenium, 42

S

sacrifice, 141
safety, 9, 289
sales, 14, 15, 224
salinity, 245, 266, 276
salt, 6, 7, 8, 12, 24, 25, 42, 47, 53, 55, 56, 57, 58, 60, 61, 144, 157, 214, 245, 246, 247, 250, 251, 253, 254, 255, 256, 257, 258, 265, 268, 272
salt production, 246, 253, 268
salts, 9, 12, 13, 16, 25, 47, 51, 52, 53, 55, 56, 58, 60, 67, 72, 246, 248, 251, 252, 253, 254, 256, 257, 260, 261
sample, 214
sand, 247
savings, 253
SBR, 187, 205, 229
scaling, 16
scattering, 148
scientists, viii, 16, 168
search, 120
seawater, 9, 244, 245, 246, 269
seaweed, 229
second generation, 44
secretion, 209, 289
security, 285
sedimentation, 202, 248
selecting, 16
selectivity, vii, 2, 3, 6, 11, 12, 16, 17, 24, 40, 50, 56, 72, 120, 121, 124, 127, 139, 142, 143, 144, 147, 163, 166, 167, 168, 169, 170, 171, 172, 173, 174, 177, 178, 179, 181, 182, 183, 184, 186, 188, 189, 190, 192, 246, 249
self-assembling, 46
SEM, 283, 284
semiconductor, 232
semiconductors, 264
sensing, 145, 161, 291
sensors, 5, 9, 147, 152, 291
sequencing, 205
series, 6, 12, 14, 24, 32, 34, 41, 44, 69, 70, 71, 72, 132, 143, 190, 219, 232, 249, 264, 266, 276, 283
serine, 250, 258
serum, 240
serum albumin, 240
services, iv
settlements, 238
sewage, 202, 229, 230, 238
Shanghai, 237
shape, 16, 22, 33, 142, 166
shear, 204, 222, 238
short-range, 212
short-term, 203, 212
Siemens, 235, 236
sign, 47
signal transduction, 291
signals, 287
signs, 53
silane, 67, 133, 134, 136, 137, 153, 154, 155, 156, 157, 158, 160
silica, 25, 123, 125, 126, 127, 132, 138, 139, 140, 142, 143, 145, 147, 151, 152, 156, 158, 159, 160, 161, 178, 183, 196
silica glass, 123
silicate, 154
silicates, 156
silicon, 127, 140, 156, 157, 159, 160
silicon dioxide, 140
siloxane, 123, 127, 140, 146, 154, 158, 159, 161, 184, 198

silver, 71, 131, 147, 154, 248, 249, 276
simulation, 15, 173, 189, 192
simulations, 15
Singapore, 163, 237
single market, 9
SiO2, 65, 132, 143, 145, 146, 150, 151, 153, 155, 156
sites, 30, 42, 44, 125, 130, 135, 145, 147, 212, 215, 279, 280, 286
skin, 6, 160, 172, 174, 175, 176, 181, 193
sludge, 201, 202, 203, 206, 207, 208, 213, 214, 217, 220, 237, 238, 239, 240, 241, 248, 249, 253
SMS, 183
SO2, 253, 254, 255
sodium, 9, 22, 30, 51, 55, 56, 57, 58, 59, 61, 62, 63, 67, 173, 178, 192, 195, 210, 211, 212, 257, 258, 261, 262, 267, 271, 274, 276, 291
sodium hydroxide, 51, 210, 261, 262, 274, 276
softener, 247
software, 169, 173
sol-gel, 25, 39, 46, 68, 119, 120, 122, 123, 124, 125, 126, 127, 129, 131, 132, 133, 134, 135, 136, 139, 142, 143, 144, 145, 146, 147, 149, 150, 151, 152, 153, 154, 155, 156, 157, 159, 160, 161, 178
solid state, 153
solid-state, 125
solubility, 15, 16, 25, 44, 55, 142, 143, 147, 149, 169, 171, 172, 184, 186, 258
solvent, 17, 25, 30, 31, 40, 55, 69, 129, 130, 134, 135, 136, 171, 172, 174, 176, 177, 178, 188, 191, 218, 223
solvent molecules, 40
solvents, 17, 29, 30, 31, 40, 61, 135, 171, 176, 188, 260
sorption, 59, 68, 166, 167, 169, 171, 172, 174, 183, 188, 191
soy, 52, 66, 249, 251, 271, 273
soybean, 273
SP, 133
spacers, 24, 63
speciation, 160
species, 21, 24, 125, 132, 134, 135, 144, 151, 154, 165, 239, 243, 244, 280
specific gravity, 202
specific surface, 45
Spectroelectrochemical, 161
spectroscopy, 145
speed, 50
spin, 129, 160
SRT, 203, 208, 213, 215, 224, 227, 235
stability, 3, 6, 7, 12, 16, 17, 23, 25, 26, 41, 43, 46, 63, 64, 69, 72, 120, 124, 125, 129, 130, 140, 143, 144, 145, 146, 147, 174, 177, 178, 179, 182, 183, 184, 194, 245, 251, 253, 289
stabilization, 125, 140, 260
stabilize, 212
stack gas, 272
stages, 11, 63, 266
stainless steel, 48, 249, 270
standard model, 216
standards, 246, 262
star polymers, 71
starch, 227
state control, 273
steady state, 212, 216, 219, 220, 221, 222
steel, 48, 50, 64, 249, 253, 254, 265, 270
steric, 36, 44, 132, 136, 174
sterilization, 264
stiffness, 173
Stimuli, 279
storage, 4, 17, 24, 61, 215, 244, 259
strain, 213
strains, 213
strategies, 14, 218, 287
streams, 2, 12, 14, 47, 48, 57, 184, 185, 188, 260, 266, 272
strength, 6, 23, 28, 30, 31, 34, 37, 38, 41, 67, 120, 140, 142, 174, 179, 181, 212, 280, 285, 289
stress, 175, 204
strong interaction, 172, 175
structural characteristics, 285
styrene, 28, 29, 31, 32, 39, 68, 129, 132, 153, 156, 187
substances, 47, 53, 139, 165, 188, 206, 239, 241, 260, 261, 262, 279, 292
substitution, 36, 37, 41, 61, 130
substrates, 123, 129, 135, 136, 137, 138, 139, 140, 141, 142, 154, 157, 283, 284, 287, 289, 290, 293
subtraction, 211
suffering, 8
sugar, 52, 249, 251, 252, 272
sugar industry, 52, 251
sulfate, 59, 60, 245, 250, 269
sulfites, 60
sulfonamide, 128, 146, 153
sulfur, 59, 178, 272

sulfur dioxide, 60, 272
sulfuric acid, 22, 28, 48, 49, 72, 131, 178, 248, 275
sulphate, 42
Sun, 156, 158, 161, 192
supernatant, 218, 248
suppliers, 189
supply, 7, 14, 15, 54, 58, 245, 262
surface area, 9, 46
surface modification, 123, 140, 141, 159
surface properties, 182, 213
surface roughness, 181
surface structure, 157
surface treatment, 249, 260
surface water, 17, 66, 235, 261, 264
surfactant, 12
surfactants, 198
sustainability, 271
sustainable development, vii, 65, 270
swelling, 26, 30, 38, 39, 41, 42, 43, 143, 174, 175, 177, 178, 179, 181, 183, 281, 285, 286, 288, 291
switching, 65, 280
Switzerland, 161
symbiosis, 40
syndrome, 261
synthesis, vii, 5, 40, 44, 52, 59, 61, 62, 125, 131, 151, 155, 183, 197, 209, 213, 244, 286
synthetic polymers, 6
systems, 2, 4, 5, 17, 18, 21, 24, 49, 52, 63, 69, 72, 125, 135, 156, 169, 180, 183, 188, 191, 198, 202, 204, 208, 214, 219, 227, 232, 240, 246, 247, 264, 266, 277, 279, 280, 281, 282, 285, 286, 287, 288, 289, 291, 293, 294

T

Taiwan, 195
taste, 251, 252
TDI, 131, 175, 191, 193
teaching, viii
technology, 1, 3, 7, 8, 10, 12, 14, 17, 18, 19, 21, 24, 50, 51, 52, 63, 64, 120, 143, 164, 176, 188, 189, 192, 202, 225, 226, 227, 229, 233, 234, 235, 236, 245, 260, 262, 268, 272, 276, 277, 279, 280
Teflon, 130, 131
temperature, vii, 1, 2, 12, 16, 25, 36, 37, 38, 55, 123, 125, 129, 131, 140, 144, 145, 146, 149, 151, 152, 157, 159, 161, 165, 166, 169, 170, 172, 181, 182, 183, 184, 220, 223, 237, 247, 279, 280, 281, 283, 284, 285, 286, 291, 292, 294
temperature gradient, 12
tensile, 142
tensile strength, 142
TEOS, 126, 128, 130, 134, 147, 151, 156, 194
tetraethoxysilane, 126, 130, 135
textile, 274
textiles, 275
TGA, 129
theory, 6, 65, 165, 209, 212, 213, 221, 238, 239
therapy, 12, 289, 291
thermal analysis, 148
thermal properties, 71
thermal stability, 3, 6, 7, 17, 23, 124, 125, 129, 130, 140, 145, 146, 147, 174, 178, 179
thermal treatment, 176, 177, 179, 181, 182
thermodynamic, 5, 171, 172, 173, 190
thermodynamics, 6, 19
thin film, 7, 151, 152, 158, 159, 160
thin films, 151, 152
three-dimensional, 31, 125
threonine, 250
threshold, 236
time, viii, 7, 9, 30, 36, 37, 38, 40, 41, 51, 56, 125, 126, 129, 132, 134, 139, 140, 143, 168, 172, 176, 203, 206, 210, 213, 223, 249, 262, 264, 279, 280, 281, 287, 288, 293
TiO_2, 132, 135, 152, 155, 156
titanium, 50, 69, 72, 135, 156, 276
Titanium, 50
title, 121, 122
TMA, 33, 35, 36, 37, 39
toluene, 188, 198
total costs, 7
toughness, 31, 124
toxic, 2, 8, 9, 178, 188, 218, 244, 246, 248, 254, 266, 268
toxicity, 203, 215, 218, 260
TPA, 35
tracking, 124
trade, 121, 143, 156, 168, 178, 189, 192
trade-off, 121, 143, 156, 168, 178, 189, 192
trans, 136, 206, 215, 216
transfer, 18, 42, 70, 71, 181, 188, 208, 260, 273
transition, 26, 40, 43, 61, 133, 147, 172, 181, 284, 285, 286, 288, 291
transition metal, 147
transition metal ions, 147

transition temperature, 15
transmembrane, 219
trans-membrane, 206, 215, 216
transparency, 124
transparent, 135
transport, 2, 3, 5, 7, 12, 13, 15, 17, 18, 24, 42, 43, 47, 54, 56, 70, 131, 139, 140, 142, 147, 152, 154, 155, 158, 164, 165, 166, 169, 171, 172, 174, 175, 182, 184, 189, 191, 195, 196, 218, 222, 243, 252, 262, 266, 274, 275, 282, 293
transport phenomena, 24, 275
transportation, 163, 165, 171
trend, 169
trial, 204
tricarboxylic acid, 270
trichloroethylene, 188
triggers, 120
tubular, 4, 9, 183, 184, 197, 224, 233
tumor, 279
tumor cells, 279
tumors, 286
tungsten, 52
turbulence, 204
turbulent, 17

U

ultra-thin, 174, 176, 193
uniform, 282
uranium, 48, 51, 120, 256
urea, 8
urine, 260, 273

V

vaccine, 240
vacuum, 57, 164, 227
valine, 250, 258
values, 16, 36, 41, 179, 217, 256, 272, 286, 288
van der Waals, 121
vapor, 10, 12, 18, 139, 140, 158, 164, 165, 167, 168, 169, 181, 195
variability, 130
variables, 207, 213
variation, 273, 288
vehicles, 145
velocity, 204, 206, 216
versatility, 18
vinyl monomers, 39
vinylidene fluoride, 185
viruses, 18
viscosity, 129, 135, 177, 206, 215, 216, 217, 219, 220, 223, 266
visible, 172
vitamin C, 267
vitamins, 250
voids, 139, 143
volatility, 164, 176

W

waste disposal, 247, 256, 266
waste water, vii, 14, 17, 24, 25, 57, 66, 235, 253, 269, 270, 277
wastewater, 8, 66, 188, 201, 202, 203, 205, 213, 225, 226, 227, 228, 229, 230, 231, 232, 233, 234, 235, 237, 238, 240, 241, 246, 247, 249, 256, 260, 265, 266, 269, 276, 277
wastewater treatment, 8, 201, 202, 205, 225, 226, 227, 228, 229, 230, 231, 232, 233, 234, 235, 237, 238, 240, 241, 246, 249, 265, 269, 277
wastewaters, 246, 247, 260
water permeability, 144
water quality, 225, 227, 228, 230, 231, 232, 233, 234, 235, 245, 246, 261, 262, 267
water quality standards, 262
water vapor, 12
water vapour, 12
water-soluble, 47, 56
wells, 247
western countries, 199
wheat, 270
whey, 52, 67, 244, 249, 250, 251, 266, 271
windows, 145, 146

X

XPS, 283, 284
xylene, 186, 198
xylenes, 188

Y

yield, 9, 203, 210, 281, 283, 284, 290, 293

Z

zeolites, 12, 143

zinc, 248, 249, 261
zirconia, 138
zirconium, 42, 136, 157
zwitterions, 250